Springer Praxis Books

More information about this series at http://www.springer.com/series/4097

J.B. Williams

The Electronics Revolution

Inventing the Future

J.B. Williams
Offord Darcy
St Neots
Cambridgeshire, UK

SPRINGER PRAXIS BOOKS IN POPULAR SCIENCE

Springer Praxis Books
ISBN 978-3-319-49087-8 ISBN 978-3-319-49088-5 (eBook)
DOI 10.1007/978-3-319-49088-5

Library of Congress Control Number: 2017941217

Cover Design: Jim Wilkie

Printed on acid-free paper

This Springer imprint is published by Springer Nature
The registered company is Springer International Publishing AG
The registered company address is: Gewerbestrasse 11, 6330 Cham, Switzerland

Contents

Acknowledgments

It would seem appropriate with the subject of this book that the Internet should have been such a vital tool during the writing. It was of enormous value in finding and retrieving hundreds of relevant articles from journals published throughout the twentieth century. In addition, many articles are published on the Internet alone.

I would also like to thank:

The enthusiasts who run the many online museums to show their collections.

Wikipedia images for many useful pictures.

The librarians of Cambridge University Library for leading me through some of the byways of their collections.

Clive Horwood for his encouragement, Maury Solomon particularly when it became necessary to split an overlong book, and Elizabet Cabrera.

Jim and Rachael Wilkie for turning the manuscript into a book with the cover design and editing.

My referees Robert Martin-Royle, Dr Andrew Wheen, Myrene Reyes and Professor Graeme Gooday for their positive comments which made me feel that the nearly 5 years' work had been worthwhile.

Last, but certainly not least, I want to thank my wife who has, without realizing it, taught me so much about history.

List of Figures

1

Introduction

On December 12, 1901, Guglielmo Marconi, a young man of mixed Italian and British parentage, was on the windswept Signal Hill at St Johns', Newfoundland.[1] He was sitting in a room in a disused military hospital at the top of the hill behind a table full of equipment. Outside, a wire ran up to a Baden-Powell six-sided linen kite which he and his assistant George Kemp were only keeping aloft with difficulty. They had already lost another kite and a balloon, and they were only using these because an antennae array at Cape Cod had blown down.

On the other side of the Atlantic, above Poldhu cove in Cornwall, almost as far west in Britain as you can get without falling into the sea, was another, much larger collection of equipment. A small power station generated 25 kW of electricity, and when a Morse key was pressed a huge arc leapt across a spark gap, dying away when the key was released. The signal thus generated was led to a set of wires held aloft by two 60 m towers. This arrangement replaced a much larger 120 m–diameter inverted cone array which, again, had blown down.

At 12.30 p.m., over the static, Marconi thought he heard something—dot dot dot. It was the Morse code for S that he had arranged to be transmitted from Poldhu. He passed the earpiece to Kemp who listened. He could hear it too. Twice more that day they heard the signal, but then the weather worsened and they had to stop. What they had achieved was to receive a wireless signal from the other side of the Atlantic, a feat that many experts thought was impossible.

Marconi now had the problem of when to announce his success, because the doubters were convinced that electromagnetic waves travelled in straight lines and couldn't bend around the curvature of the Earth. Basically they were right, but he had been lucky, as the particular range of frequencies generated by his transmitter would bounce off atmospheric layers and, hence, could be received way beyond the horizon. In the event, he waited a few days before saying anything to the press.

Later, he was to prove long-distance communication much more convincingly, and in the following decade or so the Marconi company built up a substantial business supplying wireless communication facilities, particularly to ships. When the 'Titanic' hit an iceberg

J.B. Williams, *The Electronics Revolution*, Springer Praxis Books,
DOI 10.1007/978-3-319-49088-5_1

and sank, it was the Marconi equipment installed on her which was used to call for help, and that on the 'Carpathia' which received the SOS which allowed her to come to the rescue and save many lives.

It might be thought that Marconi had mastered the technology for wireless communication, but what he was using was very crude and it was only refinement and huge amounts of power that enabled it to work at all. It fact, what he was using was purely electrical, and a blind alley. What was needed for really satisfactory equipment was the means to enlarge or amplify signals and the ability both to generate and receive them in a very narrow frequency band. This required electronics, and in 1901 it didn't exist.

In 1897, J. J. Thompson had discovered the properties of what he called 'corpuscules', and it was only slowly that the word 'electron', coined by George Johnstone Storey some 6 years before, came to be used. Electronics is one of those words that everyone knows what is meant by it, but when it comes to a definition, it slips through your fingers. As will be seen, a reasonable definition is 'equipment using devices that manipulate electrons'.

The Electronics Revolution is about how we went from absolutely nothing to the abundance of electronic items that we regard as normal today. For communication we expect mobile phones, and the Internet. Now it seems that every man, woman and child has a mobile phone in their pocket or handbag. Something that was the plaything of the few is now the norm for everyone. Modern business could not function without the instant communication of email.

For entertainment, there are radios and TVs, and music on tap for our personal use. We so expect instant TV from anywhere in the world that it is difficult to realize that it was as late as 1962 before the first live TV pictures flashed across the Atlantic, and they could only be maintained for a matter of minutes. The effect has been for us to think in terms of 'one world'. We become involved in wars or disasters in other countries in ways that would have been incomprehensible to our forefathers.

For work and play, there is computing. Home computers have become ubiquitous. Offices are full of computers, and yet more computers control whole organizations. But computers can also be in the form of microprocessors buried in appliances such as washing machines or central heating boilers. Outside, the household's car is very likely to have electronic management systems.

There is a long trail of successes, and some failures, but with one thing leading to another, often in an apparently unrelated way: wartime radar to microwave ovens, and moonshots to worldwide live TV. The intention is to follow these links to show how it all fits together. This not a 'history of technology' but one of how the technology produced all this diversity and, in doing so, how these developments created a revolution in everyday lives. For example, without radio and TV our modern democracy is barely imaginable. So often this impact of technology has been largely ignored.

How was all this achieved? Not by 'science', but by engineering and technology. There is often a great confusion between these, with the words being used interchangeably, but there is a key difference. Science is about ideas in the natural world, while engineering is about their exploitation into technology useful to humankind. Even once the science is in place, in order to introduce a major technological change there are three distinct phases: invention, development and exploitation.

The fundamental science comes first. James Clerk Maxwell predicted the existence of electromagnetic waves, and Heinrich Hertz proved he was right—that was the science. Others, such as Eduoard Branly who developed the coherer detector, or Oliver Lodge who demonstrated a wireless communication system which could ring a bell remotely, were the inventors.

At this point, though there was interesting knowledge, there was still nothing useful. It required a Marconi to come along and bring it all together. It was often said by his detractors that Marconi didn't invent anything. Besides being rather unkind about his abilities, it completely misses the point. What he was able to do was to take these various inventions and develop them into a system. He could then take this and exploit it to allow communication from a ship to the shore or vice versa. He had produced something useful to people.

Though most of the knowledge of the relevant physical laws of electricity was understood before the end of the nineteenth century, much concerning electronics had still to be learned. In some cases, it took some time to understand the devices that had been invented. With solid state physics, mostly investigated in the first half of the twentieth century, it was the search to understand that led to the invention of transistors. The fantastic bounty brought by integrated circuits and microcomputers, though, was more a matter of technology than of science.

To follow the story, we need to look at the people who made things happen. The twentieth century was characterized by the rise of the large research organizations such as Bell Labs so the developments are often hidden behind the corporate façade, but where the individuals can be identified their stories are told. Some of these people will be well known, but others may well be surprising. There are many unsung heroes who made vital contributions but received little credit because someone else was better at the publicity.

Some names are known, such as Tim Berners-Lee for the World Wide Web, and John Logie Baird for television (though his contribution was not quite what it is often assumed to be). People such as Isaac Schoenberg and Alan Blumlein, who really got the system going, are barely remembered, and what of Boris Rosing who always seems to be at the back of all the successful television developments? And who has heard of Nobel prizewinner Charles Kao whose determination gave us fiber optics on which modern communications depend?

We don't want to get bogged down in the sterile arguments about who was the first to invent something. When the time was ripe often a number of people come up with much the same idea. Simultaneous invention is quite common: Jack Kilby and Robert Noyce came up with integrated circuits at much the same time, though the concept was already known. What matters here is the turning of ideas into something useful.

At the end of all these developments we have reached a state where electronics have permeated every aspect of our lives. As so often, these things were summed up by a TV advert. An amusing one for Renault electric cars imagined a world where common items were powered by tiny engines and not by electricity. The star was a handheld card machine which had to be refuelled to work. In one way it made its point, but in another completely missed it. Without electricity there would be no electronics. Without electronics there would be no computers. Without computers credit cards would not exist so there would be no need for a card machine. In any case, the system depends on wireless and other electronic communication, so the handheld unit would not exist.

What did Marconi do when the world began to change to electronics? The record of people and organizations, when the technology on which they depend becomes obsolete, is not good. He was lucky in that he employed J. Ambrose Fleming as a consultant and so, by chance, his company came to hold one of the key patents. He had the foresight to employ Henry Round to develop the necessary devices, so the company smoothly made the transition. The great arc transmitters were phased out, and electronics took over.

NOTE

1. The whole story of Marconi's exploits is well covered at the website Marconi Calling available at: http://www.marconicalling.co.uk/introsting.htm

2

Missed Opportunities: The Beginnings of Electronics

Just because something doesn't do what you planned it to do doesn't mean it's useless.

Thomas A. Edison

One of the areas that particularly attracted the interest of experimenters in the nineteenth century was the behavior of evacuated glass tubes when excited by electricity. As the characteristics change with the amount of gas remaining in the equipment, the whole experiment depended on how well the air could be removed. Without the development of efficient vacuum pumps electronics would never have got started.

Though many people had played with these devices before, it was Herman Geissler who brought everything together. He was from a glassblowing family but had set up in Bonn in Germany as an instrument maker, which often involved working with glass.[1] In 1857, he produced a more efficient vacuum pump and used it to evacuate a glass tube into the ends of which he had sealed metal electrodes. The tube was energized with a high voltage produced by a Ruhmkorff coil, an induction coil similar to those now used to provide the spark for a petrol engine, and also as used later for generating wireless waves.

It was soon found that a very small amount of a particular gas gave rise to a discharge of a specific color. With a range of gasses, tubes exhibiting different shades could be produced. This work eventually led to gas discharge lamps, and in particular the neon lights so popular for shop signs. Geissler produced many tubes of complex shapes and colors which had quite a vogue as decorations, though his main work was used for experimental purposes by serious physicists within universities (Fig. 2.1).

Further work on vacuum pumps by Sprengel and others brought pressures sufficiently low to enable the development of electric incandescent lamps around 1879, but at the same time Englishman William Crookes managed to improve the pump still further, producing even lower residual pressures. What was discovered was that, as the amount of gas in the tube was decreased, a dark space started to appear in the glow in the tube at the cathode (negative electrode) end. As the vacuum increased this dark space would eventually fill the whole tube, but the end of the glass by the anode (positive electrode) would glow.

In an elegant paper presented to the British Association in 1879, Crookes demonstrated the properties of the particles, or whatever they were, in these tubes.[2] The anode could be

J.B. Williams, *The Electronics Revolution*, Springer Praxis Books,
DOI 10.1007/978-3-319-49088-5_2

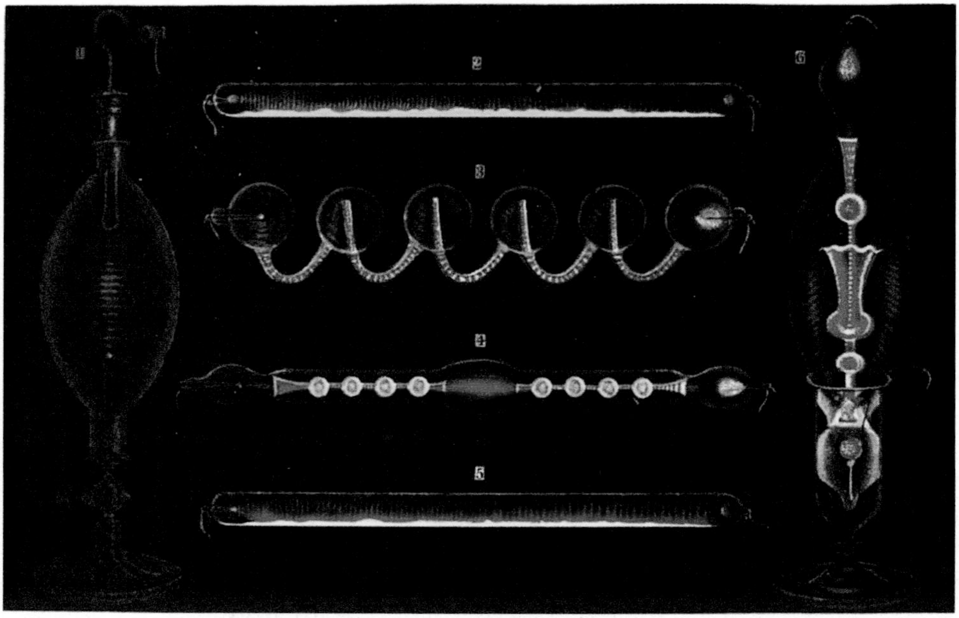

Fig. 2.1 An example of the different colors from Geissler tubes. *Source:* https://en.wikipedia.org/wiki/Geissler_tube#/media/File:Geissler_tubes.jpg

any shape he liked and the particles would go straight past and strike the end of the tube, which glowed because of phosphorescence. The particles were causing the glass itself to glow, and different glasses would produce different colors. His famous 'Maltese cross' experiment used an anode of that shape which cast a matching shadow on the end of the tube. Clearly the particles went in straight lines.

In other demonstrations he showed that the 'beam' of particles could be bent with a magnet and that it was strong enough to turn a small paddle wheel. It was all terribly interesting to the scientific community, but neither Crookes nor any of his colleagues had any idea of what they were looking at or what it could be used for apart from entertainment. Because various minerals were found to glow different colors in the 'beam' very fancy devices could be produced.

It was at this point that Thomas Edison entered this story, as in so many others. In the early 1880s he was trying to improve his incandescent lamps. One of the main problems was that the bulbs became blackened.[3] To try to prevent this he placed another electrode in the bulb and made this more positive than the filament. The theory was that, as it was known the carbon atoms from there were negatively charged, they might be attracted to the other electrode instead of the glass.

It didn't work in stopping the blackening, but he found that a current flowed. When he reversed the polarity, no current flowed. It was a very interesting result, but he had no idea what was happening, or what he could do with it. He demonstrated it to anyone who was interested, and it became known as the Edison effect. However, though many people experimented with the effect its significance was not appreciated for many years.

In 1897, J. J. Thomson took a Crookes tube and by using two lined-up apertures formed a beam of these particles.[4] He then carefully measured the deflections with both magnetic and electric fields. From this he could deduce that these particles always had the same charge-to-mass ratio, i.e., the amount of electricity they carried for their weight. They were also extremely light. Basically, he had discovered the electron, though he didn't call it that.

In the same year back in Strasbourg (then in Germany), Ferdinand Braun, a professor at Strasbourg University found a use for the Crookes tube. He built a long tube with a small hole in an aluminum plate so that he obtained a narrow beam producing a spot on the end phosphorescent 'screen'. A coil set horizontally with an alternating current in it caused the spot to move up and down in time with that source. By viewing the spot in a rotating mirror he was able to see the waveform of the signal in the coil.[5]

It was most ingenious. What he had produced was a means of examining waveforms—the oscilloscope, the essential tool of electronic engineering. Two years later, Braun's associate at Strasbourg, Jonathan Zenneck, added another coil to achieve a method of scanning the spot horizontally (a timebase). This meant that the waveform would appear on the end of the tube instead of in the rotating mirror.

The 'Braun tube' soon set people thinking about further ways in which it could be used. Its great advantage over other methods of displaying signals was that it used electrons of almost negligible mass so the beam could be moved around the screen at very high speed. This was to have a very important application in the future (see Chap. 4).

In 1899, J. Ambrose Fleming, then the professor of electrical engineering at University College, London, was made a consultant to the Marconi Wireless Telegraph Company.[6] Marconi, having pioneered wireless telegraphy, was in need of more technical expertise as his ambitions grew. This was particularly important for the attempt to transmit signals across the Atlantic and it was Fleming who designed the huge spark transmitter installed at Poldhu to enable this.

Fleming had had a complicated career as he struggled to educate himself and provide for his widowed mother. Much of his time had been spent teaching in one form or another. An important exception was when he worked for the Edison Lighting Company in London, starting in 1882 and remaining until 1885 after it had merged with Swann's company.[7] There he had known about the Edison effect and had conducted some research into it. In the 1890s, after returning to academia, he gave a number of public lectures on the subject. He was fascinated by it.

It was only in October 1904 that he had a 'sudden very happy thought' as to how he could use the Edison effect.[8] At the age of 55, he was becoming increasingly deaf and so was looking for a way of detecting the wireless waves which avoided the use of an earphone to listen to the output of the magnetic detector. The wave was oscillating backwards and forwards at a high frequency; it was an alternating wave. A normal electrical instrument could not follow this and so could not detect its presence. What he suddenly realized was that, if he passed the signal through a device showing the Edison effect, only half of the signal in one direction would go through. Now he had a signal that his meter would respond to.

It didn't take him long to set up a crude spark transmitter and then connect one of these modified light bulbs to the receiving aerial. With a suitable battery to light the lamp, sure enough his meter detected the presence of the radio signal. Realizing how important a discovery this was, he rushed out and patented it.[9] The trouble was that the terms of his consultancy with the Marconi Company meant that he had to assign the patent to them.

Fig. 2.2 An early Fleming 'valve', showing its derivation from a light bulb. The looped filament and the squiggly other electrode (anode) can clearly be seen. *Source:* Fleming J A, *The Thermionic Valve and its developments in Radiotelegraphy and telephony*

Fleming called the device the 'oscillation valve', which seems rather confusing today (Fig. 2.2). What he meant was that for an oscillation, in other words an alternating signal, the device acted like a 'valve' in that it only allowed the current to flow in one direction. Today, in the UK it is known as a 'thermionic valve', the last part of the name having stuck, though in America it became known as a 'vacuum tube'. What he had actually invented was a diode or rectifier to convert alternating current into direct current.

There were those who thought he had done very little, only using a known effect in a slightly different way.[10] This misses the point in that invention is often the taking of an idea from one area and applying it in another. What he achieved was to give the world the thought that vacuum tubes derived from light bulbs could be applied to electrical circuits with advantage. In other words, electronics had been invented.

The scene then moved to the United States where Lee de Forest was interested in the same area and had been experimenting with devices where one side was heated with a Bunsen burner. He heard about Fleming's presentation to the Royal Society in London on March 16, 1905 describing his oscillation valve. He patented a similar device but connected to a telephone receiver instead of Fleming's meter.[11] He probably thought that he had invented something different, but it was to lead to no end of difficulties later (Fig. 2.3).

By October of the following year he had had a better idea and patented a device 'for amplifying feeble electrical currents', which he called an Audion.[12] In this he added a third electrode in the glass envelope in the form of a wire grid lying between the heated filament and the current collecting plate or anode. A small change in voltage applied to this grid would cause a large change in the current between the filament or cathode and the anode. In a suitable circuit, amplification of a signal could take place. The patent was rather confused; it was unclear if he really understood what he had invented because he described the grid's movement as the cause of the change.

Fig. 2.3 Lee de Forest, the American inventor who was always embroiled in patent disputes. *Source:* http://upload.wikimedia.org/wikipedia/commons/6/65/Lee_De_Forest.jpg

Fleming took the view that the patents were deliberately confusing to disguise the fact that they were closely based on his work.[13] De Forest failed to acknowledge Fleming's contribution and this led to a feud between the two men. Also the problem for de Forest was that he couldn't use the devices in wireless receivers without infringing Fleming's patent. The ensuing patent battle between the Marconi Company in America and De Forest's company dragged on until 1916 when victory went to Fleming and Marconi.

It seems that de Forest didn't really understand how his device worked, particularly whether the electrons travelled through the small amount of residual gas or whether through the vacuum. Though he spent the next few years after its discovery trying to improve it, he was unable to make useful stable devices. As a result, he didn't really know what to do with his invention.

In Britain, the Marconi Company started to manufacture Fleming valves and use them as detectors in wireless receivers. Unfortunately, though more sensitive, they were not as reliable as the trusty magnetic detectors previously used. The filaments had a habit of burning out, so the sets were made with two vacuum tubes and a switch to change over between them should one fail mid-message.[14]

In 1911, de Forest let the British version of his patent lapse. The way was now open for the Marconi Company to make their own 'amplifying' vacuum tubes, and they produced the 'Round' valves. These were so named not because they were circular but after Henry Round who developed them for the company, though they were manufactured by the Edison Swann electric lightbulb company.

They were curious devices and difficult to make, though they worked well. They depended on a small amount of residual gas to function correctly. The problem was that during operation the gas molecules would 'stick to the walls', so reducing the performance. To overcome this, an asbestos pellet was used in a narrow extension at the top of the tube, and this adsorbed some of the gas. By heating this, more gas was released and the correct quantity could be maintained. The wireless sets incorporating the vacuum tubes had small electrical heating coils to do this, but the operators often preferred to use a lighted cigar for a little judicious warming.[15]

These vacuum tubes were satisfactory for Marconi's business where only relatively small numbers were required and the sets were attended by skilled operators, but for general use they were neither easy enough to make nor to use. Clearly some further improvement was necessary.

In America, de Forest was still struggling to improve his Audions, but they were far from satisfactory, suffering from unreliability of operation and short life. However, he thought that he could make a telephone amplifier which would be free of the patent disputes. In October 1912, he took his experimental unit to Western Electric so that the Bell Telephone engineers could look at it. They had a specific problem, in that while a relay would extend the distance of a telegraph signal there was no equivalent device for telephones. After a certain distance the telephone signal became too weak to be heard. The engineers were in great need of a way of boosting the signal so that conversations could take place right across the United States.

The Audion couldn't do this, but Western Electric engineers thought it held promise and in 1913 the company bought the rights to use the Audion for telegraphy and telephony. Harold D. Arnold set to work to improve the device. It was realized that the vacuum was not high enough and they obtained a superior vacuum pump to enable them to produce a 'harder' vacuum. After Western Electric had made numerous detail improvements they had a satisfactory device. In 1914 it enabled the company to extend the long-distance telephone system from New York to Denver and soon afterwards to San Francisco.[16]

Around the same time Irving Langmuir had finished his work on the blackening of electric lamps and started to look at the Edison effect.[17] By using the improved pump which he had developed for his work on the lamps he soon independently confirmed that the Audion worked much better with a higher vacuum. He investigated the characteristics thoroughly and soon produced greatly improved devices, which he called 'Radiotrons'. Inevitably, this led to yet another patent dispute as to who had improved the Audion, this time between General Electric and AT&T, Western Electric's parent company.

Around this time, other people were discovering some of the properties of these devices, and their uses. Edwin H. Armstrong, still a student, somehow got hold of an Audion and began to experiment with it. He found that if he fed back just enough of the output of the device to its input in a tuned wireless receiver he could achieve a much higher amplification. He had discovered 'regeneration' which was enormously valuable in improving the performance of the sets.

He also discovered that if he increased the feedback further the device would begin to oscillate, to generate a continuous signal at the frequency defined by the components around the Audion. At last a source of continuous output was available which eventually could replace all those intermittent sources like the spark transmitters. Inevitably he became embroiled in disputes with de Forest who claimed them as his invention, along with several others. Meanwhile, in Germany, a different route was taken. Two branches of the armed forces were interested in the possibilities of wireless communication. The Army was backing the Siemens Company who used Professor Braun as a consultant. The Navy supported AEG (*Allgemeine Elektrizitäts-Gesellschaft* or General Electricity Company), the other large electrical combine that depended on Professor Slaby in Berlin.[18] These two organizations fell out, and it was only after the intervention of the Kaiser that they combined their efforts, and thus the Telefunken Company was formed.

The German engineers also had access to the work of the Austrian Robert von Lieben who was trying to develop a vacuum tube based on the Braun tube. Later he developed a device with a grid, and though he claimed it was based on a different principle to the Audion he inevitably ended up in patent scuffles with de Forest. The vacuum tubes used a small amount of mercury vapor for their function.[19] At the time the need for high vacuum to make the true vacuum tube was still not properly understood.

Telefunken was aware that it needed something better. A mission to America was arranged, to visit the various companies and try to obtain more information on developments. The man they chose for this task was Frenchman Paul Pichon. In 1900, he had deserted from the French army and fled to Germany. There, he had taught French for a while, including to the daughter of Georg Graf von Arco, the technical director of Telefunken. Pichon decided to study electrical engineering and ended up working for the company.[20]

He became an international representative for Telefunken, and because he was a Frenchman it was felt that he would be less obvious about what he was up to in the charged atmosphere of 1914. He was remarkably successful, gathering examples of the latest wireless equipment but, crucially, he visited Western Electric. There he was given samples of the latest high vacuum Audions and information on how to use them. Quite on what basis he obtained them isn't clear.

Together with all his samples and information he took a ship back to Europe. It didn't look as though he was following events too closely because the one he chose docked first in Southampton. His timing was immaculate as it was August 3, 1914 when he landed, the very day that Germany declared war on France. Now he had a problem. As he was a French national he would be declared an enemy alien in Germany, and in France he was a deserter and liable to immediate arrest.

He is said to have consulted Godfrey Issacs, the managing director of the Marconi Company, as to what to do. If so, this was another example of missed opportunities as Issacs failed to understand the significance of what Pichon was carrying. Whether Issacs advised him to return to France or not, Pichon ended up in Calais and was promptly arrested. There his only hope was to get the French authorities to appreciate the importance of his samples.

Now his luck changed as his story reached the ears of Colonel Gustave Ferrié, the head of the French Military Telegraphic service (Fig. 2.4). He ordered that Pichon, with his baggage and papers, should be brought to him immediately. He understood their importance straight away. Unconcerned about patents and exactly who owned everything—there was a war on—he had the characteristics of the vacuum tubes tested, improved, and rapidly put into production. Pichon was much too valuable to rot in prison as he also knew about German developments and so was drafted into Ferrié's unit. After the war he returned to work for Telefunken. Unintentionally, he had rendered the enemies of Germany, and his company Telefunken, a vital service.

Ferrié had been in the French military since the 1890s and as an engineer he became involved in the development of wireless communication. He was responsible for the work on the Eiffel Tower station after Eiffel had handed over the tower for this purpose. Ferrié steadily increased the range until reliable communication was achieved between Paris and the forts on the German border. As a result of his work French military wireless communications were the best in the world when the War broke out.[21]

Fig. 2.4 Gustave Ferrié. *Source:* http://en.wikipedia.org/wiki/File:Gustave_Ferri%C3%A9.jpg

It took until early 1915, after some false starts, before a satisfactory vacuum tube was available. The French made a number of improvements to the American design, particularly by making it more robust and reliable. Ferrié now had the tool he needed for mobile communications, and to build up the *sapeurs-télégraphistes*, the unit to operate them. Not only were these using what had now become the conventional radio system, but also low frequencies which were transmitted through the ground using a couple of spikes driven in.

The production of vacuum tubes increased rapidly and a second manufacturer was also employed. These devices were given the mark TM (Télégraphie Militaire) and in 1916 100,000 were delivered. Production reached 1000 a day in November 1918.[22] This was not all, because the French had sent samples and information to their allies in Britain. Production began in 1916 with three manufacturers, British Thomson Houston, Ediswan and Osram (English General Electric), all electric lamp makers. These were known as 'R' valves or, more commonly, French 'R' valves.

With a regular supply of tubes (the early ones only lasted around 100 h) tremendous developments took place.[23] Sensitive receivers on the Eiffel Tower monitored German communications, giving clues to troop movements, and were instrumental in trapping the spy Mata Hari. In Britain, Henry Round developed sensitive receivers with directional antennae so that the position of ships could be determined. In May 1916, the receivers were monitoring transmissions from the German Navy at anchor at Wilhelmshaven. A 1.5° change in the direction of the signals was picked up, which suggested that the ships were leaving port.[24] This was reported to the Admiralty, who ordered the Grand Fleet to sea, hence giving them an advantage in the ensuing Battle of Jutland.

Wireless systems also took to the air; spotter planes flying over enemy lines could radio instructions back to the guns to range on the targets using a specific code to indicate where a shot fell and what correction was needed.[25] This had begun haphazardly, but rapidly became a system covering all the lines. The Royal Navy also used spotters when bombarding the Belgian ports where the Germans based their submarines. Ironically, when the United States entered the war they had no suitable equipment and were forced to copy French and British designs.[26]

As so often, war had driven the development of electronics. In 4 years it had come from the plaything of inventors to the mass communication for the military. Before the war, only the specialists in companies such as Marconi were familiar with basic electronics, but now thousands of people were aware of its possibilities. With the ending of the First World War a huge quantity of surplus equipment became available. This not only consisted of wireless sets and vacuum tubes but included other components, such as resistors and capacitors, which had also been the subject of considerable development.

With so many people now involved, there were continual advances in circuitry, finding out all the things that the vacuum tubes could do. Steadily the subject moved from the commercial and military arenas to the domestic. Now, 35 years after Edison had stumbled on his effect, something that had largely been hidden from the average person was set to impact on their lives.

NOTES

1. There are numerous sources, e.g., Dijkstra, H., Short biography of Johann Heinrich Wilhelm Geissler, available at: http://www.crtsite.com/Heinrich%20Geissler.html
2. A report of this meeting was given later in *The New York Times*, February 23, 1896.
3. Bowen, H. G. *The Edison* Effect. West Orange, NJ, The Thomas Alva Edison Foundation, p. 10; Poole, I. Vacuum tube thermionic valve history, available at: http://www.radio-electronics.com/info/radio_history/thermionic-valve-vacuum-tube/history.php
4. Bowen, p. 42.
5. Dijkstra, H. (1987) Annalen der Physik und Chemie Bond 60. Leipzig 1897, available at: http://www.crtsite.com/Annalen%20der%20Physik.html
6. Dictionary of National Biography, Fleming, Sir (John) Ambrose (1849–1945).
7. Howe, W. D. O. (1955) The genesis of the thermionic valve. *Journal of the Institution of Electrical Engineers*, 1:3. This was a lecture given to the Institution of Electrical Engineers in 1954 on the 50th anniversary of the invention of the thermionic valve.
8. Dictionary of National Biography. Fleming, Sir (John) Ambrose (1849–1945).
9. British patent number GB190424850.
10. Howe. Of course, he was an expert witness against the extension of Fleming's patent, so he might just be a little biased.
11. US patent 823402 Filed December 9, 1905.
12. US patent 841387 Filed October 25, 1906.
13. Fleming, J. A. (1919) *The Thermionic Valve and its Developments in Radio-telegraphy and Telephony*. London, New York: The Wireless Press, p. 105.
14. Fleming, p. 65.
15. Vyse, B. and Jessop, G. (2000) The saga of the Marconi Osram valve, available at: http://r-type.org/articles/art-014.htm
16. Bowen, p. 31.
17. Bowen, p. 33.
18. Archut, O., The history of Telefunken AG, available at: http://tab-funkenwerk.com/id42.html
19. Fleming, p. 135.
20. This story is well documented by G. Garratt, Why the French R valve?, *Radio Communication* February 1981, available at: http://r-type.org/articles/art-020.htm; also in van Schagen, P., DE TM-METAL lamp uit 1915, available at: http://www.veronalkmaar.nl/ham/ham-mrt10.pdf,

(in Dutch); Tyne, G. F. J. (1997) *Saga of the Vacuum Tube*. Indianapolis, IN, Howard W. Sams, pp. 192–198.

21. Lornholt, I., Eiffel Tower history, available at: http://www.e-architect.co.uk/paris/eiffel_tower_history.htm
22. Dassapt, P., Raconte-moi la Radio, the history of French radio (in French), available at: http://dspt.perso.sfr.fr/Triode.htm
23. Dassapt.
24. H. J. Round, available at: http://en.wikipedia.org/wiki/Henry_Joseph_Round
25. 'Contact', An Airman's Outings, p. 177, also available at: https://archive.org/stream/airmansoutings00bottuoft#page/n7/mode/2up
26. Katzdorn, M. and Edwin H. Armstrong, Site about the wartime activities of Edwin Armstrong, available at: http://users.erols.com/oldradio/eha7.htm

3

From Wireless to Radio

An inventor is one who can see the applicability of means to supplying demand five years before it is obvious to those skilled in the art.

Reginald A. Fessenden

I do not think that the wireless waves I have discovered will have any practical application. Radio has no future.

Heinrich Hertz

Marconi's objective had always been to set up a commercial wireless telegraphy system. This was modeled on the wired telegraphy network but dealt with the gaps that it could not easily fill, such as maritime communication. His single-minded development of commercial wireless telegraphy was very successful, and soon was standard equipment on ships. However, he and his adviser, Fleming, subscribed to the theory that the discontinuous signal of their 'spark' transmitters was essential for wireless transmission, and that it worked by a 'whiplash' effect.[1]

This mistake did not matter for Morse telegraphy transmissions which were just dots and dashes. With many of the early detectors it was, in fact, an advantage. However, if you wanted to send a more sophisticated signal, such as speech or music, it was totally inadequate. There were those who aspired to send 'telephony' and knew that, to achieve this, a continuous wave was necessary as the 'carrier'. The question was: how to generate such a signal?

Reginald Aubrey Fessenden was a Canadian by birth, but he had drifted south of the border in search of opportunities in electrical engineering. After a checkered career, including working for Edison for a time, be became a professor, first at Purdue University and then at the University of Pittsburgh. Later, he became the general manager of the National Electric Signaling Company. His interest, if not obsession, was to transmit speech by wireless.

His first attempt, in 1900 at Cobb Point, Maryland, used a spinning disk that effectively produced sparks at 10 kHz for his transmitter, but still was not a true continuous wave. The received speech was intelligible, but accompanied by an irritating noise due to the sparks. Still, he had succeeded in transmitting a message over a distance of just over a kilometer.[2]

J.B. Williams, *The Electronics Revolution*, Springer Praxis Books,
DOI 10.1007/978-3-319-49088-5_3

He realized that this was inadequate and spent the next few years looking at various ways of generating a true continuous wave.

Most of these methods derived from the work of the Dane Valdemar Poulsen on continuous arcs. An arc, as used in arc lamps for lighting, when powered by direct current but tuned with a suitable inductor and capacitor, would 'sing' at the resonant frequency determined by these components. Running this in various gases could increase the frequency it generated. Quite useful results were obtained with a number of these.

However, Fessenden's preferred method was the high-speed alternator. This requires a large number of poles and to be spun at high speed. It also must be small in diameter or the forces generated by the peripheral velocity will tear it apart. He asked General Electric to build him a suitable device; they delivered it in 1906, but it could only reach 10 kHz which was nowhere near a high enough frequency. He had it dismantled and rebuilt by the company's workshops to his redesign, and then he was able to reach around 80 kHz which was a usable frequency. He could now build a transmitter for 'telephony'.

On Christmas Eve 1906, along the American east coast, ships equipped with suitable radio receivers were astonished to hear speech and music coming from the earphones instead of the dots and dashes of Morse code. This was part of a 'concert' broadcast by Reginald Fessenden from Brant Rock, Massachusetts. He reported that:

> "The program on Christmas Eve was as follows: first a short speech by me saying what we were going to do, then some phonograph music. The music on the phonograph being Handel's 'Largo'. Then came a violin solo by me, being a composition of Gounod called 'O, Holy Night', and ending up with the words 'Adore and be still' of which I sang one verse, in addition to playing on the violin, though the singing of course was not very good. Then came the Bible text, 'Glory to God in the highest and on earth peace to men of good will', and finally we wound up by wishing them a Merry Christmas and then saying that we proposed to broadcast again New Year's Eve."[3]

It is unlikely that Fessenden fully realized what he had started. His objective for the broadcast was to demonstrate the capabilities of his system. He was interested in 'wireless telephony' in the same way that Marconi was concentrating on 'wireless telegraphy'. Despite his advances in the high frequency alternators they still couldn't really reach the frequencies that would be required of 1 MHz and beyond.

There was a sad end to Fessenden's work, in that the investors in the company tried to ease him out. After court battles he eventually turned his back on the wireless work and concentrated on other fields. As so often, the brilliant inventors are not the ones who develop the idea into a useful system. He was also very restricted by the monopolies already established by the Marconi Company, though his technology was almost certainly superior.

As far back as 1874, Ferdinand Braun (yes, him again) noted an interesting phenomenum. He found that if certain crystals were touched by a metal contact an electric current would flow more easily in one direction through this junction than in the other. Two decades later, the Indian inventor Jagadis Chunder Bose worked his way from the metal filings coherers to a single metal contact on a crystal as a way of detecting radio waves. He was working with much higher frequencies than were subsequently used by Marconi and others, but was only interested in the science and so didn't patent his ideas at the time. Eventually, in 1901, he was persuaded to do so.[4]

No one took much notice of Bose's work, and it wasn't until 5 years later that two patents for crystal detectors for radio waves appeared at almost the same time. General H. H. C. Dunwoody in the USA used carborundum (silicon carbide), while Greenleaf W. Pickard employed silicon as the crystal. Soon other materials such as galena (lead sulfide) were also being used.[5] A springy metal wire was used to press on the crystal as the other contact, and the resulting name of 'cat's whisker' was soon given to this type of detector. In use, the 'cat's whisker' had to be moved to different spots on the crystal until the wireless signal was suddenly heard in the earphone.

The detectors worked in much the same way as the diode or 'Fleming' valve by allowing only the signal in one direction to pass, hence cutting the incoming radio wave in half. By filtering out the carrier radio frequency (which the headphones would do anyway) only the required speech or music that had been 'modulated' on to it remained. It was a very simple and elegant way of detecting wireless signals which didn't require the batteries needed by vacuum tubes (Fig. 3.1).

Once these detectors became available it was now quite simple to produce a receiving set. With the increasing numbers of stations transmitting there was something to listen out for, and it wasn't long before amateurs started to be interested in building their own sets for both receiving and transmitting. In Britain, surprisingly, Parliament had been ahead of the game and in 1904 had given the Post Office, which had a monopoly on communications, powers to issue licences for wireless operation.

At the beginning, these licenses were mostly issued to Marconi and his competitors for commercial operation, but more and more were issued for 'experimental purposes' and were largely for amateurs. There was a huge rise in these from 1910 to 1912 (Fig. 3.2). In Britain, components, or even complete sets, could be bought from such places as Hamleys[6] and Gamages.[7]

A book on *Wireless Telegraphy for Amateurs* was first published in 1907 and revised many times in the next few years.[8] In April 1911 the Marconi Company published a magazine called the *Marconigraph*. This proved so popular that it grew from 16 to 52 pages in

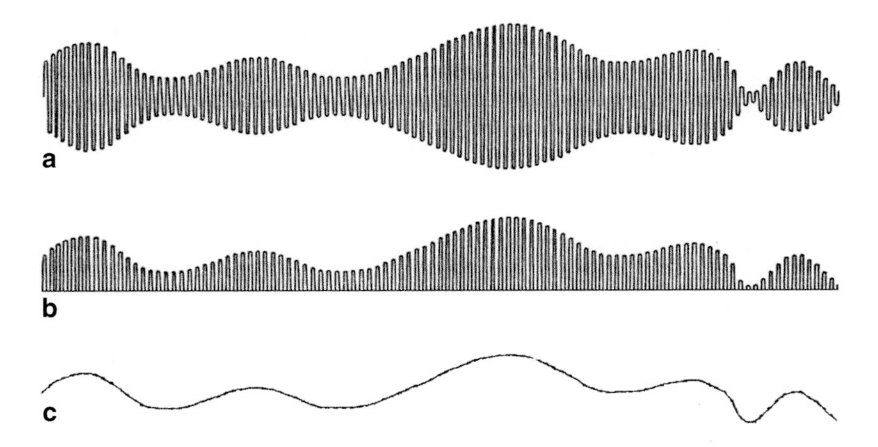

Fig. 3.1 Modulation of a radio wave. (**a**) modulated carrier wave; (**b**) cut in half by the detector; (**c**) the signal after filtering out the carrier. *Source:* Author

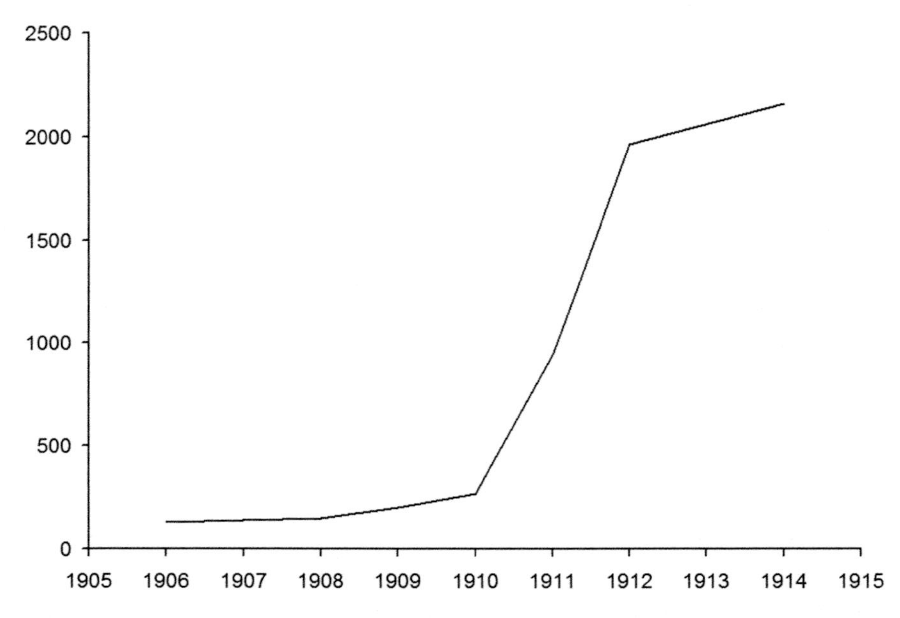

Fig. 3.2 The number of licences issued by the UK Post Office for wireless operation 1905–1914. *Source:* Author[9]

its first year. The following year it changed its name to the more encompassing *Wireless World* and its extent rose to 96 pages.[10] Interest in wireless communication was exploding as shown by the number of wireless clubs, including in 1913 the London Wireless Club, which later became the Radio Society of Great Britain. Just as it was really getting going, the First World War intervened, and on its outbreak the majority of the licence holders received a telegram ordering them to stop operating and hand in their equipment.[11]

Charles David Herrold's father was a bit of an inventor, so it was not surprising that his son should be interested in science and mechanics and study at Stanford University in California. Luck was not with him—his electrical engineering business was destroyed in the 1906 San Francisco earthquake. He turned to teaching in a college for a while before, on January 1, 1909, he set up his own 'Herrold School of Radio' in San José.[12]

To generate interest in his school, Herrold set up a transmitter to send out what amounted to advertisements. To encourage people to listen he also transmitted news and music from gramophone records. Initially, he used a spark transmitter, but the quality was not good and he soon changed to an arc transmitter of his own design. To get listeners he would go out and install crystal sets up and down the valley. From around 1910 he was transmitting on a regular schedule, well before the Radio Act of 1912 introduced licences and required all transmissions in America to have a 'call sign'.

Purists say that he wasn't 'broadcasting' because his listeners were amateurs and not ordinary members of the public.[13] This seems a bit nit-picking, as he was certainly moving over from 'communication' to 'entertainment'.

Like everyone else, he had to shut down when America entered the war in 1917. Afterwards, it took him some time to re-establish as the government introduced designated frequency bands. His was 360 m (833 kHz) which was a higher frequency than the 600 m (500 kHz) his arc transmitter was capable of reaching, so he had scrap his equipment and the station to start again.[14]

Over the next few years a few people were transmitting 'programs' but they were still amateurs doing it on a haphazard basis. Some transmissions of weather information also took place and altogether this provided enough traffic on the air to interest keen young men in building or obtaining receivers to experiment themselves. Despite this, the idea of 'broadcasting' had still to take root.

Meanwhile, Marconi had realized his error concerning the use of continuous waves for transmission, and was developing wireless telephony systems. In 1913, he bought out the French company, Compagnie Universelle de Telegraphie et de Telephonie sans Fil, in order to obtain the patent rights to the Goldschmidt generator outside Germany.[15] This was an interesting device in that it was able to multiply the basic frequency it generated inside the device. The result was that the speed at which it spun didn't need to be as high as the American GE 'Alexandersen' alternators based on Fessenden's work.

By the next year, Henry Round's work on vacuum tubes was providing results, and though they still couldn't produce the power achieved by the generators and arcs, they could more easily reach higher transmitting frequencies. In any case, the generators were not suitable on ships, Marconi's main business area, as the movement produced gyroscopic effects that were likely to damage the high speed machines. In 1914, this work produced results, and successful trials were undertaken with the Italian Navy.[16]

Then the First World War intervened, and with it the enormous developments that war so often brings. Chapter 2 dealt with the tremendous advances in the vacuum tubes, but this then drove improvements in the transmitters and receivers. Once there was the ability to amplify signals, everything became a lot easier. In addition, many other things could be achieved with vacuum tubes and most of the basic circuit configurations were invented over quite a short space of time. By the end of the war there was not only a vast amount of radio equipment being used, but a large number of people familiar with using it.

The emphasis was still on using wireless for communication point to point either by telegraph using Morse code or, increasingly, by telephony. However, the idea of transmitting a signal to be picked up by many listeners was in the air. This had most clearly been expressed by David Sarnoff who worked for the American Marconi Company. He said: "I have in mind a plan of development which would make radio a household utility in the same sense as the piano or phonograph."[17] However, most people in the industry were concentrating on the war and so his ideas were a little premature.

With the war over, the situation was different. By this time, Henry Round had succeeded in making high power transmitting vacuum tubes and in March 1919 a wireless telephony system linking Europe to America was set up. Meanwhile, the UK's Postmaster General needed to be pressured to allow amateurs back on the air; it wasn't until May that he relented, but the new licences for transmission were only issued for experimental purposes, and were not to interfere with military communications.[18] They were only allowed to use 10 W of power and use frequencies too high for high speed alternators or arc transmitters. Despite these limitations, interest in the subject grew.

In January 1920, a 6 kW telephony transmitter was installed at the Marconi works in Chelmsford for testing receivers and reception. The engineers soon became tired of reading from railway timetables to produce a test signal and turned to live music which was heard by amateurs and ships operators up to 1450 miles away. The next month they introduced a news service. They thought that the future of broadcasting lay in information.[19]

Over in Holland, the delightfully named Hanso Henricus Schotanus á Steringa Iderza had a more commercial objective in mind. He was trying to sell radio receivers, and he thought that if he transmitted speech and music there would be something for people to listen to, and then they were more likely to buy his receivers. On November 6, 1919 he started transmitting under the call sign PCGG.[20] He was pushing his luck as he only had an experimental licence, but no one stopped him. By November 1921, he was able to convert this to a full broadcast licence and he received sponsorship from the British newspaper, the *Daily Mail*, to transmit some of his output in English, as his signals could be received in England. The *Mail*, interested in technical advances, was keen to promote the use of radio.

In 1920, British Thomson Houston started manufacturing radio sets as there were now some stations that listeners could access, although it was still a very minority interest. Marconi, though mostly concentrating on their telegraphy and telephony business, decided on a publicity stunt. On June 15, 1920 they arranged for the singer Dame Nellie Melba to perform live and broadcast it from their transmitter at Chelmsford. This, too, was sponsored by the *Daily Mail*. One listener said that the transmission was 'perfectly wonderful' and that 'there must be a great future for wireless concerts'.[21]

Unfortunately, the Postmaster General (PMG) had other ideas. Though a number of companies had been given licences to make experimental broadcasts, he felt it was getting out of hand and, using the excuse that they were interfering with military communication systems, shut them all down. This position was not tenable for very long as many broadcasters were setting up in other countries, particularly America, and the growing number of set manufacturers could see the advantages of local broadcasting.

In February 1922, the Marconi Company was starting to hedge its bets though they were still unsure that the radio amateurs really constituted a market. They persuaded the PMG to issue them an experimental licence and started transmitting from one of their sites at Writtle under the call sign 2MT. The station was run by engineer Peter Eckersley, and broadcast everything from pieces performed by local people, songs, competitions and even a lonely hearts club. It rapidly built up a sizable audience and Marconi realized that they were on to something.[22]

By May, they had obtained another licence to transmit from the Marconi building in London under the call sign 2LO. With the number of amateur receiving licences beginning to climb rapidly, and numerous companies applying for broadcast licences, the Postmaster General, Mr. Kellaway, acted. His proposal was that the set makers should get together and set up a single broadcasting service. It took many months of negotiation to agree on a scheme, which was complicated by Marconi's patents and the need to have a means of financing the service. There was a slight suspicion over some of the arrangements when Kellaway ceased to be the PMG and promptly became a director of the Marconi Company.[23]

Eventually, on 14 November, financed by six electronics manufacturers, the British Broadcasting Company (BBC) was inaugurated and began transmitting on the medium wave band from the roof of Selfridges in London, reusing the call sign 2LO.[24] It reached an audience of around 18,000. The next day two further transmitters started up: 5IT in Birmingham and 2ZY in Manchester. Gradually, more were added to increase the coverage of the country.[25] In July 1925, the high power Long Wave Daventry transmitter came into service. Listeners needed a receiving licence, but that could easily be obtained from the local Post Office in exchange for a small sum of money.

The rate of take-up of these licences was extraordinary. Within about 2 years of the creation of the BBC, their number had passed the one million mark, and by the start of the Second World War there were nearly nine million licenced sets. At the beginning many of these were crystal sets, but gradually the advantages of the vacuum tube sets meant that they took over. Initially these were mostly battery powered, though in the late 1920s 'battery eliminators' appeared allowing the radio to be run from the mains where it was available. These in turn were superseded by all mains sets.

What is also remarkable is that in Britain the rise of radios mirrored that of houses wired for electricity so closely (Fig. 3.3). With so many sets either needing no power—crystal sets—or run from batteries, it is perhaps surprising, though no doubt listeners soon got tired of their limitations and migrated to mains sets when they could.

The dramatic take-up showed that there was clearly a thirst for information and particularly entertainment. Radio provided this in the home, and contributed to the idea that it could be obtained without going somewhere and partaking in a 'mass' entertainment. As a result, it began to change society. Even those who only turned it on to listen to the

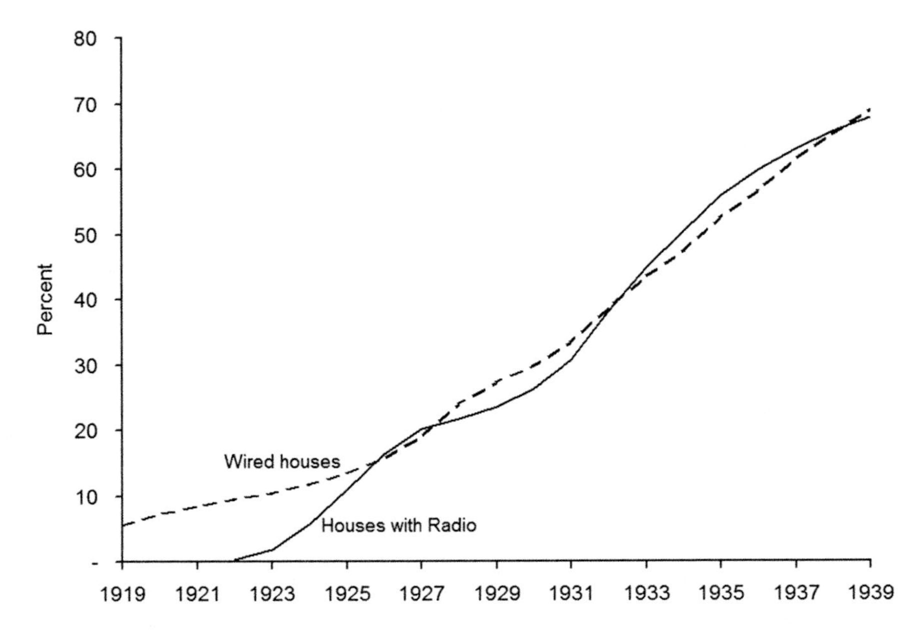

Fig. 3.3 Close correlation between the percentage of households with a radio and with electricity, though the numbers with electricity is only an estimate. *Source:* Author[26]

dance bands would naturally hear the news bulletins and become much better informed. This particularly applied to women stuck at home, who could pick up quite an education without really noticing.

While at this time most people would have called the device a 'wireless', the international word 'radio' was beginning to creep in. In 1923, the BBC issued its own guide to the programs and named the *Radio Times*. Despite the severe Director General John Reith, with his motto of 'inform, educate and entertain', trying to keep it 'highbrow', it expanded into a general service, broadcasting everything from news, to weather forecasts, to football results, music and talks, as well as panel games.[27] It was so successful that in 1927 it was taken over by the government and became the British Broadcasting Corporation.

Official attitudes took some time to change. In 1923, the King didn't allow the broadcasting of the wedding of the Duke of York (later King George VI). By the next year the King made a broadcast himself, and in 1932 he reluctantly made a Christmas broadcast. Only 4 years later King Edward VIII made his abdication speech to the nation on the radio, and it was natural for Chamberlain to announce the outbreak of the Second World War the same way. It provided a way to effectively 'speak to the nation', a thing that had never existed before.

It is often said that in the Second World War everyone in Britain had a radio and so the government could communicate that way with the people. It isn't strictly true. As Fig. 3.3 shows, only about two thirds of homes had a set. What is true, though, is that this was sufficient penetration for those without a radio to soon be told important news by their neighbors who did have one. Thus it was still able to perform the task of disseminating information to the whole country, quickly and simply.

So it was that a technology which had grown up as wireless telegraphy and later telephony as a communication medium for government, business and the few, had metamorphosed into something that served the mass of people. No one at the beginning, when those first shaky experiments proved that electromagnetic signals really could travel through the air, would have believed where it was to lead, and quite the effect it would have on everyday life. They would have been even more surprised to find that this was only the beginning.

NOTES

1. Belrose, J. S. (1995) Fessenden and Marconi: Their differing technologies and transatlantic experiments during the first decade of this century, available at: http://www.ieee.ca/millennium/radio/radio_differences.html
2. Fessenden, R., Inventing the wireless telephone and the future. Canadian Institute of Electrical and Electronic Engineers, available at: http://www.ieee.ca/millennium/radio/radio_wireless.html
3. Fessenden, H. M. (1940) *Fessenden, Builder of Tomorrows*. New York, Coward-McCann, p. 153, also available at: https://archive.org/stream/fessendenbuilder00fessrich#page/n7/mode/2up
4. GB patent 1901 18430.
5. Pearson, G. L. and Brattain, W. H. (1955) History of semiconductor research. Proceedings of the Institute of Radio Engineers, December 1955. Walter Brattain was one of the inventors of the transistor.

6. Advert in *Wireless World*, April 1913. Hamleys is now better known as a toy shop.
7. Also from an advert in *Wireless World*. Gamages was a sort of department store for boys' toys, featuring model trains and such like.
8. Howgrave-Graham, R. P. (c. 1912) *Wireless Telegraphy for Amateurs*. London: Percival Marshall & Co.
9. Data from UK Post Office, Report of the Postmaster General on the Post Office, for years 1906–14.
10. *The Marconigraph*, March 1913.
11. Postmaster General's report to the UK Parliament, 1914–1915.
12. Greb, G. B. (1958) The golden anniversary of broadcasting. *Journal of Broadcasting*, 3:1.
13. Baudino, J. E. and Kitross, J. M. (1977) Broadcasting's oldest stations: Examination of four claimants. *Journal of Broadcasting*, 21:1.
14. Greb, G. B. (1958) The golden anniversary of broadcasting. *Journal of Broadcasting*, 3:1.
15. *The Times*, September 27, 1913.
16. Coursey, P. R. (1919) *Telephony Without Wires*. London: The Wireless Press, p. 366.
17. Sarnoff, David, available at: http://www.sciencemuseum.org.uk/online_science/explore_our_collections/people/sarnoff_david?agent=sc-cc3d926d211c42adb7c376fd2218bdaa
18. Hansard. House of Commons Debate November 18, 1919, vol 121 c810
19. Marconi Calling Archive, available at: http://www.marconicalling.co.uk/introsting.htm
20. Rotterdams Radio Museum, available (in Dutch) at: http://www.rotterdamsradiomuseum.nl/index.php?option=com_content&view=article&id=71&Itemid=207
21. *The Times*, June 16, 1920.
22. Chelmsford Amateur Radio Society, 2MT and New Street, available at: http://www.g0mwt.org.uk/new-street/index.htm
23. Hansard. House of Commons Debate November 30, 1922, vol 159 cc1055-6.
24. The British Postal Museum and Archive, The Post Office and British Broadcasting, available at: http://postalheritage.wordpress.com/2009/09/29/the-post-office-and-british-broadcasting/
25. Parker, B., The history of radio: The valve era, available at: http://www.historywebsite.co.uk/Museum/Engineering/Electronics/history/ValveEra.htm
26. Number of houses wired for electricity from Hannah, L. (1979) *Electricity before Nationalisation*. London: Palgrave Macmillan, p.188, and checked from other sources, numbers of households from census figures and interpolating between them. Number of radio licences from Hansard for the first couple of years and then BBC reports thereafter.
27. The BBC Story, The BBC, available at: http://www.bbc.co.uk/historyofthebbc/innovation/20s_printable.shtml.

4

Seeing by Electricity: Development of Television

Television? The word is half Greek and half Latin. No good will come of it.

C. P. Scott, Editor, *Manchester Guardian*

Well gentlemen, you have now invented the biggest time waster of all time. Use it well.

Isaac Shoenberg following the successful demonstration of electronic television, 1934

John Logie Baird invented television. Well, that's what is often said. In reality, it is a lot more complicated. Television was the result of a series of steps, and a few dead ends, taking place over the best part of a century and involved a large number of people. First of all, it requires the pictures somehow to be broken up to produce a signal. Also essential are means to turn light into electricity, and the opposite at the other end. Then, of course, radio had to be developed to a point where it could carry the signal.

Even the word had to be invented. Once the telephone had arrived there was a feeling that the next step should be 'seeing by electricity', but the concept didn't have a name. It was only in 1900, at the IV International Electrotechnical Congress in Paris, that the Russian engineer Konstantin Persky delivered a speech in which he used the word 'television' for the first time.[1] Despite C. P. Scott's distaste for its impure etymology, the word gradually caught on, and it had become accepted by the time such a device was able to be made.

The invention of the telephone partly stemmed from mimicing the diaphram in the ear, so it was natural to think about whether examination of the physiology of the eye would lead to some way of dividing up a picture so that it could be sent out. Unfortunately, the eye works by having a vast number of sensors, and this line of thought led nowhere. A totally different approach was needed.

Alexander Bain was the son of a crofter from Caithness, Scotland, who became apprenticed to a clockmaker.[2] In 1837, he moved to London to advance his trade, but soon became interested in the new field of electricity. His first invention was a way of electrically synchronizing two pendulum clocks that were remote from each other. The next step was to use this as part of a device that could send a copy of a picture over a telegraph line. He took out a British patent in 1843, though he doesn't appear to have ever built his device.[3] It could only copy something that was made up of electrically conductive and insulating

J.B. Williams, *The Electronics Revolution*, Springer Praxis Books,
DOI 10.1007/978-3-319-49088-5_4

sections and the example he gave was a metal typeface but, having said that, it was brilliantly simple. A pendulum runs a small finger over a strip of the frame holding the partially conductive sample to be copied. After each swing a clockwork mechanism drops the frame a small amount and the pendulum's swing runs over a further strip. The process is continued until all the strips have been covered.

At the receiving end is a near identical device but here the finger on the pendulum passes over a material which changes color when electricity flows through it. As it does this point by point, it builds up a copy of the image at the other end. This is the important process we now call scanning, and it seems that Alexander Bain invented it.

Though it was some while before satisfactory means of sending images were produced, the first essential step had been taken. Here was a means of breaking up a picture and sending it bit by bit simply by scanning a strip at a time and working down the image until all of it had been covered. Bain had also understood another key point: the scanning at the transmitting and receiving ends needed to be kept in step or the result would come out a jumble. Some form of synchronizing information also needed to be sent.

In 1866, Willoughby Smith and his assistant Joseph May were involved with a cable-laying operation at sea. In setting up an arrangement to communicate from the shore to the ship while the electrical tests were conducted they needed a very high value resistor. They tried thin rods of selenium, but found the resistance unstable. In 1873, Willoughby Smith wrote to the Society of Telegraph Engineers reporting the results of the investigations into this variability. It was sensitive to light.[4]

It was a short step from this to producing a device which could vary an electric current in step with a change of light falling on it. In the early 1880s this led to a series of proposals on how to build a device for 'distant vision', all using selenium cells, which either suggested large numbers of cells, mimicking the eye, or a scanning system. The most significant of these was by the German Paul Nipkow who patented his system in 1884.[5]

Nipkow proposed a disk with a spiral of holes in it. With a selenium cell on one side and the item to be scanned on the other, the effect when the disk was spun was to produce a series of scans which worked their way down the subject. The cell output thus was a series of signals corresponding to the light intensity at the points on the scan. At the receiving end, a similar disk was placed between the viewer and the light source which varied in time with the output of the selenium cell. His arrangement for adjusting the light source almost certainly wouldn't have worked and there is no evidence he ever tried to build the device.

In 1887, Heinrich Hertz, as part of his investigations into the existence of electromagnetic waves, discovered that the size of his sparks was affected by light—being a meticulous researcher he showed that ultra-violet light caused this. He published his results and, not wishing to be distracted from his main work, effectively passed it over to others to investigate.[6]

Wilhelm Hallwachs took up the challenge and soon proved that light did affect the electric charge on a metal plate. Many others, including Arrhenius, E. Wiedemann, A. Schuster, A. Righi and M. A. Stoletow, all contributed to the background of the interaction of light and electricity.[7] Two German schoolteachers, Julius Elster and Hans Geitel, took this further and found that alkali metals such as sodium and potassium produced this effect with visible light. This led them to make a photocell.

The Elster-Geitel photocell was a simple device. In a glass envelope was a curved cathode of a film of sodium potassium alloy and a metal anode. With a voltage across the device no current would flow when it was in the dark, but the current would increase when light shone onto it. Further research showed that alkali metals further up the periodic table, such as rubidium and caesium, gave even better results. The great advantage of the photocell was that it responded almost instantly to changes of light compared with the sluggish response of the selenium sensor.

Once into the twentieth century it seemed likely that a serious attempt would be made to produce a practicable television system. It was a Russian of Swedish descent called Boris Rosing (Russians feature strongly in this story) who was the first, though he talked about 'electrical telescopy' in his 1907 patent.[8] He didn't use a Nipkow disk as the scanning arrangement in his 'camera' but a couple of multifaceted mirrors at right angles to each other and a photocell as the detector.

The receiver was a form of Braun's tube (see Chap. 2). The brightness of the spot was controlled from the output of the photocell, and the horizontal and vertical scans from signals generated by magnets and coils attached to the rotating mirrors. By 1911 Rosing was able to demonstrate a practical device. The disadvantage was that it needed three signals between the transmitter and the receiver, but this wasn't too great a drawback if it was used for what he envisaged, which is what we would now call closed circuit television (Fig. 4.1).

In practice, Rosing's device could do little more than transmit silhouette images, but at 40 pictures a second it was faster than the 12 images a second that had been found necessary to give a reasonable sense of motion.[9] Though it was recognized at the time that it wasn't a complete solution, it was a useful step along the way. Rosing had understood the need for a receiver that could respond rapidly and that an electronic solution was better than a mechanical one. What he didn't know was how to produce a transmitter following the same principle.

Almost at the same time there was a man in Britain who thought he knew the answer. He was A. A. Campbell Swinton, a consultant in London who had helped Marconi on his way. As his contribution to a discussion on this subject he proposed that 'kathode rays', as he called them, should be employed at both the transmitter and receiver.[10] Some form of Braun's

(a)

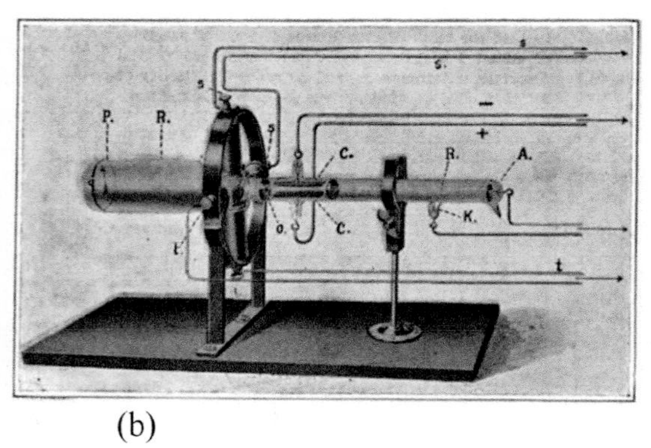

(b)

Fig. 4.1 (**a**) Rosing's mirror drum transmitter and (**b**) his Braun tube receiver. *Source:* Scientific American 23 Dec 1911 tube was needed at both ends. He based his reasoning on a simple piece of arithmetic. To get a reasonable picture there needed to be something like 150,000 points (pixels in today's nomenclature) sent at least ten times a second (in practice rather more). Thus, around 1,500,000 million bits needed to be sent every second. Even if quality is seriously sacrificed, and this is reduced to say 150,000 per second, it is still far too fast for mechanical systems. He argued that only the electron beams in cathode ray tubes could move at these speeds.

In 1911, Campbell Swinton outlined his ideas in more detail, and proposed a transmitter where light fell onto a grid of separate light-sensitive cubes which were then 'read' by the scanning electron beam of the tube.[11] The scanning signals and the intensity were sent to the receiver, which was much like Rosing's system. He admitted that it was only an idea, and that it would take a well-equipped laboratory to develop it. Unlike most of his contemporaries, he was remarkably clear about what was needed.

There the matter largely rested until after the First World War. During that time, electronics made huge strides and, with the coming of broadcast radio in 1922, interest in the possibility of television gained ground. What was now the objective was not just 'seeing at a distance' but broadcast television in a similar fashion to radio. Inventors started to see if they could produce such a system, but they hadn't taken note of Campbell Swinton's comments.

Probably the first was John Logie Baird in Britain, who in April 1925 demonstrated a silhouette system in Selfridges store in London. It was very crude with 30 lines and just 5 frames/s. Basically, he had taken Nipkow's system and made it work. The main difference was that in the receiver he used a neon lamp, invented since Nipkow's day, which could change its light level fast enough to generate the picture. As he himself said, it was surprising that no one else had done it.

In January 1926, Baird demonstrated a working system, able to show shades of grey, to the Royal Institution in London. The picture, at the receiving end, was very small, only 1.5 × 2 in. and a dull pinkish color, but a head was recognizable.[12] Of course, at this scanning speed the flicker was very bad, but it definitely worked (Fig. 4.2).

Another pioneer was American C. F. Jenkins. In June 1925 he demonstrated a system which showed a slowly revolving windmill but it was only in silhouette.[13] However, the information was carried by a radio signal between the transmitter and receiver. As scanners at each end he used elegant bevelled glass prismatic discs, but these were limited in the speed that they could be rotated, so only a picture of a few lines could be achieved.

Also in the US there appeared a serious contender—Bell Labs. In 1927, they demonstrated a system using both wire and radio to connect the two ends. It was a mechanically scanned arrangement and achieved a 50 line picture at 18 frames/s. It was competent but though it had some merit for showing a picture of the person on the other end of the telephone it was nowhere near suitable for real pictures.

Baird didn't stand still. In 1927, he broadcast live television pictures from London to Glasgow by telephone line and, in 1928, from London to New York using short-wave radio.[14] This was where his skill lay: in publicizing himself and television in general. Using the enthusiastic support he had built up with the public, in September 1929 he was able to persuade the BBC to broadcast his television, but it was only after 11 p.m. when the normal radio programs had shut down.[15] This used 30 lines but at 12½ frames/s; just enough to give a reasonable impression of movement, but the transmitter could only broadcast sound or vision. The temporary solution was to have 2 min of vision followed by the sound.

Fig. 4.2 Baird with an early television scanner showing the Nipkow disk and the dummy's head he used. *Source:* Burns R W, *The First Demonstration of Television*, IEE Electronics and Power 9 October 1975

Six months later, signals were broadcast from the new London Regional Station at Brookman's Park, which could transmit both sound and vision simultaneously. By this time, Baird had found backers and set up the Baird Television Company, which started selling the receivers, or 'televisors' as he called them, manufactured by the Plessey company.[16] Some of these were complete but around a 1000 kits were sold because many of those interested were radio experimenters. The 'televisors' were used in conjunction with a radio receiver; they were based around a Nipkow disk of about 20 in. in diameter and produced an image about 1.5 in^2. This was viewed through a lens in a rectangular opening to one side of the front of the set. Unlike most other experimenters Baird used vertical scanning instead of horizontal, which he claimed gave a better picture at this resolution (Fig. 4.3).[17]

While this was going on, others had recognized Campbell Swinton's message or come to the same conclusion. What was needed was an electronic camera based on some combination of the cathode ray tube and the photocell. First into the field was Vladimir Zworykin, another Russian, who even more importantly had been a student of, and assistant to, Boris Rosing in St Petersburg.

In 1919, Zworykin went to America and it was there in 1923 that he filed a patent for a television system.[18] It was remarkably like Campbell Swinton's proposals and was so contested that patent wasn't granted until 1938! He worked on trying to produce a system while employed by the Westinghouse Labs, but they weren't interested in his rather unpromising results.

Philo T. Farnsworth was a farm boy turned inventor. By 1926, he had talked some backers into supporting his efforts financially and setting him up in a laboratory. In 1927, he filed a patent for his television system employing an electronic tube camera, which he called an 'image dissector'.[19] The following year he was able to demonstrate a crude

Fig. 4.3 (*Left*) Baird commercial 30 line 'televisor' and (*right*) the quality of the image produced in the opening. *Source:* Garratt G R M and Mumford A H, *The history of television*, Proceedings of the IEE - Part IIIA: Television (1952) Volume: 99, Issue: 17; http://www.ambisonic.net/tv2london1.html

electronic television system which displayed 20 pictures per second and was fast enough to show movement.

Both these electronic systems and all the mechanical transmitter systems suffered from a fatal flaw. They scanned the subject and what was seen at each point was only the tiny piece of light viewed at the instant that the scanning system passed over it. The result was that the camera was very insensitive and vast amounts of light were needed to get a reasonable picture. The more lines and frames per second that were used, the worse the problem became.

What was needed was something that collected the image for the whole frame and then was read very quickly as the scanning system passed over it. In theory this was more sensitive by the number of points in the picture per frame. This could be many thousand times. The person who realized this was a Hungarian, Kalman Tihanyi. In 1926, he patented his ideas, refining them in 1928.[20] This time he took out patents in a number of countries, including Britain, and his essential improvement became widely understood.[21] The American version of his patent, after much discussion, was later assigned to RCA.[22] Tihanyi's idea was crucial in order to make a practical electronic television camera.

In 1929, Zworykin moved to RCA where the boss, David Sarnoff, another Russian émigré, really was interested in the potential. He started to develop a scheme that seemed remarkably like a development of Tihanyi's, which he called the Iconoscope.[23] Time and effort were put into not only the camera, but a complete system including a cathode ray tube receiver.

In Britain, the record company HMV was interested in widening the field of home entertainment. In around 1930 they became interested in the possibilities of television. What was clear to them was that there was little entertainment value in the low definition system used by Baird and others. They also felt that mechanical receivers had no place in the home.[24]

By 1931 they also appreciated that, apart from the camera, there were two further problems. The implications of Campbell Swinton's calculation of the number of picture elements, in what they called a high-definition system, was that it required at least a 100,000 points.[25] This meant that the radio system needed a wider bandwidth to be able to handle this rather than the few kHz of the normal radio broadcasts. This required a very high frequency (VHF) transmission of a wavelength around 50 m (60 MHz). This was new territory for broadcasters, and a considerable concern as the coverage area of the transmitter was reduced to 'line of sight', which would mean many more transmitters to cover the country.[26]

They foresaw a further difficulty: the speed requirements at the receiver. It was quickly realized that only the cathode ray tube had any real chance of achieving this, though a lot of development work would be needed to produce a product that was in any way comparable with a home cine film. By around 1932, virtually all those interested in developing television systems had come to the same conclusions.

In 1931, HMV merged with the Columbia Record company. Crucially this brought two men to the joint research team, Isaac Shoenberg, yet another Russian, and his brilliant electronics engineer, Alan Blumlein. Shoenberg, originally from the Ukraine, had studied in Kiev before being involved in setting up radio systems in Russia for what later became the Russian Marconi company.[27] He was to head the joint team. It was an inspired choice.

Shoenberg and Blumlein faced a series of problems. First, Shoenberg had to persuade his management to invest a considerable sum of money in the development of a television system. At the time, it was far from clear that a system of adequate quality was achievable, and it was his confidence in this that persuaded them. His next task was to recruit scientists and engineers to build up the teams to tackle the various parts of the problem. Part of his skill was in choosing the right people.

Shoenberg sidestepped the transmitter problem by using his contacts with Marconi to hire a VHF one. They had quite enough other problems to solve. One team was set to work to produce improved cathode ray tubes for the receiver. Another dealt with the electronic circuit problems, particularly dealing with the wide range of frequencies or the 'bandwidth' of the signals. Last, and probably the most important, was a team headed by a new recruit, Dr. J. D. McGee, to tackle an electronic camera, as he was convinced that mechanical systems would never be satisfactory.[28]

Work started in 1932 with a cleverly constructed photosensitive disk mounted inside an evacuated glass sphere. To one side was a window to bring in the light from the subject being viewed. On the same side, at an angle, was an electron gun arranged with deflecting coils to produce the scan in both horizontal and vertical directions. The disk took up an image of electrical charge corresponding to the light falling on it. The scanned electron beam would discharge it spot by spot and this signal could be read off electronically.[29]

This led to the development of the Emitron tube, which would be the basis of Shoenberg's camera. The tube was only around 5% efficient compared with the theoretical maximum, but this was an enormous improvement over previous systems, and it could work outdoors in all normal light levels and indoors without excessive lighting.

There has been some controversy as to whether this was really an American design from RCA, particularly as the early Iconoscope and Emitron tubes look remarkably similar. Also, HMV had a patent exchange arrangement with RCA and this was carried over to EMI. While Shoenberg and his team undoubtedly had access to RCA patents, (and RCA to theirs) they had already started work when Zworykin published his design in 1933.[30] The descriptions of their development by the EMI team leaves little doubt that they produced the Emitron separately, and it is backed up by the fact that it took RCA 3 years longer to get a working television system into use.[31]

Meanwhile, at the Baird Company there had been a management shake-up and Baird himself was sidelined to a laboratory. The company, under its new chief engineer Captain A. D. G. West, moved to premises in the Crystal Palace just outside London, and started working towards a high definition system.[32] To overcome the problem of lack of illumination with the transmitter they had already turned the system around. Instead of having the subject lit as normal and the sensor on the opposite side of the Nipkow disk, they put a bright light source shining through the disk and then the photocells on the same side as the person being televized.

The effect was to produce a spot of light which was rapidly moved over the subject. This was known as the 'flying spot scanner'. To get sufficient illumination they used the arc light lamp house from a film projector.[33] The person being televized had to be in the dark with just this brilliant light scanning over them.[34] To get sufficient lines in the image—and they were seeking 240—the holes became small and it was necessary to run the Nipkow disk in a vacuum at very high speed.[35] Even with this arrangement it was impossible to televize a whole scene in a studio (Fig. 4.4).

With three partners, Baird had set up a company in Germany, Fernseh AG, to develop his system there. They ran into exactly the same problems, but came up with an ingenious solution. The idea was to record the scene with an ordinary film cine camera. The film then was fed straight out of the camera into a high speed developing system, and while still wet it was scanned by a variant of the flying spot scanner and turned into a television signal.[36]

The Baird company took up this idea and produced their own version. Using aggressive chemicals involving cyanide, the time delay was reduced to 60 s, but this still meant that the sound had to be recorded on the film and played back when the film was read. It worked, but it was large and cumbersome and was inevitably fixed in one spot, the actors having to come to the camera instead of the other way round as was already the norm for filming.

Fig. 4.4 Inside the claustrophobic Spotlight Studio showing the photocell arrays. *Source:* http://www.transdiffusion.org/2003/03/13/fools_2

The Baird low definition transmissions had, however, performed their purpose: they had raised the level of interest in television. This was sufficient to persuade the UK's Postmaster General, Sir Kingsley Wood, to set up a committee under Lord Selsdon to examine whether a television service should be set up.[37] At the end of January 1935 the report was published; it recommended the setting up of a service with a minimum standard of 240 lines at 25 frames/s.[38] The systems from two companies were to be trialled: EMI, which had now formed a joint company with Marconi who would provide the transmitters, and of course Baird.

The now united EMI–Marconi team had reached this standard with a 243 line system, on which they used a clever arrangement to minimize the flicker. In one-fiftieth of a second they transmitted half the lines and then in the next fiftieth the intermediate lines. This kept the required bandwidth of the signal down but gave a picture at effectively 50 frames/s and so had less flicker. They had found that as the brightness of the receiving tubes was increased this became noticeable at 25 frames/s.

Seeing how the committee was thinking, Shoenberg took a calculated gamble. He believed that they could go further to 405 lines and that this would produce a standard which would last for some time.[39] The downside was that nearly all the parts of the system had to be redesigned to get this increased specification. There is often puzzlement as to why he did this, but in my mind the reason is simple. He wanted EMI's to be the superior system, and though at first sight 243 lines is as good as 240 there was a problem. In the electronic system some of the notional lines were lost at the end of each frame, and in practice there would have been somewhat less than 240 actual lines.[40] It was thus arguable whether it would have truly met the specification.

The BBC, which was going to run the new television service, decided on Alexandra Palace in north London as their chosen site. It was on a hill so the antenna on top of the tower didn't need to be so high to get a reasonable coverage area.[41] The building itself provided enough space for a double set of studios and equipment so that Baird and EMI Marconi could each have one. By mid-1936, the building work was sufficiently advanced for the two companies to start assembling their equipment.

The opening of the television service was scheduled for November 1935, but suddenly there was a call from the organizers of the trade show, RadiOlympia, to have a television demonstration. This was at the end of August so there was a frantic rushing around to get equipment ready and some sort of show that could be broadcast. It was called 'Here's Looking At You', and went out twice a day for 2 weeks, using the Baird and EMI Marconi systems on alternate days.

The service started to operate on November 2, 1935. The opening ceremony was broadcast on the Baird system (who had won the toss) and repeated on the EMI Marconi (Fig. 4.5). After that one system transmitted for a week before swapping over. The limitations of the Baird system soon became apparent. In addition to the problems of the fixed cameras, the system was overly sensitive to the red light used in the studio for announcements, so the three presenters, Elizabeth Cowell, Jasmine Bligh and Leslie Mitchell, had to use peculiar blue and black makeup.[42]

Though originally the trial was due to last 3 months, at the end of January 1936 the Baird system was shut down. There was really no comparison between the systems; the electronic one was clearly better even though it was at the beginning of its development

Fig. 4.5 (*Left*) The opening ceremony in November 1935, with the Emitron camera. (*Right*) Douglas Birkinshaw with an Emitron camera outside Alexandra Palace. The antennae is in the background. *Source:* http://www.teletronic.co.uk/tvera.htm; http://www.transdiffusion.org/tv/baird/fools

while the mechanical system probably was nearly complete.[43] However, the Baird equipment was superior in the film scanners for broadcasting cine films , as the electronic cameras didn't like the rapid switching between bright light and dark that the normal film shutter method produced.

The start of the service was heralded as a great success, even though there were only some 500 receivers in use, and claimed as the first high definition television system in the world.[44] Certainly it was well ahead of America where RCA demonstrated their system in 1939.

In Germany the situation was different. In addition to the Baird input to Fernseh, Telefunken had links with RCA and had taken the Iconoscope design and developed it into a usable device. The Berlin Olympic Games began on August 1, 1936, and was televized. The German systems by this time had reached 180 lines at 25 frames/s and both Farnsworth and Telefunken electronic cameras were being used.[45] However, there were very few receivers, perhaps 600, and most of these were in the hands of party officials.[46] To give the public a chance to see television, public viewing halls called 'Fernsehstube' were set up, often next to Post Offices, where for a small fee anyone could go and view the broadcast.[47] Thus the argument as to who was first revolves around quibbles about what constitutes high definition television and whether the German system was real broadcasting.

In Britain, the outside broadcast cameras captured the procession for George VI's coronation in May 1937, but television didn't make a huge impression as the sets were very expensive. By the time the system was shut down on September 1, 1939 in preparation for the war, there were only some 19,000 sets in operation.[48] In comparison, in

Germany only around 1600 sets were in use at this time.[49] After their later start, America had only some 7000–8000 sets in use when the War Production Board halted manufacture in April 1942.

This wasn't a very auspicious beginning and it wasn't clear what future television had. Few could have foreseen the rapid take-up after the war and the impact it would have by the 1950s and 1960s. This story is continued in Chap. 6.

NOTES

1. The Russian newspaper *Pravda*, http://english.pravda.ru/science/tech/18-08-2005/8778-television-0/ though others seem to agree.
2. Dictionary of National Biography, Alexander Bain (1810–1877), clockmaker and inventor.
3. It was filed on May 27, 1843 and granted November 27, 1843, though it hasn't been possible to find the number. The American equivalent was US patent 5957 from 1848.
4. Garratt, G. R. M. and Mumford, A. H. (1952) The history of television. *Proceedings of the* IEE—Part IIIA: Television, 99: 17, 25–40.
5. German patent number 30105.
6. Hertz, H. (1887) Uber sehr schnelle elektrische Schwingungen. Ann. Physik, 267:7, 421–448.
7. Bonzel, H. P. and Kleint, C. (1995) On the history of photoemission. *Progress in Surface Science*, 49:2, 107–153.
8. GB version of his patent no 27,570 of 1907.
9. *Scientific American Supplement* No. 1850 June 17, 1911.
10. Campbell Swinton, A. A. (1908) Distant electric vision, Letters to the Editor. *Nature* June 18, 1908.
11. Campbell Swinton, A. A. Presidential address to the Röntgen Society, as reported in *The Times*, November 15, 1911.
12. Garratt, G. R. M., Mumford, A. H., and Burns, R. W. (1975) The first demonstration of television. *IEE Electronics and Power*, 21:17, 953–956.
13. Garratt and Mumford.
14. Malcolm Bird, John Logie Baird, the innovator, available at: http://www.bairdtelevision.com/innovation.html
15. Baird, J. L.,Television in 1932, from the BBC Annual Report, 1933.
16. Early Television Foundation and Museum, Mechanical television, Baird Televisor, available at: http://www.earlytelevision.org/baird_televisor.html
17. Waltz, G. H. (1932) Television scanning and synchronising by the Baird System. *Popular Science*, also available at: http://www.earlytelevision.org/tv_scanning_and_synchronizing.html
18. US patent No. 2141059.
19. US patent No. 1773980.
20. Hungarian Intellectual Property Office, note on Kalman Tihanyi, available at: http://www.sztnh.gov.hu/English/feltalalok/tihanyi.html
21. GB patent No. 313456.
22. US patent No. 2158259.

23. Zworykin, V. K. (1933) Television with cathode ray tubes. *Proceedings of the Institution of Electrical Engineers, Wireless Section*, 8:24, 219–233.
24. Lodge, J. A. (1987) Thorn EMI Central Research laboratories—An anecdotal history. *Physics in Technology* 18:6.
25. In practice it needs to be a good deal more than this—a few MHz.
26. Browne, C. O. (1932) Technical problems in connection with television. *Proceedings of the Institution of Electrical Engineers, Wireless Section*, 7:19, 32–34.
27. Dictionary of National Biography. Sir Isaac Shoenberg. Some other sources say he was also a student of Boris Rosing, but his history doesn't seem to allow for this and hence it is unlikely.
28. Shoenberg, I. Contribution to the discussion on 'History of Television', at the Convention on the British Contribution to Television, April 28, 1952. *Proceedings of the IEE*—Part IIIA: Television 1952, 99:17, 40–42.
29. McGee, J. D. and Lubszynski, H. G. (1939) E.M.I. cathode-ray television transmission tubes. *Journal of the Institution of Electrical Engineers*, 84:508, 468–475.
30. Zworykin.
31. Shoenberg and McGee & Lubszynski; The EMI system was working in 1936 while it took RCA until 1939.
32. Elen, R. G. TV Technology—High Definition, available at the BFI Screenonline: http://www.screenonline.org.uk/tv/technology/technology3.html
33. In photographs of the equipment the name Kalee, a well-known brand of projector, can be seen on the light source.
34. Baird.
35. Macnamara, T. C. and Birkinshaw, D. C. (1938) The London television service. *Journal of the Institution of Electrical Engineers*, 83:504, 729–757.
36. Elen, R. G. Baird versus the BBC, available at: http://www.transdiffusion.org/tv/baird/bairdvbbc
37. *The Times*, May 1, 1934/May 15, 1934.
38. The Science Museum. Making of the modern world, available at: http://www.makingthemodernworld.org.uk/stories/the_age_of_the_mass/07.ST.04/?scene=5&tv=true
39. The strange numbers come from using odd number divisions between the frame and line frequencies. 3×3×3×3×3 = 243, and 3×3×3×3×5 = 405.
40. For comparison, the 405 line system actually has 376 complete lines and two half lines.
41. BBC, History of the BBC, Alexandra Palace, available at: http://www.bbc.co.uk/historyofthebbc/buildings/alexandra-palace
42. Elen, R. G. TV Technology 4, Here's looking at you, available at: http://www.screenonline.org.uk/tv/technology/technology4.html
43. EMI went on to produce the more advanced Super Emitron and CPS Emitron tubes for the cameras, giving even better pictures. See McGee and Lubszynski.
44. The Science Museum. Making of the modern world, available at: http://www.makingthemodernworld.org.uk/stories/the_age_of_the_mass/07.ST.04/?scene=5&tv=true
45. Genova, T. and Bennett-Levy, M. Television history—The first 75 years, 1936 German (Berlin) Olympics, available at: http://www.tvhistory.tv/1936%20German%20Olympics%20TV%20Program.htm

46. Uricchio, W. Envisioning the audience: Perceptions of early German television's audiences, 1935–1944, available at: http://www.let.uu.nl/~william.uricchio/personal/SWEDEN1.html
47. Pendleton, N. A brief history of German prewar television, available at: http://www.earlytelevision.org/german_prewar.html
48. The Science Museum. Making of the modern world, available at: http://www.makingthemodernworld.org.uk/stories/the_age_of_the_mass/07.ST.04/?scene=5&tv=true
49. Institute of Media Studies. History of television, available at: http://im.com.pk/ims/2009/02/history-of-television/3/

5

Seeing a Hundred Miles: Radar

Freya had as her most prized possession a necklace, Bressinga-men, to obtain which she not only sacrificed, but massacred her honour. The necklace is important because it was guarded by Heimdall, the watchman of the gods, who could see a hundred miles by day and night.

R. V. Jones

In the 1930s, the UK's Air Ministry had a standing reward of £1000 (nearly $5000) to anyone who could kill a sheep from a distance of 100 yards (91 m).[1] In America, the U.S. Army's Aberdeen Proving Ground later also offered a similar reward, but here the target was a tethered goat. To get the prize the claimant had to demonstrate the task in front of official witnesses, but the real setback was that electromagnetic energy had to be used—in the phase popular in the newspapers, it had to be a 'death ray'. The animals were never in any danger, as there were no takers.

In 1924, there had been a brief flurry of activity when an Englishman, Henry Grindell Matthews, claimed to have made such a device. He got the press very excited but, when challenged to demonstrate the device, always blustered his way out. He tried to play off the British, French and Americans against each other,[2] but eventually they all got tired of being given the run around. Periodically, the press reported inventors who claimed to have created death rays, so it was largely to protect themselves against the charge of not taking these things seriously that the military offered these prizes. The task was much simpler than what was really required, but would show if someone had something that could be developed. The powers-that-be were skeptical, but you never knew.

In the early 1930s, Britain had begun to wake up to the threat of attack by air. Though there had been bombing raids in the First World War the tremendous development in aircraft performance was now a considerable worry. In 1932, Stanley Baldwin, amongst other things the Chairman of the Imperial Defence Committee, made the statement in Parliament: "I think it is well also for the man in the street to realise that there is no power on earth that can protect him from being bombed. Whatever people may tell him, the bomber will always get through…".[3] By this he didn't mean they would all get through but that some of them always would. The current idea was that the only defence was offence, but fortunately not everyone was so pessimistic.

J.B. Williams, *The Electronics Revolution*, Springer Praxis Books,
DOI 10.1007/978-3-319-49088-5_5

By early 1935, this concern had hardened into a desire for action and the UK's Air Ministry set up a Committee for the Scientific Study of Air Defence under the chairmanship of H. T. Tizard (later Sir Henry). Up to this time the only ways of detecting hostile aircraft were by eye or ear, neither of which had proved very useful. A much more effective method was needed.

The first task was to decide definitively whether there were any possibilities in these 'death rays'. Robert Watson-Watt, the Superintendent of the Radio Department at the National Physical Laboratory, was asked to comment.[4] In a memo he poured cold water on the idea, showing that it would require many megawatts of power even if the pilot and the engine weren't protected by the metal of the aircraft. As a throwaway line he mentioned that detection was a much more practical option.

The committee jumped at this, and on February 12, 1935 Watson-Watt sent a memorandum outlining proposals for a system of aircraft detection using radio waves. He and his team had calculated the power and wavelengths needed, together with suggestions as to how range and position could be obtained. The right man had been asked, as some years before Watson-Watt had worked on trying to track thunderstorms using radio methods, which is a closely related problem.

Such was the enthusiasm that, on February 26, a demonstration was arranged to show that aircraft really did reflect radio waves. For simplicity the existing 50 m (6 MHz) transmitter used for the BBC Empire service at Daventry was used. Two receiving aerials were placed across the beam 10 km away and the results of two receivers were fed to an oscilloscope display in antiphase. The result was that the signal directly from the transmitter canceled out and the blob on the display remained in the center. Signals from another source, such as those bouncing off the Heyford bomber that was flying at a height of 2000 m to and fro in the main beam, would be seen as a vertical deflection. This was clearly visible as the plane flew around and there was no doubt it was being detected.

This is often cited as the 'birth of radar'. Unfortunately, it was not. First, there was no information on range or position, only that there was a reflection from the aircraft that could be detected. This wasn't new. Heinrich Hertz had shown that radio waves were reflected from metal objects as part of his experiments way back in 1886.

In 1904, Christian Hülsmeyer demonstrated his Telemobiloscope at the Hohenzollern Bridge, Cologne, successfully detecting ships on the Rhine up to about 3 km away. The equipment used the spark transmitter and coherer detector of that time and so was unable to give much more information than that something was there. Though he obtained patents, he could get little interest in it.[5] He was too far ahead of the capabilities of the technology.

Marconi was also aware of radio reflections from metal objects and in 1922 he said: "It seems to me, that it should be possible to design apparatus by means of which a ship could radiate or project a divergent beam of these rays in any desired direction, which rays, if coming across a metallic object, such as another steamer or ship, would be reflected back to a receiver screened from the local transmitter on the sending ship, and thereby immediately reveal the presence and bearing of the other ship in fog or thick weather."[6] Perhaps he knew it was still beyond the current technology as he didn't pursue the idea.

However, by the mid 1930s, there had been considerable improvements in the technology, both of the vacuum tubes and in the other components. Watson-Watt had a clear view of what was possible with the technology available. He was convinced he could make an aircraft detection system work, and that it needed to be done in a hurry. He had what he called the 'cult of the imperfect'—'Give them the third best to go on with; the second best comes too late, the best never comes.'[7] Many people would disagree, but he was almost certainly right in the circumstances of the time.

The Daventry test might have given more confidence in the calculations, but it was really a political stunt to convince the British government to disgorge some money to do some real development. In this it was successful, and 2 weeks later three people left the Radio Research Station at Slough for the wilds of Orfordness on the east coast of England. This remote site was chosen so that work could proceed in complete secrecy.

Within 5 weeks, on June 17, 1935, the three-man team had succeeded in getting a clear echo from a Supermarine Scapa flying boat at a range of 27 km. A transmitter was used to send pulses rather than a continuous signal. To obtain the range of the aircraft was merely a matter of measuring the time taken for the pulse to come back from the target. The distance it had traveled there and back was thus twice the range. As the electromagnetic wave travels at the speed of light, the timing gave a direct measure of distance. By setting the horizontal trace of an oscilloscope going when the transmit pulse was sent the echo would be seen further along the trace. The distance could simply be read off the suitably calibrated screen.

The next task was to determine the horizontal (azimuth) and vertical (height) position. Instead of moving the antenna, Watson-Watt proposed a system with a fixed 'floodlight' transmitting antenna and a pair of crossed receiving antennae. By comparing the signals in these two antennae and determining the direction from which the signal came using a device called a radio goniometer, the azimuth bearing could be found. By having more than one pair of receiving antennae stacked vertically a similar arrangement could be used to determine the vertical bearing.[8]

Thus, by the end of 1935 the team was able to determine the position of an aircraft in three dimensions at a considerable distance. The progress was sufficient for the government to sanction the building of five radar stations to cover the approaches to the Thames estuary. With the considerable increase in work, during the first half of 1936 the enlarged team moved to Bawdsey Manor not far away. Over the door was the motto: 'PLUTOT MOURIR QUE CHANGER' (Rather Die than Change) which seemed particularly inappropriate for the work going on inside.

Watson-Watt was well aware that the task was not just to produce a radar system but to supply information to Fighter Command for air defence. In September 1936 an air exercise was arranged and the radar stations were to provide information. In reality, only the station at Bawdsey was ready and that one had been rushed out and not properly tested. A row of VIPs, including the Commander-in-Chief of Fighter Command, Air-Vice-Marshal Sir Hugh Dowding, watched the three operators at their oscilloscope displays while nothing happened—until faint blips were seen at the same time as the incoming aircraft were heard![9] This could have been a complete disaster, but Watson-Watt talked very fast and mitigated the damage. The problem was found in the transmitter and better results were obtained before the exercise was over.

Somehow, despite the problems, on the basis of this exercise Watson-Watt was able to persuade the Chief of the Air Staff to decide that 'the RDF system was already proved', and so the next stage of building more stations should go ahead. If he hadn't managed this, the system would certainly not have been working by the beginning of the war and probably not even in time for the Battle of Britain.

Fig. 5.1 A Chain Home radar station with three transmitter masts and four receive masts. *Source:* http://planefinder.net/about/the-history-of-flight-radar/

The radar stations used four giant steel lattice towers 110 m high to house the antennae for the transmitters which used a wavelength of 23 m (13 MHz).[10] The wavelength had been moved from the original 50 m to keep away from commercial transmissions. The receiving aerials were on four wooden towers 75 m high. It was crude, but it worked (Fig. 5.1).

The next stage was to build the air defence system. First the information was fed to a 'Filter Room' where information from more than one station was combined into a single coherent picture of what was happening. This could then be fed to the Fighter Control Room for action to be taken. This whole operation took a large amount of work to get right, but the system kept on being refined and improved.

One example of the detail was working out the path a fighter should take to intercept the incoming bombers. It was known as the 'Tizzy Angle' after Sir Henry Tizard who first proposed it.[11] The development of Operational Research greatly assisted the whole process of refinement of air defence.

By 1937, the Manor station could detect planes 160 km away.[12] A second air exercise using three of the original five radar stations was sufficiently successful for an order to be placed for 20 more stations. As the political situation in Europe deteriorated, these were built in a great hurry, and when war came in September 1939 there were some 19 stations on the east coast and six more on the south. In the nick of time Britain had a working air defence system.

In Britain, it was thought that they were the only country which had worked on radar, but it wasn't so. Because of the international situation every country wanted to keep their work secret.[13] Some experiments took place in the U.S., France, Holland, Russia, Japan and Germany. Many of these played with systems little different from Watson-Watt's crude Daventry demonstration which had been proven to not really be capable of obtaining range information.[14]

The country that made the most progress was Germany, where Dr. Rudolf Kühnold of the Navy's Communications Research Laboratory NVA, worked with the commercial company GEMA.[15] They eventually produced a short wavelength shipborne system known as Seetakt, which was useful for determining the range of other ships at sea and hence useful for gunnery. There was also a longer wavelength land system called Freya which could detecting only the range and horizontal position of aircraft. In 1939 there were only eight of these, while a few ships had been fitted with Seetakts.[16]

Also entering the fray was the radio manufacturer Telefunken. The company produced a high frequency radar with a wavelength of 50 cm (600 MHz) with a dish antenna that could move to track an aircraft, known as a Würzburg. It was originally intended to help guide anti-aircraft guns, but later was used to track bombers and assist fighters to find them. It wasn't ready until 1940. Behind both GEMA and Telefunken as a consultant was the microwave expert Dr. Hans Eric Hollmann without whom German radar would not have made so much progress.[17] Hollmann's 1935 book *Physics and Technique of Ultrashort Waves* was very influential both inside and outside Germany.

One major difference of approach was in the choice of wavelength. Watson-Watt went for a relatively long wavelength. Partly this was on the mistaken idea that the reflections would be greatest if the size of an aircraft was around a quarter of a wavelength but, more importantly, he knew that this was well within the capabilities of the available technology and so could be made quickly with minimal technical risk.

Although there has been some criticism of the Chain Home (CH) system, unlike other countries Britain had a working air defence system when she went to war.[18] This was down to Watson-Watt's persistence. He may not have invented radar, but the country had much to thank him for in getting a working system rather than waiting to have a better one. In view of the difficulties that other nations encountered trying to generate sufficient power at the shorter wavelengths, his approach was well justified.

In any case, it didn't rule out further improvements which were being furiously undertaken, leading to radar systems in ships and planes. For example, one development was of a coastal radar working on a wavelength of 1.5 m (200 MHz), which was added to the CH stations to detect low-flying aircraft. Known as the Chain Home Low (CHL), though it had only half the range of the 100–200 miles of the CH stations it usefully filled a gap and was ready in 1940 when the main German bombing assaults occurred.

What Watson-Watt and Tizard, and those around them, had realized at an early stage was the importance of having a system to use the information and supply it correctly to the fighter planes and so form an effective air defence. The motivation was to save having to have fighters on permanent patrols, for which there certainly weren't the resources. Hence Britain was able to defend herself in the Battle of Britain with the limited numbers of pilots and aircraft available.

Despite having the technology, the Germans were more interested in offence, and had given little thought to air defence, so it took them some time to put it into place once the Allied bombing started. The importance of air defence was also starkly demonstrated in December 1941 when the Japanese attacked Pearl Harbor. The Americans had good radar sets and were able to track the incoming aircraft at long distances, but there was no effective system for assessing the information and reporting their results so that action could be taken.[19] The result was catastrophic.

In Britain, the search was now on for equipment that would work at ever-shorter wavelengths to achieve small size of antennae and accurate detection of small objects such as submarine periscopes. Just as the war began, research programs were put in place at the Physics Departments of Oxford and Birmingham Universities to develop components that would make microwave radar possible, ideally around 10 cm wavelengths.

There were a whole range of devices that might be useful starting points but the main team concentrated on the klystron. However, two young men, John T. Randall and Henry A.H. Boot, started work on another device, a Barkhausen-Kurtz detector.[20] To test this they needed a microwave source, so they thought they would just make one. Despite the rest of the team working on this problem they went their own way. Their thinking was that it might be possible to combine the best features of the klystron with its cavities and the magnetron which was a type of vacuum tube where an applied magnetic field caused the electrons to rotate.

What they almost certainly didn't know was that considerable work had been undertaken in other countries on variants of the magnetron and most had been abandoned. They decided that the easiest way to make such a device was to drill seven holes in a copper block, one in the middle and the others spaced around with narrow slots connecting them to the central one. The cathode to emit the electrons went through the center and then ends were put on so that it could be pumped down to achieve a vacuum (Fig. 5.2).

Virtually all previous versions of the magnetron had been inside a glass envelope like a normal vacuum tube. Randall and Boot's version had the immediate advantages that it was (a) easily cooled and (b) a magnetic field could be applied across the device with only small gaps, hence increasing its strength. It was delightfully simple and robust.

On February 21, 1940, when they finally switched it on, the bulbs used as loads kept burning out. Eventually they realized the device was generating a staggering 400 W of power at about 9.8 cm wavelength. This was so much better than other lines of investigation that work was immediately switched to developing it further. GEC's research laboratories were brought into the project and it was converted into a practical device.

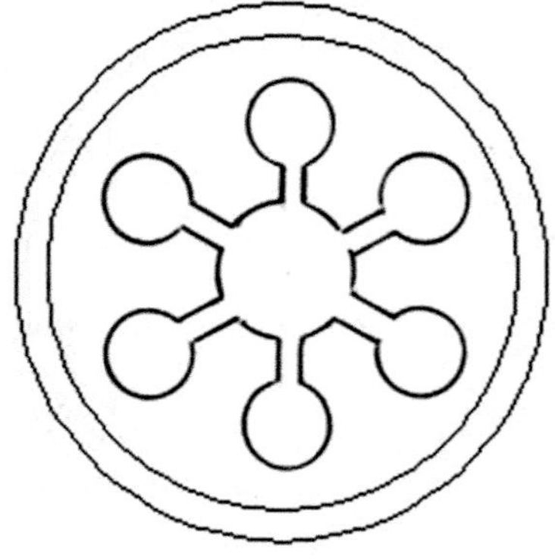

Fig. 5.2 (*Left*) Randall and Boot's experimental cavity magnetron which needed to be pumped down to obtain a vacuum. An electromagnet applied the field vertically through the device. (*Right*) The form of a 6-chamber magnetron like the original prototype. *Source:* Readhead P A, *The Invention of the Cavity magnetron and its introduction into Canada and the USA*, Physics in Canada November/December 2001

Interestingly, the treated cathode that they used derived from the French work at CSF by Maurice Ponte which was on the way to producing a useful magnetron, although not as powerful as Randall and Boot's.[21] Pointe had to stop work and flee when France fell in May 1940.

By September 1940, devices which could produce pulse powers of 10 kW had been made. One such device was taken to America as part of the Tizard mission, which went to exchange secrets so that access could be had to American resources and particularly American manufacturing strength.

When the performance of the device was explained, the Americans were staggered. Alfred Loomis, the chairman of the Microwave Committee of the National Defence Research Committee, said they were getting the chance to start now 2 years ahead of where they were and were getting infinitely more from the British than they could give them.[22] One thing they did give was the name 'radar' which had been invented by Samuel M. Tucker and F.R. Furth that year.[23]

A vast amount of work on both sides of the Atlantic was carried out, leading to the rapid introduction of superior radars that worked in the centimetric band. This was vital in the ongoing battle in the Atlantic against German submarines.[24] The first generation 1.5 m sets were largely negated by a detector that could warn of the aircrafts' approach. With the centimetric radar a suitable detector came too late to have a significant influence.

With the centimetric sets the British, and later their American allies, had leapfrogged the Germans who were only able to scrabble behind when they captured an airborne centimetric radar using a magnetron in a crashed bomber whose explosive charge had failed to destroy the device. By the war's end a huge number of different types of cavity magnetrons had been produced by the Allies. Some of these had peak powers as high as 2 MW and wavelengths less than a centimeter.

After the war, military radar continued to develop as the possible threat came from missiles. With their greater speeds it was necessary to have even more powerful radars that could spot incoming objects at huge distances if there was going to be any hope of interception. Britain had its Early Warning system, but even then it was only supposed to give a 4-min warning. America built a Ballistic Missile Early Warning System (BMEWS), many of the chain of stations in inhospitable places in the Arctic, but even then the warning was limited to 10 min.

It was obvious that this technology would be useful for civilian Air Traffic Control. At first, modified military radars were used both for checking the positions of aircraft and also helping them land. In Britain, air traffic in the 1950s increased at around 15% per year meaning that it doubled every 5 years.[25] From almost zero at the end of the war it had reached around ten million passengers a year by 1960. In the US, it started from a higher base, and despite slower percentage growth it reached around 50 million in the same period.[26]

Radar had arrived in the nick of time because the old methods of visual sightings and radio contact couldn't cope with numerous aircraft in the same space at the same time.[27] Only a proper layout on a sophisticated radar screen could provide the controller with a 'picture' of what was happening, particularly as the jets came into service that were faster than the older planes.

A further development was of Secondary Surveillance radar.[28] This was separate from the primary radar; a pulse was transmitted in the direction of the aircraft, which responded by transmitting a code to say who it was and also other information, particularly its height. These data were added to the display of the position markers provided by the primary radar, giving the controller an even fuller picture.

Without these developments the huge growth in air travel just wouldn't have been possible. By the end of the century 200 million passengers were using Britain's airports alone—in 2016 286,075,280 passengers passed through US airports.[29] It would not have been safe to have the numbers of aircraft needed in the sky at the same time, and certainly not landing and taking off at the rates that were needed. Without it, all those flights would not have happened. Something that started as a means of detecting hostile aircraft ended up keeping track of friendly ones.

The idea of using radar to track the weather was not new. Watson-Watt had started his career trying to use radio to track thunderstorms, but by the Second World War it was noticed that rain showed up on the radar screens.[30] With the war over, researchers started to develop this observation using surplus military radars, and were able to give an indication of the rainfall rate and hence detect severe storms.

Around 1960, the Doppler Effect was introduced to the radars to gain information on the speed of movement of storms. This is a well-known phenomenon which causes the pitch of a siren to appear higher as it approaches and lower as it moves away. Frequency or phase shifts in the returning signal of the radar could be used to determine the speed of an object moving towards or away from the radar set.

This was developed extensively in America where the need was to be able to give a warning of storms that were likely to develop into tornados. Gradually, as computing power was added, more and more information could be gained which greatly assisted weather forecasting. By the end of the century, the plots from a number of different sets could be combined to produce a weather map which showed the position of clouds producing rain or other precipitation. It became one of the pillars of weather forecasting.

As the wavelengths became shorter and shorter, and simple cheap components for working at these became available, another use presented itself: measuring the speed of vehicles. Here the distance and position information weren't required, only the speed, and the Doppler radar was ideal for this. With suitable computing power an output directly in kilometers per hour could be obtained.

From this came the 'speed gun', beloved by police for catching motorists who disregard the traffic signs. This was automated into a 'speed camera', which took a picture of vehicles going at speed to show their number plates as evidence. While they are credited with reducing accidents they are hated by the possessors of fast cars! Finally, came the active speed indicator which merely indicated the car's speed if the driver had let it drift a little high. This was generally effective in villages where most motorists accepted its advice to slow down without complaint.

Radar thus grew from a simple observation that radio waves could be reflected from objects to something that has a multiplicity of uses in the modern world. It guards our borders against intruders, ensures that planes keep to the invisible lanes in the sky, warns of rain to come and calms the impatient motorist.

NOTES

1. Hanbury Brown, R. and Watson-Watt, R. (1994) Father of radar. *Engineering Science and Education Journal*, 3:1, 31–40. He cites A. P. Rowe who was the secretary to the Tizard Committee and so presumably knew about this.
2. *The Guardian*, May 29, 1924.
3. Hansard. House of Commons Debate, November 10, 1932, vol 270 c 632.
4. Originally his name had been Watson Watt but he later hyphenated it. This form has been used for simplicity.
5. German patent DRP165546 and British patent GB190413170.
6. Marconi, G. (1998) From a speech at a joint meeting of the IRE and the American Institute of Electrical Engineers on June 20, 1922, quoted in R. W. Simons and J. W. Sutherland, 40 years of Marconi radar from 1946 to 1986. *GEC Review*, 13:3.
7. Zimmerman D. (2010) *Britain's Shield: RADAR and the Defeat of the Luftwaffe*. Stroud: Amberley Publishing, p. 79.
8. He applied for a patent for this in September 1935 though for security reasons it was not finally published until July 1947. Patent GB593017.
9. Hanbury Brown was one of the operators and documents this in his article.
10. There was plenty of expertise in this sort of construction after the building of the Grid. The number of towers was later reduced to three.
11. Rowe, A. P. (2015) *One Story of Radar*. Cambridge: Cambridge University Press, p. 45.
12. Buderi, R. (1998) *The Invention that Changed the World*. Ilford: Abacus Publishing, p. 71.
13. A rare example of information being published were reports of the anti-collision system fitted to the French ship 'Normandie', e.g., 'Feelers for ships', *Wireless World* 38, June 26, 1936, p. 623.
14. Brown, L.,(1999) *Technical and Military Imperatives: A Radar History of World War 2*. London: CRC Press, pp. 40–48 and 72–90.
15. NVA—Nachtrichtenmittel-Versuchsanstalt; GEMA—Gesellschaft für elektroakustische und mechanische Apparate mbH.
16. Clark, G. C. (1997) Deflating British radar myths of World War II. Research paper, available at www.radarpages.co.uk/download/AUACSC0609F97-3.pdf
17. Hollmann, M. Radar world, available at: http://www.radarworld.org/hollmann.html
18. For example, Clark above. He appears to have not quite understood the capabilities of the CH system, though he does admit that having an air defence system was important.
19. Hanbury Brown.
20. This section relies heavily on Randall, J. T. and Boot, H. A. H. (1946) Early work on the cavity magnetron, *IEE*, 9:1, 182–183, and also Redhead, P. A. (2001) The invention of the cavity magnetron and its introduction into Canada and the USA. *Physics in Canada*, November/December 2001, 321–328.
21. Neel, L., La vie et l'oeuvre de Maurice Ponte, Academie Francaise, available at http://www.radar-france.fr/eloge_Ponte.htm (in French).
22. Redhead.
23. Buderi, p. 56.
24. Rowe, p. 162.
25. CAA, *Aviation Trends*, Q2 2009.
26. FAA, Regional Air Service Demand Study, available at: www.faa.gov/.../eastern/planning_capacity/media/DVRPC%20Task%20B%20&%2
27. History of NATS, available at: http://www.nats.co.uk/about-us/our-history/

28. Mansolas, Y. and Mansolas, A. A short history of Air Traffic Control, available at: http://imansolas.freeservers.com/ATC/short_history_of_the_air_traffic.html
29. Office of the Assistant Secretary for Research and Technology, U.S. Department of Transportation.
30. IEEE Global History Network. Radar and weather forecasting, available at: http://www.ieeeghn.org/wiki/index.php/Radar_and_Weather_Forecasting

6

The 'Box': Television Takes Over

Television is a bomb about to burst.

Grace Wyndham Goldie, shortly after her appointment to the Television Talks department of the BBC in 1948

It will be of no importance in your lifetime or mine.

Bertrand Russell in reply

On June 7, 1946, television in Britain came on the air again, showing the same cartoon as when it had shut down 7 years before. The presenter, Jasmine Bligh, reintroduced herself to viewers with the words: "Do you remember me?"[1] It was as though there had just been a slight interruption. The government (as usual) had set up a committee under Lord Hankey to investigate what to do with television after the war, but in the ensuing austerity period it was inevitable that the existing system should be resurrected. Hankey did propose that the system should be rolled out to other parts of the country, and higher resolution systems should be investigated.

While there were enthusiasts, many, like the philosopher Bertrand Russell, still felt that there was little future in television. This included a substantial proportion of the BBC's management who were hardly interested in the few thousand television viewers when their radio audience was measured in millions. The television staff at Alexandra Palace were dismissively known as 'the fools on the hill'. However, the next day—8 June—the infant television service was able to show the Victory Parade in London. It was a small foretaste of what it was capable of doing.

Before the war the production of television sets had only reached about 8000 a year, and little more was expected.[2] Of the total of 19,000 sets produced it was unlikely that all of them remained in working condition, so when the service reopened the audience was very small, probably around 15,000, and all in London. The situation was not improved by the shortages of materials which meant that the manufacturers could only make around 2000 sets per month which, though greater than prewar, was nowhere near enough to meet the demand.[3]

J.B. Williams, *The Electronics Revolution*, Springer Praxis Books,
DOI 10.1007/978-3-319-49088-5_6

As the service personnel were demobbed, they returned to civilian life with a gratuity in their pocket. Many of the married women were still in work and the conditions of full employment meant that people had money to spend.[4] Considerable savings had been built up during the war when so much had been controlled or rationed, but now it was over those restrictions were still in place. What could the people buy with their money? There was a rush for electric fires and a few other items that escaped the controls, but the main thrust went into leisure. Holiday destinations were crowded, cinema attendance hit a peak in 1946, and dance halls and football grounds were full.

While television had started as a 'rich man's toy' it rapidly lost that image. Once the shut-down in the disastrous winter of 1947 was out of the way, things began to pick up. The first fillip was from the televising of the wedding of Princess Elizabeth to Philip Mountbatten in November 1947 followed by the London Olympics in August 1948. It was during that year that sales began to accelerate: 3500 in January, 6300 in August, 11,800 in September and 13,000 in December.[5] This was greatly helped by the halving of purchase tax on radios and televisions in June 1948.[6]

Steadily, the social mix of viewers was changing to one much more representative of the population in the London service area as a whole. While still only in the hands of a small minority, it was the aspiration of around half. The opportunity to stay at home for entertainment was a huge draw.[7] This was even though small black and white screens were competing with large picture color films in the cinemas. Despite some claiming it was educational and widened interest, the BBC's research found that, given equal resources, it was the family with the lower educational level that was likely to have the television set first.[8]

Its popularity was surprising given that, in 1948, the entire day's schedule could consist of children's programs from 4 pm to 5 pm, a shutdown until 8.30 pm to allow the children to be put to bed, followed by a play until 10 pm. Then there was 15 mins' news in sound only.[9] To watch it the light needed to be turned out as the picture wasn't very bright, and then there was the problem of where to put it. There was an element of 'social snobbery' about obtaining a set, be it bought, rented or on hire purchase.

Following the war there was a baby boom. It didn't take all those new parents very long to discover the wonderful ability of television to child mind. It wasn't that the children just watched their programs; they devoured everything, and absorbed a considerable amount of knowledge without really noticing. With children seeing television at their friends' houses 'pester power' was a considerable factor in persuading other parents to obtain a set. As more people had sets, this pressure increased.

Countering this were the difficulties that the program makers were laboring under. This included the growing hostility of those who thought that it might reduce attendances, such as the film and theater industries and the owners of sports rights.[10] Some of this was serious; football matches and other sporting events could not be televized, and theaters would often not allow their casts to perform in front of the cameras. Other restrictions were just petty, such as British Movietone refusing to allow the BBC to have a copy of its film of the re-launch of the television service. Even Disney would no longer let them show its cartoons.

Another curious restriction was that television didn't show the news because the Director-General thought it would be inappropriate. This was only partially altered in 1948 when Newsreel, a weekly bulletin of filmed reports, was finally screened. It took another 6 years before a proper news program was shown, and even then the newsreader, Richard Baker, wasn't allowed to appear on screen in case he 'sullied the stream of truth'.[11]

To extend the limited coverage of the Alexandra Palace transmitter another was opened in 1949 at Sutton Coldfield to cover the Midlands area adding the potential of another nine million viewers to the 12 million in London. In 1951 and 1952, three more were added at Holme Moss to cover the northern area, Kirk o' Shotts for Scotland and Wenvoe for the Bristol channel area, virtually doubling the population covered.[12] Though there were still large gaps, almost 80% of the population could receive a signal. The numbers of viewers really began to climb, roaring past the one million mark.

In 1951 the government raised the Purchase Tax on sets from 33⅓ to 66⅔%.[13] The following year, the new Tory government wanted to cut production to reduce imports. They placed restrictions on production, trying to limit this to two-thirds of the previous year. However, despite the restrictions some 30% of sets were bought on hire purchase so the minimum initial deposit was raised to 33⅓% and the repayment had to be made within 18 months.[14] The measures didn't achieve entirely what was intended, but did slow the growth.[15]

On June 2, 1953, Queen Elizabeth II was crowned in Westminster Abbey. Television cameras were there with the young Queen's approval. Some 20 million people watched the event,[16] but as there were only around two million sets, this averaged ten people to a set. In many houses up to 20 people crammed themselves into a small living room to peer at the small screen. It was almost as big a festivity as the coronation itself, and quite a social event. It brought the country together in a way neither the cinema nor radio could. There was something magic in watching an event such as this live.

It is commonly stated that this was the making of television in Britain. The growth curve suggests it was already on its way. The government's restrictions had slowed the growth before this, and the reduction of Purchase Tax to 50% shortly before the coronation helped to accelerate it again.[17] This is sufficient to explain the slight change of pace in the acquisition of television sets. People had already decided that they were desirable things to possess.

The typical price, compared to most people's earnings, was very expensive.[18] It is hardly surprising that, despite hire purchase and rental deals, it took time for people to be confident enough to get a set. One hazard for the intending owner was the poor reliability of the sets. The nightmare was that the 'tube would go' as the cathode ray tube was very expensive to replace. On top of this a television licence had to be purchased annually. Despite these problems sets were still being sold at an average rate of around 100,000 per month.

What the coronation coverage did achieve was to persuade those in authority to take television more seriously, because when the audience figures were analyzed it was found that the number watching the television coverage had been greater than those listening on the radio.[19] The radio audience was starting to decline. The effect on the cinema was even

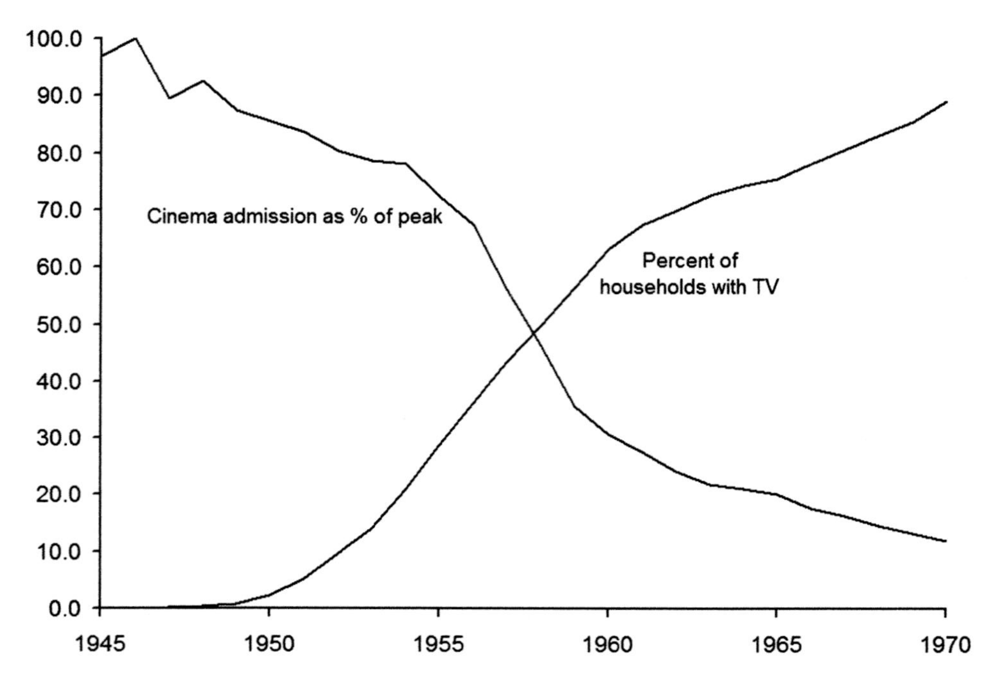

Fig. 6.1 Cinema attendance in the UK declined as more homes obtained televisions. *Source:* Author[20]

more severe as can be seen from Fig. 6.1. The match between the rise of one and the decline of the other is too precise to be anything other than cause and effect.

The coronation also gave a boost to the international connection of television systems. This was not a trivial problem, as in addition to getting the signal to and from the various countries, the standards were different. The French used 819 lines and the Germans 625 lines and there were other incompatibilities. However, in 1952 the BBC's engineering department had managed to convert French signals to the British system so they were able to extend the principles of the converters to the other systems. As a result, a million people in France, Belgium, the Netherlands and West Germany saw the coronation live. The Americans had to wait for a telefilm to be flown to the United States as there was no way of transmitting the signal across the Atlantic.[21]

In America, television had effectively started from much the same low point after the war. However, the building of transmitters was far more rapid and without the financial and governmental restraints imposed in Britain. It was also free of a licence as it was funded by advertisements. The result was that it took off even faster. In 1950, British television had reached about 2% of homes whereas in the United States it was 9%.[22] This had grown to 50% by 1954, but it took Britain a further 4 years to achieve this growth. Other countries were even slower off the mark. Where the US had 13 million sets in 1951 and Britain one million, the Soviet Union had 50,000 and France 30,000.[23] Other countries had yet to start a full service though many were experimenting.

In the UK, once television started to take off so did a movement to remove the BBC's monopoly on television broadcasting. It was largely supported by politicians who were advocates of free enterprise, and television set makers who stood to benefit from increased sales.[24] Ranged against them were supporters of the BBC and those who were concerned about the change in standards that might result. In particular, they did not want to follow the American method of unrestricted sponsorship and adverts controlling the programs.

The Television Act of 1954 set up an Independent Television Authority (ITA) to regulate the industry and award franchises for fixed terms.[25] There were separate contractors for different parts of the country. With strict controls on what could and could not be done in the way of advertising, it was a reasonable compromise.

On September 22, 1955 the first commercial station, Associated-Rediffusion, started broadcasting in the London area.[26] Gradually, one by one over the next 5 years the regional independent stations opened. In the beginning the existing television sets had to be modified to receive the new service but the new ones were designed to receive the appropriate channels.

At first viewers quite liked the adverts, but it rapidly became apparent that these were of very variable quality from quite fun and memorable to the downright awful.[27] Though the companies found it difficult at first, the momentum of the industry was such that an ITV franchise was soon described as 'a licence to print money'.[28] The effect on the BBC service was to force of the organization to produce programs that people wanted to watch rather than what was thought to be good for them, though all the time was the worry of maintaining standards.

Once a television set was present in the majority of homes, its impact became marked. It had two rather contradictory effects. The first was to bring families together. People went out less to the cinema, but also less to football matches and the pub. This particularly applied to the man of the house who was more often at home.[29] This resulted in a growing trend towards a family-centered society rather than one based on collectiveness. The effect wasn't just in leisure but contributed generally to the growth of individualism.

The contradictory effect was that, particularly at big national events such as the coronation, it brought the whole country together. Because it was live, the viewers were all 'there' watching what was happening, and though they were merely sitting in their own homes they felt a part of the event just as much as those who lined the streets. Often they could feel more involved because they had a much better view and understanding of what occurred rather than the bystander who only saw a glimpse of what was happening.

Television also informed in ways that weren't expected. The act of seeing pictures of what was really happening had an effect much greater than reading about it or listening to a description could. In the US the news footage of the struggle for civil rights brought pressure for change, while the pictures that came out of the Vietnam War soon turned the majority of the country against it.[30]

In Britain, though it took the politicians some considerable time to understand its power, television began to affect the outcome of elections. It was as though the average person lived simultaneously in two environments, the normal everyday direct experience, and a second interconnected one which linked them to national—and later international—matters. It was the beginning of 'one world'.

In 1964, the BBC was able to add another channel—BBC2. Recognizing that the 405 line system was now getting rather obsolete, they used what had become the European standard of 625 lines. This seems to have derived from conversion of the American 525 line 60 Hz standard to the European 50 Hz mains. Though the standard is normally thought to have originated in Germany, there could well have been Russian input into it.[31]

The next step was obviously color. The principle of building a color picture from only red, green and blue had been known for a long time. In the 1930s Baird and Zworykin both experimented with color systems, but the difficulties meant that only black and white was introduced. It was not too complicated to make a camera with prisms behind the lenses to pass the incoming view through color filters and to three of the new smaller tubes. The problem lay in producing a receiver.

In America in 1950 the Columbia Broadcasting System (CBS) persuaded the authorities to let them broadcast a system which used the obvious three colors transmitted separately.[32] As this used the UHF band as opposed to the black and white signals on VHF, the lack of compatibility didn't seem to matter. However, it did in the marketplace. CBS couldn't break the cycle of not having enough sets so the advertisers weren't interested which led to not enough money to develop the service, and so round again. In late 1951 they gave up the struggle.

Meanwhile, RCA developed a system based on the work of Frenchman Georges Valensi. He had patented a scheme that was compatible with the existing black and white sets.[33] The idea was simple. The red, green and blue signals were combined to produce a 'luminance' signal which was broadcast in exactly the same way as the black and white service and gave the same picture on existing sets. In addition, there were 'chrominance' signals red minus the luminance and blue minus luminance, which were also transmitted. At the receiver these were recombined to produce the red, green and blue signals.

With RCA providing much of the impetus, the National Television System Committee (NTSC) worked to produce a color television standard that was compatible with the existing black and white. The result, which began to appear in 1954, was known as the NTSC system. With expensive receivers the system was slow to take off, and it took until 1964 before 3% of homes watched in color.[34] There was no hurry as the programs could still be watched on all the existing black and white sets that were in nearly every home.

The solution to the receiver problem was rather elegant. The screen of the cathode ray tube was made with a sequence of dots of colored phosphors red, green and blue. Three electron guns were aimed to pick up the appropriate colored dots. However, it was just not possible to make electron optics good enough to do this reliably, so a metal plate with a grid of holes was placed a short distance from the phosphor screen. This ensured that the beam from the electron gun fell exactly on the correct dot. By having the three guns at slightly different angles the beams could be arranged to fall on the appropriately colored dot. This concept required considerable detailed development, but the principle eventually allowed practical color cathode ray tubes to be manufactured at acceptable prices.

In the late 1950s, the BBC in the UK started experimenting with a color television system. At first they modified the NTSC system for the British mains frequency, but the shortcomings of the system, where the color could be quite variable, led them to abandon it. It was colloquially known as 'Never Twice the Same Colour'. The French produced a system called SECAM which overcame these problems; although it was rather complex, they went on to use it. However, in Germany another system, called PAL, seemed more attractive.

PAL stood for phase alternation line, which meant that on alternative lines the phase of the color signals was inverted. This had the effect of effectively cancelling the errors in the color and produced a better stable picture. Britain and the rest of Western Europe, apart from the French, standardized on this using the 625 line system.[35] As BBC2 was already transmitting a 625 line signal, this was the obvious place for a color service which started in 1967 (Fig. 6.2).

Over the next few years, BBC1 and the ITV channels all moved to UHF and introduced color transmissions.[36] Despite the receivers being much more expensive than the black and white sets, the take-up was even more rapid than that for black and white sets. Despite the much later start it only ran about 3 years behind the US. In 1982 another channel was added to the UK service, with the unimaginative name of Channel Four. By this time 97% of households in Britain had a television though there were more sets than this, equivalent to 112% of households.[37] In other words, some 15% of homes had more than one set. Having started in the living room they were migrating into the kitchen and particularly teenagers' bedrooms. This was a trend that would continue. By the end of the century virtually every household had at least one television, but there were half as many sets again as that needed. Many families had two or even three.

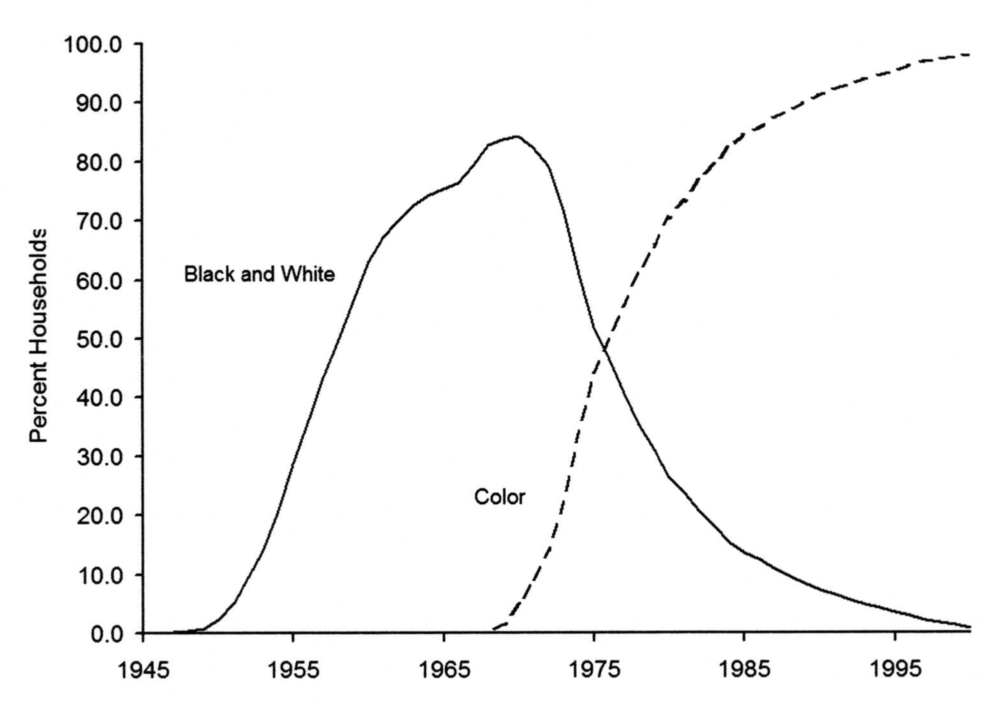

Fig. 6.2 Percent of households with black and white and color television in Britain, 1945–1995. *Source:* Author[38]

There were still three more technological steps that impacted on television. The first was the conquest of space which brought communication satellites. This made it possible to transfer a television signal from anywhere in the world to a studio ready to broadcast. As a result, events on the other side of the world could be watched at the moment they were happening. News became global and the world became a village. With such knowledge it became more and more difficult to hold populations in thrall.

Satellites had another impact in direct broadcasting. Now it was possible to transmit large numbers of channels, and the few that everyone used was multiplied into the hundreds. Needless to say, the quality of much of the content was not great, and some lived entirely on repeats, but it was now possible to cater for many minority tastes. Satellites and the possibilities they opened up are explored further in Chap. 20.

For many years it had been the objective to reduce the size of the television set while the screen became bigger and bigger. Unfortunately, with a cathode ray tube the rules work the other way. A larger screen needs a longer 'neck' behind to accommodate the electron gun and deflection systems. Clever arrangements produced sharper angles of deflection, but there was a limit to this and a big screen set was always rather chunky. The arrival of plasma and then LCD displays broke these constraints and led the way to smaller, lighter sets that could be mounted in the wall if required. These displays are explored further in Chap. 12.

One of the other things that had made flat screen televisions possible was the ever onward march of digital technology. Computing and its derivatives began to creep in everywhere. It therefore seemed logical to transmit even the television signal as a digital instead of an analogue quantity. Apart from having to change the set and the presence of even more channels it didn't have an enormous effect on the viewer.

Television had thus come from the dream of a few inventors to something that was central to everyday life. It was the main source of news—the newspapers and radio had largely relinquished that role—and the first place to turn when world events unfolded. Billions watched the Olympics and royal weddings, and every day viewers could see the troubles from around the globe. Television pictures of disaster and people in distress could result in great generosity as fabulous sums of money were collected to aid famine and earthquake victims. Like all developments there were good and bad aspects, but there was no going back. Television had become such a matter of everyday living that hardly anyone could imagine life without it.

NOTES

1. The story of BBC television—Television returns, available at: http://www.bbc.co.uk/historyofthebbc/resources/tvhistory/afterwar.shtml.
2. Radio Manufacturers' Association, UK television receiver production in the 1930s, available at: http://www.terramedia.co.uk/reference/statistics/television/tv_set_production_1930s.htm.
3. Fisher, D., Take-up of radio and television in the UK: The early years, available at: http://www.terramedia.co.uk/media/change/radio_and_tv_takeup.htm.
4. Addison, P. (1985) *Now the War is Over: A Social History of Britain, 1945–51*. London: Jonathan Cape, p.144.
5. *The Times*, December 4, 1948 and November 19, 1949.

6. *The Times*, June 12, 1948.
7. 'Mass observation', quoted in Kynaston, D. (2008) *Austerity Britain: Smoke in the Valley*. London: Bloomsbury, p.17.
8. BBC, Audience Research Department, 1952 available at: http://www.terramedia.co.uk/quotations/Quotes_Anon.htm#TVhomes.
9. *The Observer*, November 14, 1948.
10. BBC, The story of BBC Television—The 'Cinderella' service, available at: http://www.bbc.co.uk/historyofthebbc/resources/tvhistory/cinderella.shtml.
11. BBC, The story of BBC Television—Memories of Ally Pally, available at: http://www.bbc.co.uk/historyofthebbc/resources/tvhistory/memories.shtml.
12. *The Times*, August 13, 1952.
13. *The Times*, April 11, 1951.
14. *The Guardian*, January 30, 1952.
15. *The Guardian*, July 15, 1952.
16. BBC, The story of BBC Television—TV's crowning moment, available at: http://www.bbc.co.uk/historyofthebbc/resources/tvhistory/coronation.shtml.
17. *The Times*, April 15, 1953.
18. See contemporary advertisements, e.g., History World available at: http://www.historyworld.co.uk/advert.php?id=1311&offset=0&sort=0&l1=Household&l2=Television.
19. BBC, The story of BBC Television—TV's crowning moment, available at: http://www.bbc.co.uk/historyofthebbc/research/general/tvstory12.
20. Cinema figures from Terra Media, UK Cinema admissions, available at: http://www.terramedia.co.uk/reference/statistics/cinema/cinema_admissions.htm; Television licence figures from BBC reports (available from Wikipedia, Television licencing in the United Kingdom (historical), available at: https://en.wikipedia.org/wiki/Television_licensing_in_the_United_Kingdom_(historical); and Terra Media, UK television licences, available at: http://www.terramedia.co.uk/reference/statistics/television/television_licences.htm; and households extrapolated from censuses.
21. Alexandra Palace Television Society, June 2, 1953—Coronation day, available at: http://www.apts.org.uk/coronation.htm.
22. Television History. The first 75 years television facts and statistics - 1939 to 2000, available at: http://www.tvhistory.tv/facts-stats.htm.
23. *The Times*, December 14, 1951.
24. Given, J. (2011) The Commonwealth of television: EMI's struggle for a new medium. *Media History*, 17: 3.
25. It received its royal assent on July 30, 1954.
26. BFI Screenonline, Timeline, available at: http://www.screenonline.org.uk/tv/timeline/index.php?y=1955&t=2&et=&st=.
27. Kynaston, D. (2010), *Family Britain 1951–1957*. London: Bloomsbury, p. 665.
28. Lord Thomson of Scottish Television, quoted in Parliamentary Business, Chapter 3: British Television, available at: http://www.publications.parliament.uk/pa/ld200910/ldselect/ldcomuni/37/3707.htm.
29. Kynaston, D. (2010) *Family Britain 1951–1957*. London: Bloomsbury, p. 671.
30. Cyber College, The social impact of television, available at: http://www.cybercollege.com/frtv/frtv030.htm.
31. A discussion on this is on UK Vintage Radio, 625 line television origins, available at: http://www.vintage-radio.net/forum/showthread.php?t=83653.
32. Color television, available at: https://en.wikipedia.org/wiki/Color_television.
33. British Patent 524443 which is the UK version of his original French patent.

34. Color televisions, available at: http://www.tvhistory.tv/Color_Households_64-78.JPG. It wasn't until 1972 that sales of color receivers exceeded those of black and white sets. See, TV History, Television Sales, available at: http://www.tvhistory.tv/TV_Sales_70-77.JPG.
35. The Russians and their satellites in Eastern Europe also used the SECAM system.
36. Elen, R.G., TV Technology, 8 Britain in colour—and UHF, BFI Screen Online, available at: http://www.screenonline.org.uk/tv/technology/technology8.html.
37. Terra Media, UK Television receivers in use, available at: http://www.terramedia.co.uk/reference/statistics/television/tv_sets_in_use.htm.
38. Television licence figures from BBC reports (available from Wikipedia, Television licensing in the United Kingdom (historical), available at: https://en.wikipedia.org/wiki/Television_licensing_in_the_United_Kingdom_(historical); and Terra Media, UK television licences, available at: http://www.terramedia.co.uk/reference/statistics/television/television_licences.htm; which are then from 1970 onwards, blended into the figures from the General Household Survey. The number of households is extrapolated from censuses.

7

Spinning Discs: Recorded Music

My idea is that there is music in the air, and music all around us; the world is full of it, and you simply take as much as you require.

Edward Elgar

Before the later part of the nineteenth century, if you wanted music you either had to play it yourself or find someone else to do it for you. Generations of young ladies were encouraged to learn the piano as an 'accomplishment', but really to provide their families with a source of entertainment. Music only existed 'live'. The exceptions were the chiming of clocks, and barrel organs which could only play one tune, or at best a few short ones.

The breakthrough was the use of rolls of stiff paper similar to a Jacquard loom which could easily be changed. This is usually ascribed to the Black Forest firm of Michael Welte und Soehne in 1887, and it also gave the great advantage that the rolls could be duplicated much more easily than barrels with their complicated sets of pins.[1] Before long, the same idea was transferred to the piano to produce what was commonly known as the pianola.

At first, the mechanical system could only operate the notes and any 'performance' involving dynamics and phrasing had to be supplied by the operator via a series of controls. Later, even this was overcome and the rolls could reproduce the playing of a real pianist. Some rolls were recorded by famous musicians such as Paderewski, and huge music libraries were built up in the early years of the twentieth century, when pianolas were quite popular.

After the First World War they went into decline, and by the 1930s had almost disappeared. Their demise coincided with the rise of radio and with another instrument which was able to reproduce music—the gramophone. This had the advantage that not only the player but also the 'record' was much easier, and hence cheaper, to produce. To find its origins it is necessary to go back to 1877.

Thomas Edison had been beaten to the telephone by Alexander Graham Bell, but he realized that the crux of the device, the vibrating diaphragm, could have other uses. He used it to move a stylus which indented a groove on a layer of tinfoil wrapped around a cylindrical drum. This was turned by a handle, which both rotated the drum and moved it laterally by returning the stylus to the start of the groove and winding the handle again.

J.B. Williams, *The Electronics Revolution*, Springer Praxis Books,
DOI 10.1007/978-3-319-49088-5_7

Fig. 7.1 Emile Berliner. *Source:* https://en.wikipedia.org/wiki/Emile_Berliner#/media/File:Emile_Berliner.jpg

The stylus would follow the groove, and vibrate the diaphragm in sympathy.[2] The result was a recording of the nursery rhyme 'Mary had a little lamb' although it was a very feeble and distorted version of the original.

Edison paused long enough to name the device the 'Phonograph' before losing interest as he was trying to make a telephone repeater and it didn't work well in this role. It took a few years before Bell and Charles Tainter substituted a wax cylinder for the tinfoil coated drum and this led Edison to return to it and improve it further. Though the cylinders didn't last long, the machines found a use for recording dictation and providing some 2–4 min of low-quality music reproduction.

The great drawback of the cylinder was that it could not be reproduced—each one had to be recorded separately (or possibly a couple recorded at the same time). However, one man thought that there had to be a better method. He was Emile Berliner, an American who had already made his name in the development of the resistance microphone which had turned the telephone into a practical device (Fig. 7.1).

Berliner was originally from Hanover, but had gone to America in 1870 to avoid being drafted into the Franco-Prussian war.[3] His drive and inventiveness took him from doing odd jobs to being a researcher for the American Bell Company. Having become financially secure from the proceeds of his various patents he decided to leave and set up as an independent inventor.

His particular interest was in being able to record and replay speech and music. He looked at the devices made by Edison and Bell and Tainter, and felt they were not the solution. He also needed to make something quite different in order not to infringe their patents. The first part of his solution was to make the recording on a flat disk instead of a cylinder. A disk is much easier to handle and store than a cylinder—it takes up less space—but crucially it opens up the possibility of making multiple copies. This was the real advantage.

The other half of his approach was to cut the groove which recorded the sound from side to side instead of the hill and dale of previous inventors. This wasn't just to be different. He argued that the stylus would find it more difficult to cut the deeper grooves and hence not reproduce the sound correctly. This was overcome in his side-to-side movement. Another advantage was that the constant depth of the channel could be used to guide the stylus so it no longer needed to be moved by a mechanical screw on the player. In the previous designs, a period of low indentation would cause it to lose the track. This meant a simpler player.[4]

Berliner began to experiment with different types of disks. For the master, he settled on a zinc disk with a wax coating. This could then be put into an acid bath for a controlled time and the zinc partially etched away hence making a substantial master. He then plated this to produce a negative version which could be used to produce 'stampers' to press copies of the original.

The material for the final disks caused him a considerable amount of trouble. He tried plaster of Paris and sealing wax before turning to the new substance, celluloid. None of these were really satisfactory. The next attempt was to use a hard rubber, which was reasonably satisfactory, before he finally came up with shellac. This, though breakable, produced a durable and practicable record which could be played many times.

Berliner protected his ideas for what he called the 'Gramophone' by taking out patents in the US, Britain, Germany and Canada.[5] As he came up with further improvements he also protected these. The next step was to try to set up companies to exploit the ideas. However, while demonstrating his ideas in Germany he was approached by a toy manufacturer, Kammerer and Reinhardt, to make his machines. So the first records and players were produced in his native Germany as toys. These used 5-in. records and a tiny player.

In America he set up a Gramophone company in partnership with Eldridge Johnson who contributed a clever spring-wound motor, hence producing a playback machine that ran at a constant speed. Outside the US, he founded the Gramophone Company in England, Deutsch Gramophon in Germany, and also the Gram-o-phone Company in Canada.

In America he made an agreement with an experienced salesman, Frank Seaman, to distribute his machines and records. At first this seemed to go well, but Seaman had his own agenda and wanted to control the business himself. While still selling Berliner machines he set up the Zonophone Company to distribute and sell his own machines and records.[6]

Berliner reacted, not unnaturally, by refusing to supply Seaman any further. However, Seaman managed to get an injunction to prevent Berliner operating. He was also able to dispute the validity of the Berliner patents and gained a judgment in another of those disputed patent cases that seemed to happen around the late nineteenth century.

In disgust, Berliner went to Canada where he could still operate, assigning his US patents to Johnson and leaving him to continue the fight. Eventually Johnson triumphed, and set up the Victor Talking Machine Company with Berliner still in the background. Though there was some competition they now had a successful business as their shellac records rotating at 78 rpm became the *de facto* standard. The sales really began to grow.[7]

By the early twentieth century the Columbia Record Company, which was a merger of the cylinder phonograph interests, was giving up the fight and starting to produce disks. In 1908, they introduced double-sided disks which up to then had only been recorded on one side. In America, and the other countries where these companies operated, the sales of records and players increased dramatically once records of reasonable quality and low-cost players were available. By 1913, it was estimated that a third of British homes had a gramophone.[8]

It was Britain that was the source of the famous 'His Masters Voice' logo. Francis Barraud had painted a picture of his old dog looking into the horn of a cylinder phonograph. When the manager of the Gramophone Company saw the picture he asked if it could be changed to show one of their disc machines instead. Barraud modified the picture and sold it to them together with the slogan 'His Masters Voice'. The Berliner-associated companies made very good use of this as a trademark, which was shortened to HMV but accompanied by the picture of the dog and a gramophone.[9]

Even though the records only lasted for some 4 min, the ability to reproduce the work of real singers and musicians made them very popular. The companies persuaded famous singers such as Caruso and Dame Nellie Melba to record for them. By 1914, the Gramophone Company was selling approaching four million records a year and it had collected many competitors such as Columbia Records.[10]

All this was achieved without any electronics to amplify the signals. Recording took place in a small studio with the musicians huddled around the mouth of a giant horn (Fig. 7.2). This usually projected from a curtain, which was there partly to deaden any sound from the recording apparatus, though this was unlikely to be a problem with the insensitivity of the system. More importantly, it was there to hide the important recording equipment which the companies wanted to keep secret.

On the narrow end of the horn was a diaphragm that vibrated with the sound, moving a stylus which cut a groove in a wax master record which was slowly inched sideways as the recording progressed. Thus it was the opposite way round to the playback where the stylus and horn moved as it worked across the record. The recording horns were simply too big and heavy to do anything else.

Fig. 7.2 The composer Edward Elgar recording with an orchestra. The horn can be seen projecting on the right. The tape wound around it is to dampen resonances. *Source:* http://www.charm.rhul.ac.uk/history/p20_4_1.html

The wax recording was then plated to produce a copper master which was a negative. From this was made a 'mother' which was a positive and hence a durable copy of the original vulnerable wax master. The mother was then used to make negative 'stampers', used to press the actual shellac records. It seemed a complicated process, but it only had to be done once and then large numbers of saleable copies could be made of popular songs such as 'Ma he's making eyes at me', 'Yes we have no bananas', 'Tea for two', 'Blue moon', and many more.

When J.P. Maxfield and H.C. Harrison at Bell Labs, using their experience with telephone and public address systems, produced an electronic means of recording in 1925 the poor quality of mechanical recording became very evident. The electronic system was licenced out by Western Electric for other record companies to use.[11] The changeover was rapid.

In Britain, the Columbia Record Company—now British controlled and also owning the American Columbia—was searching for a way to avoid paying the royalties on the Western Electric system. In 1929, they employed a bright young electronics engineer, Alan Blumlein. He and his team were tasked with finding a method of recording that avoided the patents.[12]

They produced a moving coil cutter with electronic adjustment to control its characteristics, which performed even better than the Western Electric system's moving iron head with its mechanical filtering. Now really good recordings could be made. At the same time, electronics could be applied to the players to give better reproduction and, particularly useful in halls, greater loudness.

The coming of electronics turned out to be a mixed blessing. While it enabled better recording systems, it also brought two serious competitors: talking films and radio. These soon started to have a drastic effect on the fortunes of the record industry, particularly in 1930 when a slump set in. Sales and profits fell sharply which soon led to a merger between Columbia and the Gramophone Company with the new company called Electrical and Musical Industries (EMI). The search for other business areas led them into television which was where Alan Blumlein was to have yet another impact (see Chap. 4).

Blumlein was to make yet another contribution to recording. The story goes that he and his wife were at a cinema and were disconcerted by the sound coming from a single set of speakers which could be a long way from the actor on the other side of the screen. He claimed to have the answer. In 1931 he invented 'binaural sound', later known as stereo.[13] The system uses two channels which are recorded separately (in the two sides of the same record groove) and manipulated in such a way that, if the listener is in the correct place between the two loudspeakers, the sound appears to come from a particular position. Hence, when applied to a cinema film the sound followed the actor across the screen. The system only started to appear on records in the late 1950s. (Blumlein was to go on to work on the development of the H2S airborne radar during the Second World War. Unfortunately, he was in a Halifax bomber trialling the system in 1942 when it crashed in Wales, killing all the crew. He was only 38.)

One device that helped save the record industry was the automatic player, which was introduced in the 1930s, and after 1940 was known as the juke box.[14] This could play a record from a large number stored inside. A customer inserted his money in the slot and then choose a song or other piece of music. The machine would then pick the record from its store, place it on a turntable, lower the stylus and start to play it, all automatically. They were popular in bars, cafés and restaurants and absorbed a considerable proportion of the industry's output in the 1940s and 1950s.

Throughout the 1930s the search was on for a type of record that could play for much longer than the 78. What was required was a much finer groove so that they could be packed much closer together and hence give a greatly extended track length. Also if the rotational speed was reduced the time would be increased. Two innovations made this possible. The first was a superior material for the disc, Polyvinyl Chloride (PVC), commonly known as 'vinyl'. The second was the use of synthetic gems, usually sapphire, for the stylus which could then be formed to a much finer point.

In 1948, Columbia introduced a long playing record (LP) using the new vinyl rotating at 33 rpm (actually $33\frac{1}{3}rpm$ but commonly known as 33). This used a smaller groove one-third of the size of those on the existing 78s, so on a 12-in. disk each side lasted for 25 min. In addition, the new material lowered the surface noise and produced a sound with less hiss.[15]

The following year RCA Victor—RCA had bought out the Victor Company—introduced a rival. It used a similar microgroove and rotated at 45 rpm. It was much smaller in size at 7-in. diameter but still played for the 4–5 min of the 12-in. 78.[16] Far from being competition for each other the two formats both blossomed, with the LP being used for albums and classical recordings while the 45s carried the popular 'singles'.

In the UK, these formats only appeared in 1952, but the availability of multispeed record players soon made them popular. By the end of the decade they had virtually replaced 78s. While the fortunes of the record industry had been improving since the war, it was the arrival of these formats and the pop music boom that drove sales. This was entangled with the youth culture of that period which is examined further in Chap. 9 (Fig. 7.3).

Waiting in the wings was another recording technology. Its genesis lay back in 1898 when Valdemar Poulsen, of arc transmitter fame, invented his Telegraphone.[17] His idea was that in a telephone system where the receiver is an electromagnet and a diaphragm, if the diaphragm was replaced by a moving steel wire it would become magnetized with a copy of the telephone speech. By running the wire back under another electromagnet connected to a telephone receiver the recorded signal could be played back.[18]

The system worked, but he and his associates never managed to build a satisfactory telephone answering machine, which was what they were attempting. Later the focus changed to a dictation machine, but though this was more satisfactory it was never able to compete with the phonograph cylinder devices. What had seemed like a good idea fizzled out. It needed to wait, once again, for the electronics to catch up.

In the early 1930s, the German company AEG made a serious attempt to develop a magnetic recorder.[19] The objective was to use a new plastic-based tape with a magnetic iron oxide coating developed by the chemical company I.G. Farben. This resulted in a device they called the Magnetophone which was a high-quality recorder for use in radio broadcasting stations. The German broadcasting authority, Reichs Rundfunk Gesellschaft, became a major user and installed Magnetophones in nearly all its studios so that programs could be recorded and then replayed at a convenient time.

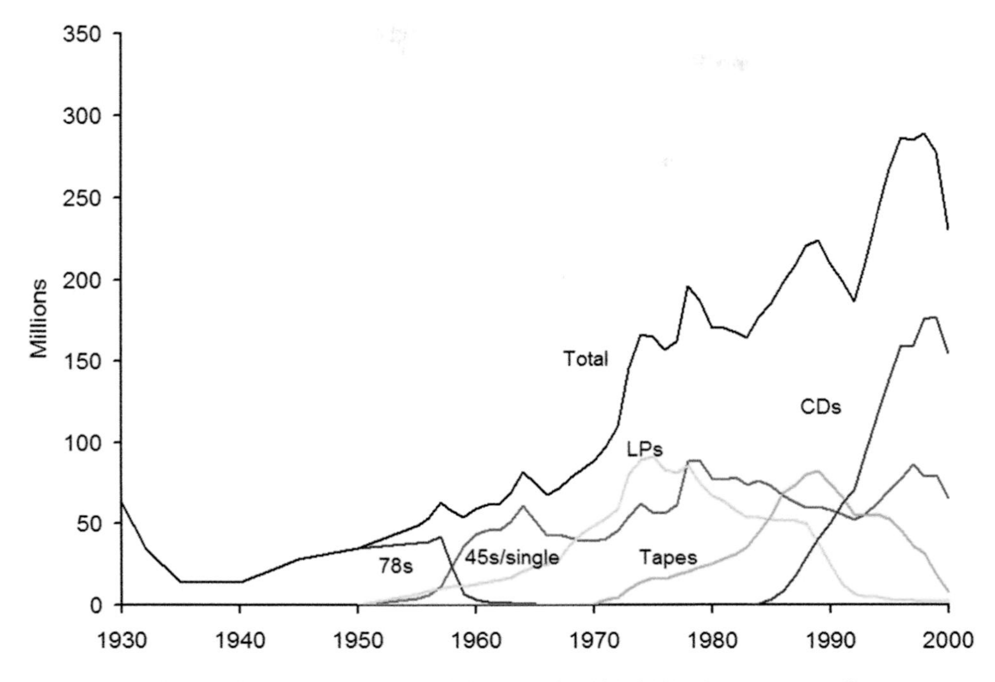

Fig. 7.3 Sales of recorded music in the UK, 1930–2000. *Source:* Author[20]

During the Second World War this technology was used in virtually all the stations controlled by the Germans in the occupied countries. The result was that magnetic recording became fairly standard in European radio stations. In America, with its bias towards live programing, there was little interest in the devices.

At the end of the war the victorious Allies were on the look out for any useful technology and soon discovered the Magnetophones. The technical details were studied and examples of the machines were sent to England, France and the US. These soon formed the basis of a renewed interest in tape recording for use in broadcasting, and the development of industries in those countries.

As the quality of these machines was higher than gramophone records, tape was soon being used to make the original recordings as it could easily be edited so small mistakes could be removed. The masters for making the records were derived from this edited master tape. The tape itself was soon available from a number of suppliers such as 3M who had taken all the information about the German tapes and produced their own versions.

As the 1950s progressed, the prices of tape recorders began to fall, and the quantities sold to rise. Though some manufacturers tried to turn them into a mass-market item the basic drawback was the tape-reel system. To play a tape the spool had to be loaded on to one spindle and an empty spool on to the other. The tape had to be threaded through the heads and started into the slot on the take-up spool winding it on until it held. The whole process was rather fiddly. In addition, the tape would have to be wound back before it could be removed from the machine.

Manufacturers started to look for a more convenient format. RCA tried a cassette where the feed and take-up spools were all enclosed and it could he dropped into the machine. While they produced the cassette they were too late with the players and the whole project fizzled out.

In Hasselt in Belgium, in a subsidiary of the Dutch Philips company, the director Lou Ottens and his team were working on a compact cassette.[21] They knew that others were trying to produce their own formats and that this was not a matter of who had the best technical solution but who established the standard in the marketplace. To achieve this they did a deal with the Japanese manufacturer Sony, and were also prepared to give other manufacturers a free licence. The result was that after its launch in 1963 it rapidly became the standard.

Though progress was made with reducing the noise or hiss from the tape it was still noticeable. One man, Ray Dolby, came up with a practical solution to this. He was an American who had come to study for a PhD at the University of Cambridge. In 1965 he set up his own company, Dolby Laboratories, in London and worked on this problem. His first system Dolby A was for professional tape systems, but in 1967 he produced Dolby B which was a simplified system for commercial tapes.[22]

The Dolby B system boosted the higher frequencies when recording, but progressively more as the signal decreased. On playback this process was reversed, thus giving an unaffected reproduction. The clever part was that it boosted the signal up away from the tape noise, thus producing an output with considerably reduced noise. When applied to the cassettes it made a marked improvement in performance. Within a few years most prerecorded tapes were using the system and it was being built into playback devices.

The advantage of the tape over records was that the user could record their own music, speech or whatever sounds they liked. One of the main uses was to record programs from the radio or make copies of other tapes or records. This was initially a considerable worry to the record industry but was never really a very serious problem. Tapes were also useful for recording speech such as a talk or a conversation and this led to them being employed in telephone answering machines. A short one was used for the outgoing message and a longer one to receive incoming ones.

By the late 1970s, the manufacture of consumer electronics such as tape machines had largely been taken over by Japanese companies with Sony in the fore. It was in that company that a revolutionary product evolved. The concept was to do away with the recording function and make a small portable cassette player that used headphones. Crucially, the device had to be small and light enough to be carried about by the user. It was launched in the summer of 1979.[23]

Despite a slightly shaky start, the 'Walkman' was enthusiastically taken up by young people, justifying the faith of the Sony chairman Akio Morita. No longer was music only available from a fixed record player or tape player, but it could be taken around with the listener. It became a personal player rather than a collective one, thus changing the nature of the market for recorded music.

It was not long before the other manufacturers caught up and the market for personal players was born. With the advent of miniature headphones that were pressed into the ear the players soon shrank to the point where they were little bigger than the cassette. A measure of the success of the approach was that in 1989, after 10 years in production, Sony had produced more than 50 million Walkmans. By 1992 this had reached 100 million and 150 million in 1995. In the late 1980s and early 1990s tapes outsold all other forms of recorded music, but there was another competitor on the horizon.

Interestingly, it came from the same source. At Philips, Lou Ottens was well aware that it is better to make your own product obsolete rather than wait for someone else to do it. He was thus interested in a more satisfactory replacement for the vinyl LP. He took a look at the work Philips had been doing on video discs where the information was recorded on a metalized plastic disk and read with a laser. This had the advantage that there was no surface contact with the disk and it would last much longer.[24]

During the development it was realized that there were considerable advantages if the signal was recorded digitally. The sound was broken up into very short sections and each analogue level was converted into a digital number. This stream of digital information could be stored on the disk. At the playback the ones and zeros recorded on the disc could be read by the laser and converted back into a varying analogue signal which would be an accurate image of the original.

By 1979, Philips were able to demonstrate this system, but they were well aware that in order to launch a system so different from the conventional records they needed partners and try to establish an international standard. The obvious company to turn to was Sony, who were already experimenting with digital recording, and with whom they had collaborated on the tape cassette. By later that year the two companies had set up a joint development team to establish the standard for the new disk and system.

Within a year they had agreed on a disk 12 cm in diameter instead of the 12 in. of an LP. This would hold 75 min of music, importantly, as Ohga of Sony insisted, large enough to hold all of Beethoven's Ninth Symphony. The other more technical aspects of the specification were also settled and they agreed that digital error correction would be needed to overcome the slight flaws in the disk and produce a near perfect sound.

The production of a marketable product was another matter. The concept was pushing the technology in a number of areas such as semiconductor lasers, micro lenses, and the electronics which was only practicable in large-scale integrated circuits. This was in addition to the problems of the disks themselves. Also they had to convince the record companies to adopt the new standard. Here they had an advantage as Philips owned Polygram and Sony were joint owners of CBS/Sony. Even so, it took considerable persuasion to get them to move away from the LP format which had stood them so well for so long.

By the end of 1982, the two companies, and their associated record companies, were able to launch the system. Sony and Philips each produced players and the first few titles were also available on the new shiny Compact Digital Audio Discs. Despite initial criticism, the system began to be accepted. The high quality of the reproduction and the small convenient size of the discs soon convinced all the competitors to adopt this standard.

Sony were obviously going to try to produce a small portable player equivalent to their Walkman. In 1984, they achieved this and, like the original Walkman, it caused the market to take off. Sony had gambled by introducing the product at a low price and it took 18 months before the volume had driven costs down sufficiently for it to become profitable.[25]

The success of the format was such that in the early 1990s the sales passed that of tape and LPs and continued to rise. It was the most successful format of all for holding music in a physical form. The take-up of players was very rapid; in Britain some 20% of households had at least one CD player in 1990 rising to 77% a decade later.[26] This was a remarkable change seeing that many (probably most) people already had a good deal of the music on LP or tape. The superiority of the format meant that the users were prepared to rapidly renew their collections.

Of course, turning music into a digital format opened up other possibilities with the advent of computers. As used on CDs, the problem was that it generated very large amounts of data. What was needed was some means of reducing that. In Germany, professor Dieter Seitzer of the University of Erlangen was interested in the possibility of sending high-quality music down telephone lines. To do this, the digital signal would have to be very severely compressed without degrading the reproduction.

It was one of his students, Karlheinz Brandenburg, who took up the baton and worked towards a practical system. In 1989, he and his collaborators took out a patent in Germany.[27] The International Standards Organization also set up a subgroup to produce standards for compression of video and audio signals. It was called the Motion Picture Expert Group and the standard for audio they produced was heavily based on Brandenburg's work. They called the new standard MPEG (from the name of the group) audio layer III, better known as MP3.[28]

Though the development of the standard took some years, it was finally launched with the MP3 name in 1995.[29] Gradually interest increased and files that could be run on personal computers began to appear. They were very efficient as they were 12 times smaller for the same music when compared with a CD. It was, however, the appearance of small players with memories sufficient to save large numbers of records that changed the game once again. The music tracks could be downloaded from the internet or transferred from a CD.

Though these devices were beginning to have an impact before the end of the century, as can be seen by the reduction in CD sales, it was after the millennium with the advent of Apple's iPod that they really took off.[30] Music had thus lost touch altogether with the physical format that was used to bring it to the listener. It was just an ethereal collection of ones and zeros that could be transported weightlessly to provide entertainment whenever it was required.

Elgar's idea that there was music all around has come to pass, though not in the way he was meaning. The world is now full of it—too full, some would say—and people can have as much of it as they require. A whole series of inventions and a pile of technology has transformed the situation from where music only existed when played live to one where each person can listen to it when and where it is desired.

NOTES

1. The Pianola Institute, History of the Pianola—Orchestrions, available at: http://pianola.org/history/history_orchestrions.cfm.
2. Beardsley, R. and Leech-Wilkinson, D. (2009) A brief history of recording. Centre for the History and Analysis of Recorded Music, available at: http://www.charm.kcl.ac.uk/history/p20_4_1.

3. The Library of Congress, Special presentation: Emile Berliner, available at: http://memory.loc.gov/ammem/berlhtml/berlemil.html.
4. Harvey, W.F. (1912) Dr. Berliner: Master inventor. *The World's Work*, Volume 24, p.675.
5. First US patent 372,786 and UK patent 15,232 1887.
6. 'Phono Jack', Emile Berliner, available at: http://phonojack.com/Berliner.htm.
7. Estreich, R. Emile Berliner, available at: http://www.bobsoldphones.net/Pages/Essays/Berliner/Berliner.htm.
8. History timeline, available at: http://www.emimusic.com/about/history/timeline/.
9. Nipper, available at: http://en.wikipedia.org/wiki/Nipper.
10. EMI Archives, 1887–1919, available at: http://www.emiarchivetrust.org/about/history-of-emi/.
11. Beardsley, R. (2009) High quality recording and reproducing of music and speech. Centre for the History and Analysis of Recorded Music, available at: http://www.charm.rhul.ac.uk/history/p20_4_1_3.html.
12. Alan Blumlein, available at: http://en.wikipedia.org/wiki/Alan_Blumlein.
13. British patent 394325. The application was in December 1931 and it was granted in 1933.
14. Bakker, G. The evolution of the British recorded music industry. BIS, Economics paper No 6, Department of Business, Innovation and Skills, also available at: http://webarchive.nationalarchives.gov.uk/20121212135622/http:/www.bis.gov.uk/assets/biscore/economics-and-statistics/docs/10-781-bis-economics-paper-06.
15. Vinyl Record Company. History of vinyl music records, available at: http://www.vinyl-record.co.uk/Pages/VinylRecordHistory.htm.
16. 45rpm.org. A history of the gramophone record, available at: http://www.45-rpm.org.uk/history.html.
17. Clark, M. and Nielsen, H. (1995) Crossed wires and missing connections. The American Telegraphone company, and the failure to commercialize magnetic recording. *Business History Review*, 69:1, 1.
18. Gavey, J.(1901) Poulsen's Telegraphone. *IEE Journal*, 30:151, 969.
19. Morton, D.L. (1993) 'The Rusty Ribbon': John Herbert Orr and the making of the Magnetic Recording Industry, 194–1960. *Business History Review*, 67:4, 589.
20. Sources for data: from 1973 to 2000, ONS, Social Trends 40: 2010 edition Chapter 13: Lifestyles and social participation; Before 1973 the figures are less secure. 1950–1973 based on production figures from Gouvish, T. and Tennett, K. (2010) Peterson and Berger revisited: Changing market dominance in the British popular music industry, c.1950–80. *Business History*, 52:2, 187–206, Table A1; UK recorded music production (millions), 1955–1980 which are compared with the overlap period from 1973 to 1980 to produce ratios between producing and sales. Before 1950 the figures are derived from Bakker, G. (2011) Adopting the rights-based model: Music multinationals and local music industries since 1945. *Popular Music History*, 6:3, 307–343, Figure 5, and assuming that the real value of records didn't change during the period back to 1930. Before around 1973 the figures should be treated as only a general guide.
21. Lou Ottens, Godfather of the cassette and CD, available in Dutch at: http://www.mfbfreaks.com/geschiedenis/lou-ottens-peetvader-van-cassette-en-cd/.
22. Dolby, R. (1971) A noise reduction system for consumer tape recording. Audio Engineering Society, Paper of the Convention 1971, available at: http://www.dolby.com/uploadedFiles/Assets/US/Doc/Professional/Ray-Dolby-B-Type-Noise-Reduction.pdf.
23. Sony History, Chapter 6, Just try it, available at: http://www.sony.net/SonyInfo/CorporateInfo/History/SonyHistory/2-06.html.

24. Riezenman, M.J. IEEE Milestone Honors, Creation of Compact Digital Audio Disc Player, IEEE 6 Feb 2009, available at: http://theinstitute.ieee.org/technology-focus/technology-history/ieee-milestone-honors-creation-of-compact-digital-audio-disc-player328.
25. Sony History, Chapter 9, Opposed by everyone, available at: http://www.sony.net/SonyInfo/CorporateInfo/History/SonyHistory/2-09.html.
26. General Household Survey.
27. Deutsch patent DE 3912605.
28. Bellis, M. The History of MP3, available at: http://inventors.about.com/od/mstartinventions/a/MPThree.htm.
29. MP3 Licensing, The History of MP3, available at: http://www.mp3licensing.com/mp3/history.html.
30. iPod history, The history of mp3, available at: http://www.ipodhistory.com/.

8

The Crystal Triode: The Transistor

It has today occurred to me that an amplifier using semiconductors rather than vacuum is in principle possible.

William B. Shockley, laboratory notebook, 29 December 1939.

After the Second World War, electronics was beginning to run out of steam as the limitations of thermionic vacuum tubes were being reached. Their shortcomings were their size, power consumption and reliability. There was what was called 'the tyranny of numbers', a phenomenon where systems became so complex that the losses from failures and downtime started to exceed the benefits. This applied to early attempts to build computers, but also to telephone systems. In America, Bell Labs, the research arm of the American Telephone and Telegraph organization, was keen to find a solution.

Despite a cast of thousands, spread across numerous companies, one man and one organization stood out from the crowd. The man was William Shockley and he, of course, worked for Bell Labs. His scientific contribution was enormous, dominating the early stages of development, and was often the key driver in all the main semiconductor devices. His character flaws, surprisingly, also contributed to the foundation of Silicon Valley.

Despite the vacuum tube diodes totally displacing cat's whisker crystals (see Chap. 3) as rectifiers in radio receivers, there were still people who wanted to know how they worked. The interest was increased with the discovery that copper oxide coupled with copper as well as selenium could also be made to rectify electrical signals. Despite the development of quantum mechanics and considerable advances in solid state physics, by the outbreak of the Second World War these devices were still not fully understood.

What changed everything was the coming of radar and in particular the quest for ever-higher frequencies (see Chap. 5). Though the magnetron had solved the problem of generating the required short wavelengths, there was still a problem detecting them as this was beyond the practical range of vacuum tubes. It was then that attention returned to cat's whisker detectors as they were found to work at these short wavelengths.

The problem with these, recognized by every owner of a crystal set, was that they were difficult to use and it was necessary to fiddle about with the wire on the surface of the crystal until a good spot was found. In Britain, the need for the devices was so great that

J.B. Williams, *The Electronics Revolution*, Springer Praxis Books,
DOI 10.1007/978-3-319-49088-5_8

commercial versions were produced in which they were pre-adjusted to work. However, with a knock or simply with time they would often lose the spot and cease working. The solution was to make them into a cartridge that could simply be replaced.

It was found that tapping the device during manufacture improved its performance and a special wax filling helped protect against vibration.[1] This was roughly the position when the Tizard mission went to America and disclosed the cavity magnetron. Furious work began on both sides of the Atlantic to make better crystal diodes.

It was soon discovered that the variability was due to impurities in the material, now silicon, and also its polycrystalline nature. When very pure silicon was produced (a difficult task) it was found not to work very well and what was required was a controlled amount of impurity, usually aluminum. Now, at last, reasonably repeatable and reliable diodes could be made. They were turned out in vast numbers, using either silicon or the related element germanium, for radar systems during the war years.[2]

Working for Bell Labs' radio department at Holmdel, New Jersey, in 1940 was a bespectacled Pennsylvanian of Dutch descent called Russell Ohl.[3] His interest was in very short wave radio communications and this had led him to study the properties of silicon. He realized that the purity of the silicon was important, and had some melted in a crucible, the theory being that the impurities would migrate to one end leaving a slab of purer material. Cutting pieces from this ingot he set about testing their properties to find which were suitable for making point contact diodes. However, one gave very strange results. He didn't think much about this and put it on one side.

Some time later he returned to investigate this rogue piece of material and discovered that, when a light was shone on it, a voltage was produced at the connections on the ends. This dumbfounded the researchers at Bell Labs, but it was then noticed that the piece of silicon had a crack. Further investigation showed that the material was different each side of this crack. What had happened was that when it had been cut from the ingot it contained material that was pure at one end and impure at the other with a sharp barrier between the two in the middle.

It was also found that the slab of silicon performed as a rectifier without the need for a point contact. Ohl applied for patents on the device as a photocell and also as a rectifier but it is clear that quite what he had was not understood at the time.[4] What was appreciated was that there were two sorts of silicon, one with excess electrons and one with a deficit. These were due to the different impurities present.

The presence of small quantities of phosphorus or arsenic gave the excess electrons and they called this n type. If the impurities were boron or aluminium there was a shortage of electrons (known understandably as holes) and this was named p type.[5] The boundary between these two types in the sample was known as the p–n junction. It was only after the war that the full significance of this discovery was understood and its considerable impact felt. In the meantime, Bell Labs kept very quiet about it.

At the end of the war Mervin Kelly, Bell's director of research, was keen to return to investigations of solid state physics. He was convinced that this would be an important area for the future of the company. He knew exactly who he wanted to head the group, and made sure that he got him back from his war work. That man was William Shockley (Fig. 8.1).

Shockley had been born in London before the First World War, but his parents were both Americans involved in the mining industry. They returned to California and he went to school in Hollywood where he was so competitive that he stood out in classes full of the children of pushy film industry people. His results were good enough for him to go to

Fig. 8.1 John Bardeen, William Shockley, Walter Brattain at Bell Labs in 1948. *Source:* https://en.wikipedia.org/wiki/William_Shockley#/media/File:Bardeen_Shockley_Brattain_1948.JPG

California Institute of Technology (Caltech) to study physics where he had no difficulty in getting a degree.

He then went east, and in the depths of the Depression enrolled on a postgraduate course at the Massachusetts Institute of Technology (MIT) determined to learn quantum mechanics. His thesis for his doctorate was calculating how quantum waves of electrons flowed in sodium chloride, otherwise known as common salt. This seemed an abstruse subject, but it taught him a great deal about the actual workings of solid state physics.

Needless to say, once he had his doctorate, Kelly at Bell Labs snapped him up. In 1938, he was part of a small group carrying out fundamental research on solid state physics with the intention of eventually finding new materials or methods which would be useful in the telephone business. In America, research, while wanting to understand the basics, always kept one eye out for producing something that would be useful. By contrast, the many European physicists who had developed quantum mechanics were only really interested in the pure theory.

Before they could get very far war intervened, and William Shockley, like many of his colleagues, was snapped up by the military to help win it. He became involved in Operational Research on antisubmarine tactics and on high altitude radar-assisted bomb sights. In 1945, he produced a report estimating the possible casualties from a land invasion of Japan; he suggested that this could lead to 1.7–4 million American casualties and influenced the decision to drop the atomic bomb on Hiroshima.[6]

Once back at Bell Labs his multidisciplinary group included not only physicists, but engineers, chemists and metallurgists. Among them were experimental physicist Walter Brattain, who had worked at Bell Labs for many years, and new recruit theoretical physicist John Bardeen. The two men, so different in many ways, were soon to become firm friends and worked well together.

Shockley set the group to work on something he had tried before the war. This was to make a device similar to a vacuum tube, but in a solid piece of semiconductor material. He argued that a vacuum tube was a stream of electrons traveling from cathode to anode which was controlled by the electric field from the grid, so why couldn't an electrode on the surface of a piece of germanium or silicon affect the electrons flowing through a channel in it?

Though both Shockley and Bardeen calculated that it ought to work using the current theories—it didn't. This led to many months of work with Bardeen largely doing the theoretical work and Brattain the practical experiments. Bardeen developed a theory about what was happening at the surface of the material, the so-called surface states. The experimentation was mainly directed at understanding these states.

By late 1947 they were beginning to see some progress. It seemed sensible to use the well-tested point contact to probe the piece of germanium, but they found that they could break down the surface layer with an electrolyte and make another connection via this. These crude three terminal devices did show some gain, but were not really satisfactory. It took only a couple more steps before they realized that what was needed were two point contacts very close together.

Just before Christmas 1947, Brattain made a very crude device by sticking a strip of gold foil around the edges of a plastic wedge. He then cut a very fine slit in it at the point. This was then held down on the slab of germanium by a spring. First they checked that the two points worked as rectifiers and then came the moment of truth. Sure enough, when connected to suitable voltages the device amplified a signal. They had invented a solid state amplifier, but it worked by an entirely new process. It didn't depend on the electric field like a vacuum tube (Fig. 8.2).

At first sight it is surprising that this device, which became known as the point contact transistor, hadn't already been invented. With all the work that had been done on point contact diodes it seems a logical step to try another point contact. In fact, there were stories

Fig. 8.2 The crude first point contact transistor. *Source:* https://en.wikipedia.org/wiki/History_of_the_transistor#/media/File:Replica-of-first-transistor.jpg[7]

that, many years before, radio operators sometimes used another whisker and a battery to get more sensitivity. Nothing seemed to have come of this.

However, in France it did. Two German radar engineers, Herbert Mataré and Heinrich Welker, who had been involved in the race to catch up when the Allies' lead in centimetric radar had been discovered, found postwar work in 1947 with the French company CSF Westinghouse[8] which had a contract from the Ministry of Posts and Telephones to develop point contact diodes. These diodes were successfully put into production, but Mataré had worked on a dual diode device during the war and wanted to investigate it further. It didn't take him very long to bring the two points close together and, early in 1948, produce a device that could amplify a signal.

The French PTT publicized the device and named it the 'Transistron'. Due to Bell Labs' secrecy there is no way Mataré could have known of their device so it is a clear case of simultaneous invention. Though this device was patented in France and used by the PTT it didn't have much commercial success despite being, from some reports, rather better than the American devices.

Meanwhile, Bell Labs were pouring resources into producing a commercial device, and patenting everything around it. Shockley stunned Bardeen and Brattain by wanting a general patent starting from his field effect ideas. That way his name would also go on it. The patent lawyers investigated and soon framed one around just Bardeen and Brattain's work. This annoyed Shockley intensely.

Worse was to follow. He tried to patent his ideas for the field effect device but the patent agents discovered that, back in 1930, a German/Polish immigrant called Julius Lilienfeld had described and patented very similar devices so Shockley's ideas were unpatentable.[9]

This spurred Shockley into a furious period of intensive work. He was not very impressed with the point contact transistor and was convinced that he could come up with something better. For a considerable time he worked secretly—he was going to ensure that no one could cut him out this time. When he did disclose his ideas he made sure that those involved worked on specific investigations controlled by him. It was noticeable that Bardeen and Brattain were excluded from this.

Shockley wanted to start from Ohl's p–n junction. His proposal was that, if another junction was formed so that there was a p–n–p or n–p–n three-layer sandwich with a sufficiently thin central layer, the device should produce amplification. The investigations of his research group around this area suggested that such a thing might work. The problem was how to make it.

After some false starts the filing for the patent on the point contact transistor went in on June 17, 1948.[10] At last Bardeen and Brattain were free to publish what they had done. The momentous announcement was a low-key letter to the editor of the journal *Physical Review*.[11] However, on June 30, Bell Labs held a large-scale press conference to announce the transistor. By this time the devices had been made into small cylinders not unlike the cases for the point contact diodes. They were demonstrated amplifying the speaker's voice, which impressed the audience.

Such was the demand that a prototype production line was set up to supply experimental devices. Unfortunately, these were not very satisfactory as the characteristics varied wildly and they were very noisy. In addition, the devices were unstable. As someone put it, 'the performance of the transistor was likely to change if someone slammed the door'.[12]

By mid-1949 more satisfactory devices were being made and in 1950 were put into production by Western Electric.

Although Bell Labs licenced the manufacture to other companies the transistors were never made in really large quantities. The total output in June 1952 was about 8400 a month, most of those from Western Electric.[13] The users were the military and particularly AT&T themselves, though some were used in hearing aids. The problem was that they had to be made by hand and this meant that it was not possible to get the price down to the levels that had now been reached for vacuum tubes. Thus, only users who could not tolerate the power consumption of these were likely to switch to transistors. Also there was something better on the horizon.

Shockley wasn't going to make any mistakes this time and his patent application went in shortly after Bardeen and Brattain's.[14] However, despite progress it wasn't until April 1949 that he had something that worked. Morgan Sparkes, one of his team, fused a drop of p-type germanium on to a slab of n-type, forming a p–n junction. He then managed to cut a slot so that there were two sections of p type sticking up from the slab. It was rather like a point contact device but with p–n junctions instead of point contacts. It was crude but it proved that the concept was correct.

One thing that had become clear was the need for very pure germanium in single crystals with controlled amounts of impurities. Another of the researchers, Gordon Teal, developed a way of growing single crystals. A seed crystal was dipped into a crucible of molten germanium and then slowly drawn out. A single crystal rod formed as it cooled.[15]

The next step was to gently change the impurities in the melt half way through the process so that a p–n junction was formed. Slices cut across this junction formed junction diodes. The next step was to do the same but quickly add a second impurity (or dopant) to return the material to the original type. In the early summer of 1950 they started with n-type germanium, added some p-type and then n-type, thus producing an n–p–n double junction. However, the central player was not narrow enough and the devices cut from the ingot, though they worked, only did so at low frequencies much worse than that achieved by the point contact devices.

It took until January 1951 to get the central region narrow enough. Now the devices, when cut in narrow slices, could amplify signals up to 1 MHz—hardly useful for radios—and this was still lower than the point contact devices. In all other respects they were far superior. The noise was very low and they were robust and stable. The manufacturing technique was called the grown junction method.

In July, Shockley published a paper describing the transistors and giving results of the tests on those they had made.[16] This was the prelude to a symposium on the device. Once the patent was granted in September (it had only taken 3 years since filing), the company started granting licences to make them to more or less anyone who paid the $25,000 fee. In the following spring they held a training course for the 40 companies, 26 American and 14 foreign, who found the entry fee.

This might seem an extraordinary action for the company to earn just $1 million after all the effort they had put in, but they were under pressure. The military, who had contributed some of the costs, were keen that the knowledge was widely disseminated so that they had more than one supplier. In addition, AT&T was under threat of an anti-trust action which meant that they were in danger of being broken up for being a monopoly. Licencing the transistor technology would help in their defence.

It might seem that only Bell Labs were in the game, but it wasn't so. In June, even before Shockley's paper, General Electric announced prototype transistors. More importantly, they were made by a different process. They took a thin slice of n-type germanium and placed small blobs of indium on each side and then heated it. The indium melted and dissolved into the germanium producing p-type regions. By stopping at the right moment a thin central (base) layer could be left, and a p–n–p transistor produced.

This was a simpler way to make transistors and RCA soon adapted the process to mass produce what were now called alloy junction transistors. While they still had to pay the fee to Bell Labs they had stolen a march on them and were arguably ahead. The subject had now changed from research to manufacturing, and here Bell Labs no longer held all the cards.

Now that the work on the junction transistor was more or less complete Shockley could move on (as a theoretician he was not involved in the problems of manufacture). His interest returned to the field effect device, the analogue of the vacuum tube, which had eluded him ever since he had started to investigate solid state devices. Now the solution to the surface states problem was simple. If you took a channel of, say, n-type germanium and put p-type layers each side of it, then a reverse voltage would pinch off the channel preventing current flow.

Shockley worked through the theory and published a paper in November 1952.[17] At this stage only a few experiments had suggested that such a device was possible, but one hadn't been made. It took two of his co-workers, Dacey and Ross, most of the next year before they were able to make such a device and publish the results.[18]

It wasn't until 1960 that field effect devices using an insulating layer and a metal gate region were made by Kahng and Attalla at Bell Labs.[19] This led to a number of companies making these devices, but they suffered from contamination in the oxide insulation layer which took many years to fully eliminate. Though these devices were not of great interest at the time, it was to be these Metal Oxide Semiconductor MOS devices that were going to have enormous impact in the future (see Chap. 11).

Shockley had thus been involved in all three types of transistor. He was in charge of the team that produced the point contact transistor and undoubtedly contributed to the line of development that led to its discovery. He invented the junction transistor and drove the team to research and make it. He eventually worked out how to make a field effect transistor and got his team to fabricate it, even though it was unpatentable. Therefore, he was the key person that had driven the whole invention of the different types of transistor, justifying Kelly's faith in him. However, he still had a contribution to make to the industry.

By 1954, Shockley and his team at Bell Labs had managed to produce a better process for making transistors.[20] The technique was to diffuse the doping impurities into the basic slab of germanium. The process could be controlled to produce the thin base layer required to achieve a high gain from the transistor. In addition, devices made by this process could reach higher frequencies, opening up the possibilities of using them in radios.

Down in Dallas was a small company called Texas Instruments that had already decided to get into the transistor business. One stroke of luck for them was that Gordon Teal from Bell Labs wanted to return to his native state and came to work for them. Their ambition was to mass produce devices and they looked around for a suitable vehicle to do this. They lit on the idea of a small portable radio and, after considerable struggles, assisted a company called Regency to design and make one in late 1954. Of course, it used four of their transistors and was small enough to fit in the hand.

They claim it was the first transistor radio, but probably they were unaware of an earlier product. A German company called Intermetall had produced one a year earlier, based around point contact transistors made using the expertise of Herbert Mataré who had joined the company. Needless to say, once the idea was out then other companies latched on to the it and within a few years the transistor radio or 'trannie', mostly made in Japan, became the essential accompaniment for all teenagers.

Also in 1954, Teal and the team at Texas Instruments succeeded in making silicon transistors. He was still wedded to his grown junction method, but they got them into production and sold them to the military at high prices like $100 each, even though this wasn't the ideal method of manufacture.

Bell Labs were still working on developments and their next step was to try to produce silicon transistors by the diffusion method. This was much more difficult for a number of reasons, not least its much higher melting point which meant it was difficult to find any suitable material for the crucible. The reason for this attempt was that silicon devices were potentially superior with lower stray currents when the device was supposed to be turned off and also they would work at higher temperatures. By 1955 they had succeeded. This, they realized, was the way forward: silicon as the material, and diffusion as the process.

In the mid 1950s, Shockley had what was described as a mid-life crisis.[21] He felt blocked at Bell Labs as other men were promoted past him. His management style, as well as his treatment of Bardeen and Brattain and their subsequent protests, had been noted. He looked around for something else, trying university teaching and returning to government in Washington. Eventually he realized what he wanted to do, which was to run his own company. It was the only way he could benefit from his discoveries as Bell Labs didn't give him any royalties on his patents. In late 1955, after a week's discussions with the entrepreneur Arnold Beckman, Shockley Semiconductor Laboratories was set up.

Shockley wanted to return to California and the site chosen was the San Francisco Bay area close to Stanford University where there would be some mutual advantages. Shockley set about recruiting people, which with his reputation was not difficult. Leading people in the industry were only too pleased to be asked by the great man to join him. It was noticeable that very few came from Bell Labs, despite his attempts. Perhaps they knew too much about him.

At first all went well with the stated objective of making silicon transistors by the diffusion method, though progress was slow. However, Shockley kept sidetracking his people to work on his latest thoughts, and got progressively bogged down in trying to manufacture one of his previous ideas, the four-layer p–n–p–n Shockley diode. This was a device like a pair of embracing transistors. When a voltage was applied it passed no current until a particular threshold was reached and then switched to more or less a short circuit. Removing the current switched it off again. Shockley had visions of selling large numbers for telephone exchanges, but the diode couldn't be manufactured with consistent characteristics and there was no way AT&T was going to use such an untried device.

In 1956, there was a great distraction as it became known that Shockley, Bardeen and Brattain had been awarded a Nobel Prize for the work on the transistor.[22] For a while this masked the problems in the company where the talented crew were becoming increasingly disturbed by Shockley's management style. He had some very odd ideas, but the constant distractions and paranoiac behavior reached a point where they could be tolerated no longer.

Despite attempts by Beckman to resolve matters, the firm was Shockley's baby and that meant that those who were not happy only had the option to leave. The biggest exodus was in September 1957, and consisted of eight people, among them Gordon Moore, Robert Noyce and Jean Hoerni.[23] Rather than dispersing they managed to find financial backing from the Fairchild Camera Corporation and set up just down the road from Shockley. They had chosen a good moment as the Soviet Union launched its first Sputnik satellite and the American government poured resources into closing the 'missile gap'.

By late 1958, less concerned with the theory and concentrating on practicalities, they had diffused silicon transistors in production. These were sold at high prices and quite soon the company was into profit. Though in some ways crude, the process was quite sophisticated with masks being used so that the areas where the substrate was doped with impurities were well controlled. They were so successful that their theoretician, Jean Hoerni, had the luxury of time to sit in a corner and think.

That was a good investment. Hoerni came up with a new process which involved creating a protective layer of silicon dioxide on the surface of the substrate and etching holes through it to dope the various areas and create the transistors. Not only were the resulting devices more stable but the process made manufacturing easier as the connections were all on the one surface. Hence this was known as the planar process. It was to allow more than one device to be made on the same piece of silicon and lead to yet another revolution as described in Chap. 11. A patent was filed on May 1, 1959.[24]

Other fugitives from Shockley Semiconductor and departures from Fairchild Semiconductor led to the setting up of ever more companies in the San Francisco Bay area and the growth of 'Silicon Valley' as companies spun out one from another over the next few years. For example, some 40 companies were formed by ex-employees of Fairchild alone.[25] Shockley had been responsible for the foundation of a whole industry, if not quite in the way he had envisaged.

Though Shockley couldn't make a go of the four-layer diodes, General Electric were also interested in them, and went in a different direction. As a company with heavy electrical engineering interests they were keen to find a controlled form of rectifier to join the power semiconductor rectifiers that they were already making. They recognized that if they had a third connection to the next layer up from the cathode it only required a small current into it to switch on the device. In fact, it only required a short pulse which was very useful in actual circuits.

In 1957, working from their silicon transistor process using combinations of diffusion and alloying they successfully made large devices which could switch several amps of current and withstand the mains voltage. As a demonstration, engineer Bill Gutzwiller bought an electric drill and used a prototype silicon-controlled rectifier to adjust its speed by altering the point on the ac cycle where the devices switched on.[26]

General Electric rapidly brought these devices into production as they had many uses in their business and for others in the industry. At first they were called silicon controlled rectifiers, SCRs, but the general name in the industry became thyristors. They could be made in large sizes to handle huge powers dwarfing the capabilities of the early transistors. They soon found their way into variable speed fans, light dimmers and automatic washing machines, and a host of other uses. By the late 1970s, thyristors were a $100 million business despite falling prices for the devices.[27]

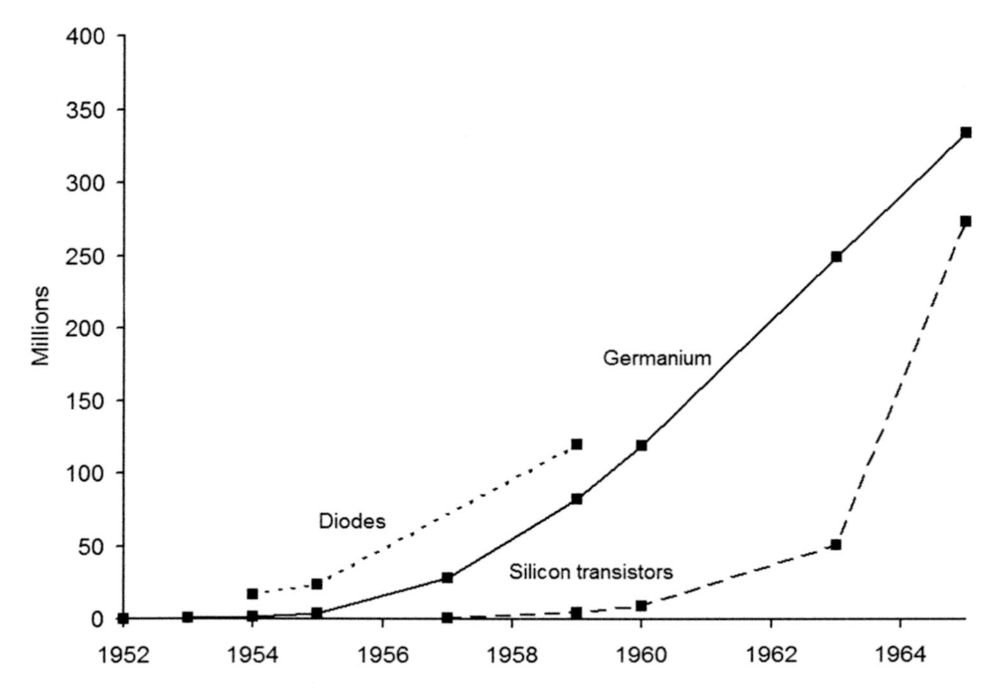

Fig. 8.3 Growth in annual sales of transistors. Silicon took over from germanium in the late 1960s. *Source:* Author[28]

The basic thyristor was to produce a whole family of specialist devices often with even more layers, such as the triac, a version of the thyristor that could cope with current in either direction.[29] This simple adaptation of adding another connection had turned the four-layer diode from Shockley's failure to something with a vast range of uses.

Though slow to start, the industry really took off in the late 1950s and particularly the 1960s. In the US, the numbers of silicon devices exceeded germanium in 1965 though it took the Japanese manufacturers, who mostly made consumer items, longer to switch over. This was driven by the falling price of silicon devices which reached that of germanium at around this time (Fig. 8.3).[30]

As mentioned above, the major impact in the domestic sphere was the appearance of the transistor radio. Most of the other advances were hidden as the steady replacement of tubes by transistors with the advantages that the transition brought in decreased power consumption, size, and failures, could not be seen directly. As shown by the takeoff in sales, the electronics industry took to them rapidly and this imparted a new dynamic. Products that had previously been impractical were now possible. However, the real revolutions were still to come.

NOTES

1. Bleaney, B., Ryde, J.W., and Kinman, T.H. (1946) Crystal valves, electrical engineers. *Journal of the Institution of Electrical Engineers, Part IIIA: Radiolocation*, Volume: 93: 5, 847–854.
2. Seitz, F. (1995) Research on silicon and germanium in World War II. *Physics Today*, January 1995, 22–26.
3. Riordan, M. and Hoddeson, L. (1999) *Crystal Fire: The Invention of the Transistor and the Birth of the Information Age*. New York: WW Norton, p.88. This chapter borrows heavily from this book.
4. US patents 2402661, 2402662, and 2443542.
5. Group four elements such as carbon, silicon and germanium have four electrons in their outer shell making their valency four. They can form crystals where each atom locks to four others forming a solid structure with effectively no free electrons. Introduction of atoms of group 5, such as phosphorus and arsenic, into the structure of the crystal means that there is a spare electron at each introduced atom which means that it can conduct a current of electrons hence being n type. However, if group 3 elements such as boron and aluminium are introduced instead, then there is a shortage of electrons, hence forming holes and the material being p type.
6. William Shockley, IEEE Global History Network, available at: http://www.ieeeghn.org/wiki/index.php/William_Shockley.
7. This is a later replica.
8. Adam, C. and Burgess, M. The first French germanium semiconductors, available at: https://docs.google.com/viewer?a=v&pid=sites&srcid=ZGVmYXVsdGRvbWFpbnx0cmFuc2lzdG9yaGlzdG9yeXxneDoxYzcxNzgxNzFjZTgxODI1; and Riordan, M. How Europe missed the transistor, IEEE Spectrum, available at: http://spectrum.ieee.org/semiconductors/devices/how-europe-missed-the-transistor; The CSF Westinghouse company (Compagnie des Freins et Signaux) was independent of the American Westinghouse company.
9. US patents 1745175, 1900018, 1877140.
10. US patent 2524035.
11. Bardeen, J. and Brattain, W.H. (1949) The transistor, a semi-conductor triode. *Physical Review*, 74:2, 230–231.
12. Riordan, M. How Bell Labs missed the microchip. IEEE Spectrum, available at: http://spectrum.ieee.org/computing/hardware/how-bell-labs-missed-the-microchip.
13. Production and delivery figures for transistors, *Electronic*s, June 1952.
14. US patent 2569347.
15. This was based on research in 1917 by a Polish scientist called Jan Czochralski. These devices are still known as Czochralski crystal pullers.
16. Shockley, W., Sparks, M. and Teal, G.K. (1951) p-n junction transistors. *Physical Review*, 83:1.
17. Shockley, W. A unipolar "field-effect" transistor. *Proceedings of the IRE*, 40:11, 1365–1376.
18. Dacey, G.C. and Ross, I.M. (1953) Unipolar "field-effect"transistor. *Proceedings of the IRE*, 41:8, 970–979.
19. Sah, C.T. (1988) Evolution of the MOS transistor—from conception to VLSI. *Proceedings of the IEEE*, 76:10, 1280–1326.
20. Seidenberg, P. (1997) From germanium to silicon—A history of change in the technology of the semiconductors. In Andrew Goldstein and William Aspray (Eds) *Facets: New Perspectives on the History of Semiconductors*. New Brunswick: IEEE Center for the History of Electrical Engineering, 1997, pp. 35–74. available at: http://www.ieeeghn.org/wiki/images/3/3a/Seidenberg%2C_1997.pdf.
21. Riordan and Hoddeson, *Crystal Fire*, p. 225.

22. Bardeen had left Bell Labs for academia to work on superconductivity. He was later to win a second Nobel Prize for his work on the theory of this, lifting him into a very select club together with Marie Curie, Linus Pauling and Fred Sanger.
23. Moore, G.E. (1998) The role of Fairchild in silicon technology in the early days of "Silicon Valley". *Proceedings of the IEEE*, 86:1, 53–62.
24. U.S. Patent 3,025,589.
25. Atherton, W.A. (1984) *From Compass to Computer: History of Electrical and Electronics Engineering*. London, Palgrave Macmillan, p. 250.
26. Burgess, M.P.D. General Electric history: Semiconductor research and development at General Electric, available at: https://sites.google.com/site/transistorhistory/Home/us-semiconductor-manufacturers/general-electric-history.
27. Morris, P.R. A history of the world semiconductor industry, p. 60.
28. Data from: Electronics, Production and delivery figures for transistors, *Electronics*, June 1952; Riordan and Hoddeson, *Crystal Fire*, pp.205/6; Seidenberg, P. *From Germanium to Silicon*; Morris, P.R. *A History of the World Semiconductor Industry*, p. 49.
29. Holonyak, N. (2001) The silicon p-n-p-n switch and controlled rectifier (Thyristor). *IEEE Transactions on Power Electronics*, 16:1, 8.
30. Morris, P.R. *A History of the World Semiconductor Industry*, p. 101/2.

9

Pop Music: Youth Culture in the 1950s and 1960s

The young always have the same problem—how to rebel and conform at the same time. They have now solved this by defying their parents and copying one another.

Quentin Crisp

At the beginning of September 1956 the UK suddenly became aware of 'Rock and Roll'. The reason was that 'disturbances' were taking place in movie theaters, and outside them in the street, leading to a considerable number of youths finding themselves in front of the magistrates. Their offence was to be 'jiving' in the cinemas and then being ejected. When they continued their dancing and singing in the street the police attempted to move them on. When this failed a number were arrested.[1] These problems continued for the next 2 or 3 weeks.

The apparent cause of these difficulties was the showing of the film 'Rock Around the Clock' featuring the band leader Bill Haley. This was all rather curious as the film was first screened on 23 July and had already been shown in some 300 theaters when the disturbances started.[2] The fact that these spread so rapidly was a classic example of a copycat phenomenon but, with hindsight, they were doing little harm, though there was some damage in the cinemas. It was showing that the older generation no longer understood the younger.

The title song had first appeared as a record at the beginning of the previous year when it briefly entered the pop music charts before disappearing again. Its use in the film 'Blackboard Jungle' (a much more violent film despite the censor's cuts) in September reignited interest in the song and it shot to the top of the chart and stayed there for 5 weeks, but remained in the charts for much of the next year.[3] It returned many times subsequently, finally achieving the status of one of the most widely sold and hence influential records ever produced.

The year 1955 was a turning point for youth culture though it wasn't that obvious at the time. It was the point where the number of teenagers began to rise sharply after an almost continuous fall since the end of the First World War (Fig. 9.1). The post-war baby boom—which actually began in the middle of the war—was beginning to have its impact. Commercial television was launched, and sales of records were increasing as the new microgroove 45s were becoming more popular. Portable transistor radios were starting to appear.

J.B. Williams, *The Electronics Revolution*, Springer Praxis Books,
DOI 10.1007/978-3-319-49088-5_9

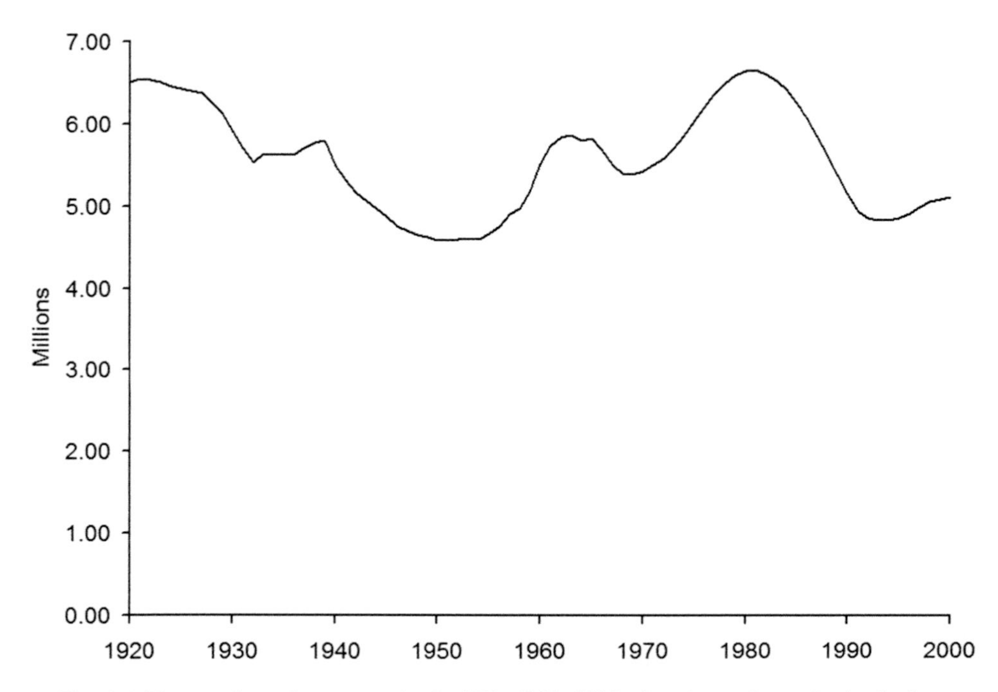

Fig. 9.1 The number of teenagers in the UK, 1920–2000, showing a sharp rise in the late 1950s. *Source:* Author[4]

The core of this new youth culture was pop music which increasingly became separated from the more traditional middle-of-the-road taste of the older generations. In Britain, it was driven by five key influences: the juke box, the portable record player, Radio Luxembourg, the transistor radio, and the electric guitar. All these, of course, derived from the development of electronics.

With the increasing prosperity of the 1950s more young people were employed on wages some 50% higher in real terms than before the war, and even those who weren't could look forward to generous amounts of pocket money from more prosperous parents.[5] Teenagers were no longer smaller versions of their parents; they matured earlier and so wanted to differentiate themselves. One trend was to dress differently—the 'Teddy boy'. This was supposedly based on Edwardian fashion and hence the name, but it had its own set of rules.

Devotees wore suits, often in various colours with 'drainpipe' trousers, thin 'bootlace' ties and 'brothel creeper' shoes with crepe soles. The hair was longer with an accented quiff at the front held in place by lashings of 'Brylcream'. At the rear it met in a vertical line known as a DA, Ducks Arse, because of its similarity to the said animal.[6] The girls, in print dresses often with splayed skirts held up by numerous petticoats, were somewhat overshadowed.

They tended to gather in the coffee bars with their Gaggia espresso machines that had sprung up, and what really attracted them was the juke box to play their favorite music. For a coin in the slot, and a press of a button, the whole group could hear the desired record.

As the recording format changed from the big cumbersome 78s to the smaller 45s the number of tunes that the juke box could hold was increased. By sharing the expense, the group could listen for hours.

What drove them out to these places (apart from not wanting to be with their parents), was that at home the means of playing records was embedded into the radiogram, that large piece of furniture that often they were not allowed to touch. Even when they were, it was in the family sitting room and this was not the place to bring friends to listen to music that wasn't going to go down well with any adults within earshot.

In the early 1950s, Birmingham Sound Reproducers (BSR) decided to make a low-cost record player deck that could handle all the record sizes, 78s, 45s and 33s, and was also an autochanger. This meant that it could be loaded with a small stack of records and set in motion. The records would drop onto the turntable one at a time and be played automatically, the pickup placing itself on the beginning of the record and removing itself at the end.[7]

This was a great advance and it was soon picked up by a furniture maker called the Margolin company which had an interest in music players. They produced a portable player under the name Dansette. This was an enormous success and soon there was a complete range of different sizes and styles but all featuring various BSR decks—some without the autochangers—and built-in loudspeakers at the front in a case covered in brightly-colored leatherette.[8] This was completely different from the conventional radiograms that looked like a piece of furniture.

The player consisted simply of the record deck, an amplifier with its mains power supply, and one or more loudspeakers. These sounded out through a special gauze at the front of the case. The cabinet was made of plywood and covered with the gaudy fabric which, while cheap to produce, gave a finish that was attractive to the young people.

The great thing about these players was that they could be in a bedroom because all that was needed was a pile of records and a plug. They could be moved from house to house if needs be (though some were fairly heavy), but that didn't matter. It meant that the gang could sit in someone's bedroom and listen to records without disturbing the household—except, of course, when it got too loud which is usually did. Other manufacturers, such as Bush, HMV and Murphy, soon piled in but it was the Dansettes which captured the bulk of the bottom of the market and became the icon for the young people.

In the 1950s, the BBC was still being true to its Reithian tradition to 'educate, inform and entertain', with the least emphasis on the last of these when it came to much that would interest the younger generation. In Britain it had a monopoly of radio broadcasting, so teenagers looked elsewhere, and that was to Radio Luxembourg.

The station, in the Grand Duchy of Luxembourg, went back to the beginnings of radio when the Anen brothers set up their first transmitter.[9] It had grown spectacularly between the wars as an independent station when most others had been subsumed into national broadcasters. During the war it had been taken over by the German government and used for propaganda broadcasts. When they retreated, the Americans used it for black propaganda.[10]

With the war over Radio Luxembourg returned to its peacetime mix, transmitting in a number of languages at different times of the day. The English programs were in the evening because it was only with the coming darkness that reception was possible over much of the country. This wasn't improved by the switch from the long to medium wave on the famous 208 m. Even with a very powerful transmitter the reception still suffered badly from fading where signals taking different paths in the atmosphere would sometimes add together to give a loud output, but at others nearly cancel out.

Despite these problems, the station was popular because it worked hard to produce programs that people wanted to listen to. Quite early on it produced the 'Top Twenty', featuring those records in the current charts. That, coupled with the sense of being a bit subversive by listening to something that wasn't quite approved of, made it attractive to the teenagers.

Luxembourg was a commercial station with its programs supported by sponsors. However, in 1955 when commercial TV started up in the UK, some of these sponsors transferred and Radio Luxembourg became more and more dependent on the record companies. This meant that their output was fewer quizzes and band shows, and more playing of records. Far from discouraging the audience this made it even more attractive to the youth.

With this came the cult of the disc jockey, the DJ. Many of them went on to make careers in this form of entertainment. These programs became a staple diet for teenage listeners and were later taken up by pirate radio stations, and eventually even by the BBC in their Radio One.

There was, however, a problem for the average teenager who wanted to listen. The main radio in the home was usually the radiogram, firmly grounded in the living room and, as mentioned above, was a large item of furniture. Sometimes a smaller set could be found that could be transferred to the bedroom and coaxed into receiving Radio Luxembourg, but what was really wanted was something small and portable.

Enter the transistor radio. As mentioned in Chap. 8 the small pocket transistor radio appeared in America in around 1954/55. There they wanted something to fit in the shirt pocket. To do that the transistor needed to be very small and that meant a tiny loudspeaker which sounded 'tinny' and didn't produce a very loud sound. There was a further problem in that the transistors were struggling to reach the frequencies required so the battery voltage had to be raised to 22.5 V. This was far from ideal.

In Britain, where people were not in the habit of going around in shirtsleeves, there was no need for the device to be that small. The result was that British radios were more the size of a handbag. The advantage was that the loudspeaker could be larger and hence the sound was louder and of better quality.

The first transistor radio in Britain was the PAM, produced by the Pye company in 1956, which fitted this profile at 9¼ × 3½ × 7¼ in and weighing just over 2½ kg with the batteries fitted.[11] It used eight transistors rather than only four.[12] The transistors had improved sufficiently that it could run from four torch batteries which was much simpler and with a longer life than the cumbersome High Tension and Low Tension batteries needed by a vacuum tube set. It was expensive at the time, but it showed the way forward.

It didn't take long for competitors such as Bush, Ultra, Roberts, GEC, Ever Ready to catch up, usually using transistors from Mullard, a subsidiary of Philips. Prices began to fall, and in 1961 two million sets were sold.[13] They tended to be quite similar in format, were easily portable, running either from torch batteries or a 9 V pack, and weighing only a few kilograms.

These radios were ideal for teenagers to carry about and gather round. At this stage, listening to music was a collective event, and preferably as loud as any adults within hearing distance would tolerate. Only later, with the advent of smaller radios, mostly from Japan, with in-ear headphones did music become a solitary practice and lead to Walkmans and MP3 personal players.

BBC TV, like its radio, wasn't really keen on pop music. It had a few shows over the years featuring hit songs but performed by resident bands and singers rather than the original artists. It wasn't until 1957 that they produced a program for teenagers called 'Six Five Special' after its time of broadcast.[14] This featured live performances where the band or singer mimed to their records in front of an audience.

In 1958, ITV responded with a livelier program called 'Oh Boy'. and gradually both channels began to take the rock and roll scene more seriously. Various attempts were made before the long running shows of 'Top of the Pops', 'Juke Box Jury' and 'Ready Steady Go' took to the air.

About this time Elvis Presley crashed on to the scene, scaring the older generation with his sexy movements. He didn't become known as the King of Rock for nothing. From the Deep South in the US, he was the great exponent of the form. A hit such as 'Don't You Step on my Blue Suede Shoes' showed the link with the teenage fashions of the day. It soon became apparent that a song like 'Heartbreak Hotel', blending the angst of teenage love with a rock beat, was a sure-fire winner.

So that he could emulate his idols, what the aspiring teenage rock star—and it was usually he rather than she—really wanted was an electric guitar. This was strung with special strings with steel cores. Underneath them at the bottom of the finger board was a pick-up which consisted of a coil of wire and metal cores under each string. When the string vibrated a small voltage was set up in the coil. Its output was fed to an amplifier and a very loud instrument ensued. It was exactly what was wanted.

The pick- up could be fitted to an ordinary acoustic guitar, but there was no need for the sounding box and the solid body guitar became the norm. Fitting it with four heavier strings, instead of the normal six, produced a far better bass sound. Here then, with a couple of guitars, a bass and the addition of a drum kit, was the line-up for the classic rock band.

There were some problems. The cable carrying the signal from the guitar to the amplifier was very vulnerable. With the player typically jumping about it was subjected to a lot of flexing and easily broke. A lot of time was spent repairing the fragile coaxial cable. This was necessary because of the small signal available from the pick-up.

The amplifier thus needed to have more amplification than that used for a record player. At this time these still used vacuum tubes because the available transistors were not powerful enough. It was normal to fit the loudspeaker into the same cabinet as the amplifier and this brought an interesting problem; the sound would cause the whole amplifier to vibrate. Now vacuum tubes are relatively large devices and don't much like being shaken about.

The first device when vibrated would often change its characteristics, thus altering the signal. This would be amplified and come out of the loudspeaker causing a 'howl round'; an ear piercing shriek. In practice, it was necessary to try many vacuum tubes until one was found that was sufficiently free of this 'microphony' for the whole thing to work properly.

Nevertheless, the rock bands loved their amplifiers, saving up for ever more powerful ones. The people who didn't like them were the adults within hearing distance, particularly the parents of the budding musicians. However, many of them settled down to play the instruments properly. Armed with their Bert Weedon guitar tutor called 'Play in a Day' (which was rather optimistic), they set to work. The follow-up 'Play Every Day' was rather more realistic, and as he was a famous English guitarist with hits to his name they wanted to emulate him.[15]

Many of the guitarists, such as Eric Clapton and Brian May, who went on to be the core of future bands, learned their craft from Bert Weedon. Among them also were the three instrumentalists, John Lennon, Paul McCartney and George Harrison who formed the Beatles, a group utterly dependent on electric guitars. These were the instruments that shaped the future of pop music.

While pop music didn't create the phenomenon of the teenagers, it certainly had a great deal to do with the way that this subculture developed. It was facilitated by that set of electrical and electronic technology that could deliver it to them. The records themselves and their production system and the jukeboxes to play them were the basis, but radio and TV to disseminate them were also important. Then there were the ways to listen the music—the portable record players and transistor radios. This also brought the desire to create the music which was served by the electric guitars and their amplifiers and electric organs. At the time, these had been very much collective occupations. It was only later that music became a personal pleasure with individuals walking the streets wired for sound.

NOTES

1. *Manchester Guardian*; September 4, 1956; *The Times*, September 4, 1956.
2. *The Times*, September 15, 1956.
3. Smith, M.J. Number Ones, the 1950s, available at: http://onlineweb.com/theones/1950_ones.htm.
4. Computed from ONS Live Birth statistics in Vital Statistics: Population and Health Reference Tables, with allowances for neonatal and deaths under 1 year. No allowance has been made for further deaths and net migration though the errors are unlikely to be large.
5. Addison, P. (2010) *No Turning Back: The Peacetime Revolutions of Post-war Britain*. Oxford, Oxford University Press, p. 179.
6. Addison, P. *No Turning Back*, p. 97.
7. Birmingham Sound Reproducers, available at: http://en.wikipedia.org/wiki/Birmingham_Sound_Reproducers.
8. Lambert, J. History of the "Dansette" with Samuel Margolin, available at: http://www.dansettes.co.uk/history.htm.
9. History of Radio Luxembourg and its English service, available at: http://www.radioluxembourg.co.uk/?page_id=2.
10. Black propaganda is false information and material that purports to be from a source on one side of a conflict, but is actually from the opposing side. It is typically used to vilify, embarrass, or misrepresent the enemy.
11. Dudek, E.J. Vintage technology, available at: http://www.vintage-technology.info/pages/history/histpamtr.htm.
12. Parker, B. The history of radio, available at: http://www.historywebsite.co.uk/Museum/Engineering/Electronics/history/TransistorEra.htm.
13. 60s Transistor radio, available at: http://www.retrowow.co.uk/retro_collectibles/60s/tranistor_radio.html.
14. Nickson, C. Rock 'n' Roll on TV, available at: http://www.ministryofrock.co.uk/TVRockAndRoll.html.
15. Bert Weedon, available at: http://en.wikipedia.org/wiki/Bert_Weedon.

10

From People to Machines: The Rise of Computers

I think there is a world market for maybe five computers.

Thomas Watson, chairman of IBM, 1943

Computers in the future may weigh no more than 1.5 tons.

Popular Mechanics, forecasting the relentless march of science, 1949

In the eighteenth and nineteenth centuries there was a great demand for tables, partly mathematical ones such as logarithms and trigonometrical functions, but also important ones for navigation. These were all compiled by hand by poor drudges called computers and, not surprisingly, they were full of errors. These occurred not only at the calculation stage, but in transcription and printing. Where text has some rationale, it allows the copier to apply intelligence; a meaningless jumble of numbers is much more prone to mistakes.

It was to this problem that Charles Babbage began to apply his mind in the early 1820s. He was a polymath with a wide range of scientific interests and was later to become the Lucasian professor of mathematics at Cambridge, a position with many illustrious holders stretching from Isaac Newton to Stephen Hawking.[1] His approach was to mechanize the process of calculation right through to printing.

He set to work on what he called his Difference Engine. This calculated a class of functions by the method of differences. In this, after the first few terms are calculated by hand, the rest can be obtained merely by a series of additions, making them much easier to calculate either by hand or machine. His Engine was designed to undertake the repeated additions and print the answer of each term.

The design was vast. Had it been completed, it would have required an estimated 25,000 parts and weighed around 15 t. It would have been 2.5 m high × 2.13 m long × 0.9 m deep (8 ft high × 7 ft long × 3 ft deep).[2] Only partial sections were ever completed and then he moved on to what he felt was a better idea. He was later to produce a revised design which was not built at the time but made from the drawings at the end of the twentieth century by the Science Museum in London, where it resides.

J.B. Williams, *The Electronics Revolution*, Springer Praxis Books,
DOI 10.1007/978-3-319-49088-5_10

What he had moved on to was the Analytical Engine which was a more ambitious and general-purpose machine (Fig. 10.1). First, it was able to calculate all four arithmetical functions: addition, subtraction, multiplication and division. These were performed in a central unit which he called the 'mill'. The numbers to be performed on were brought from a series of stores, and the result saved in another store.

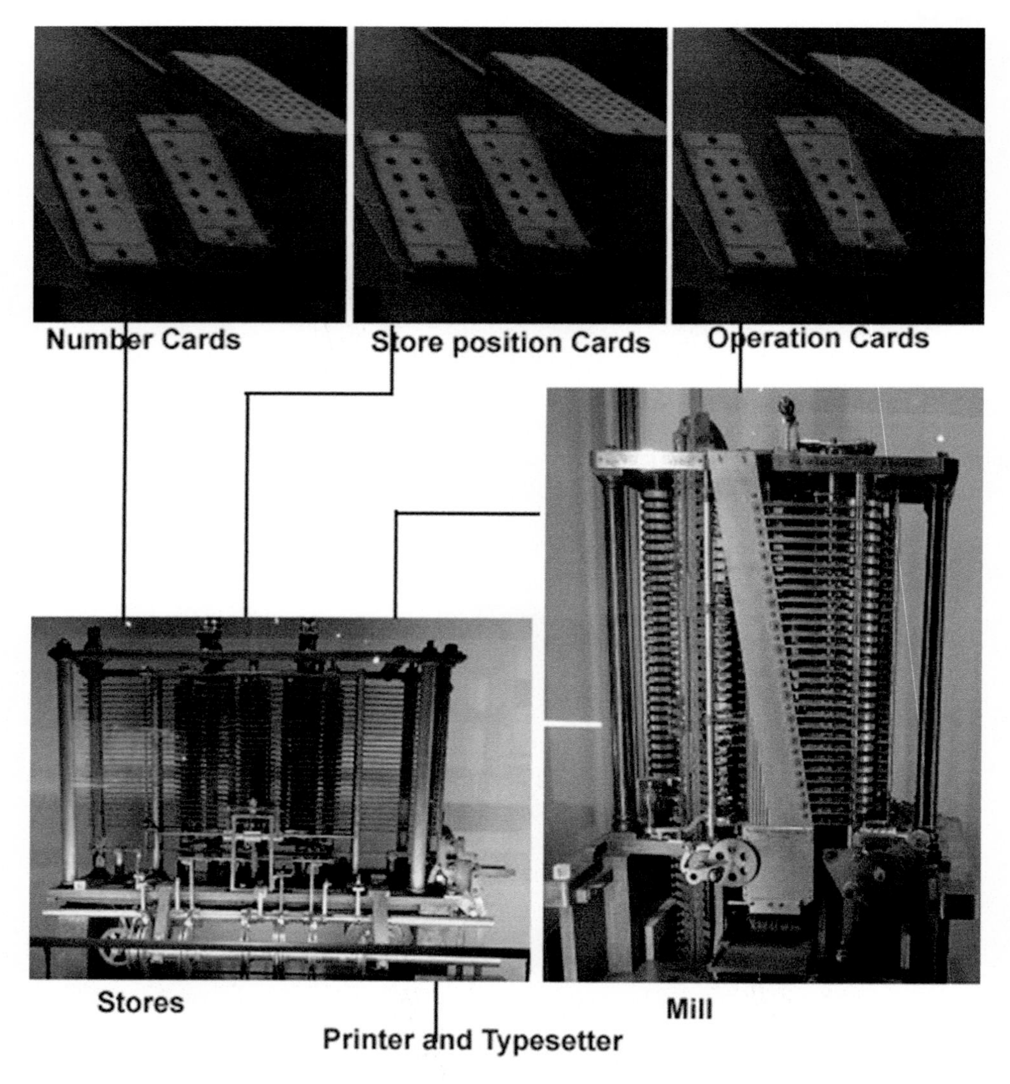

Fig. 10.1 Babbage's Analytical Engine with pictures of parts in the Science Museum, London. The machine used three sets of cards for mechanical convenience, linked together to supply variable numbers, control of which stores to use and the operation required from the mill. *Source:* Author

The whole thing was controlled by a series of punched cards, an idea borrowed from the Jacquard looms which were capable of producing a very wide range of patterns automatically. This made it a very flexible calculator, but the ideas went further. It was capable of jumping backwards or forwards a number of steps in the series of cards depending on the result of the calculation. If a result was not zero, for example, it might go back a number of cards and then repeat the steps until the value was zero and then move to the next card.

It was this ability to act on results that were not known to the operator that distinguished it from everything that had gone before. The change might not seem to be very great but it is what separates a computer from a calculator and makes it capable of a vast range of possibilities. It now means that a program can be written to perform a task and the machine itself can make the necessary decisions without the intervention of a human.

At this point, a most unlikely character entered the story: Ada, Countess of Lovelace, who was Lord Byron's daughter though he never had contact with her. Her mother, fearing the contamination of a poet as a father, brought her up to be the opposite and she became a competent mathematician. While quite young she met Charles Babbage and became a strong supporter of his work.

Babbage was not a great communicator of his ideas, but an exception was a lecture he gave in 1842 at the University of Turin. A young Italian engineer present, Luigi Menabrea, who was later to become the Prime Minister of Italy, wrote up a 'Sketch of the Analytical Engine' in French. Ada Lovelace spent the best part of a year translating this into English and, encouraged by Babbage, added notes which are considerably longer than the original document.[3]

The document clearly explains the ideas of the Analytical Engine and, importantly, how to produce the cards to make it undertake a task, or as we would say today, write a program. In the last note is a procedure (a program) to calculate the 'Numbers of Bernoulli', and it is on this that her reputation as the very first programmer rests. While there is some controversy as to how much of this came from Babbage and exactly what was her contribution, there is no doubt that she understood the process and could explain it.

More importantly, between them they had defined the essentials of a computer and how to program it. Ada had appreciated that it had uses beyond the mere calculation of numbers, a point which Babbage seems to have missed.[4] Unfortunately, she died young but Babbage, and later his son, tried to build the analytical engine but got no further than some sections of it such as part of the mill and a few stores. Despite this lack of success and the low level of interest in the ideas, it is almost certain that it would have worked but, being mechanical, would have been slow in operation.

Here the matter rested until the 1930s, when the brilliant young British mathematician, Alan Turing, published a paper describing a universal machine to compute numbers.[5] It would appear that, at the time, he was unaware of Babbage and Lovelace's work and reinvented the idea.[6] The machine was to have a tape with a series of instructions on it and was capable of going forwards and backwards and making decisions on what to do next depending on the data. The problem with his paper was that it was largely incomprehensible to lesser mortals.

Meanwhile, in Germany, a civil engineer named Konrad Zuse was getting frustrated by the tedium of the calculations he had to undertake and looked for a way to do them automatically. In 1936, he began work on a series of devices to achieve this. The first, called Z1, was all mechanical and not very reliable, but it introduced a key innovation—it used binary instead of decimal which is much easier to implement.[7]

Over the next few years he progressed to the part mechanical/part-relay Z2, and the all-relay Z3. He was working under considerable difficulties as the Nazi regime was uninterested in his work and, once the war started, he was bombed out and his machines and drawings were destroyed. Despite this, he was able to conceive a commercial device, the Z4. He only got back to this work seriously in 1948 and the first one was installed in the ETH in Zurich in 1950. In it he had gone back to his mechanical memory which was 64 words each of 32 bits. The rest used relays, with the instructions coming from paper tape. In this one he introduced the essential conditional branching which is where a computed result causes the program to jump somewhere other than to the next instruction.

In America, Howard Aiken also wanted a better way to undertake calculations and, aware of Babbage's work, thought he could design a device using relays. The Harvard Mark I, as it became known, was built by IBM to his concepts but not completed until 1943. It was only really put to work after it was moved to Harvard in 1944 where it worked 24 h a day for the US Navy for the rest of the war. Though an impressive machine it didn't really advance the technology, but the publicity it brought and Aiken's lectures advanced the cause of computing.

In Britain, with a war on and survival at stake minds were concentrating on other things. Alan Turing had been seconded to work on deciphering enemy coded signals, and had come up with the electromechanical 'bombes' to help crack the German Enigma codes. However, the Lorenz SZ40 signals used by Hitler and the senior generals proved an even greater problem.[8] A section under Max Newman worked on finding a solution, with Turing as a consultant. Their first attempt was a relay-based system but it proved unreliable.

At this stage Tommy Flowers of the Telephone Research Labs at Dollis Hill became involved. He was convinced that the correct approach was to use vacuum tubes. What he knew, that others didn't, was that they became quite reliable if never switched off or disturbed, but the codebreakers were skeptical. In a great act of faith, the management at Dollis Hill backed him and he and his team designed a special-purpose machine to break the wheel settings needed to crack the code. The machine was designed and built in an enormous hurry between February and December 1943 and was then installed at the Bletchley Park codebreaking centre. It was named Colossus and was followed by even more advanced versions.

Though the machines had high-speed tape readers, they were used to input the signal to be analyzed and not the program which was set up on plug banks and switches. Thus the state-of-the-art had moved on but it was still not a fully functioning computer as we would know it. What it had done was to show the way forward, and that was by using electronics with its enormous increase in speed. The problem was that it was all to remain top secret for many years. Its only influence was on the people, such as Turing and Max Newman who knew how it worked.

Across the water, the United States Army had a problem with the length of time it took to calculate ballistics tables to enable them to accurately fire their guns. At the Moore School of Electrical Engineering at the University of Pennsylvania, Dr. John W. Mauchly and Dr. J. Presper Eckert proposed a solution. In June 1943, the Ordnance Department signed a contract for them to research, design, and build an electronic numerical integrator and computer—ENIAC.[9]

It was to take until the autumn of 1945—too late to aid the war effort—to complete the enormous computer which had 19,000 vacuum tubes, and weighed in at 30 t. In February 1946 it began serious work. It was virtually all electronic, and though it was more general purpose than Colossus it was still programmed by switches and connectors which made changing from one task to the next very time-consuming, although it was very fast when it ran. Though in some ways an advance, the lack of tape or another simple method of changing the program was a step backward.

It was beginning to be realized that the ideal arrangement was for the program as well as the variable numbers all to be stored in a memory. Then the whole thing could run at electronic speeds. This is why switches and plugs were used in Colossus and ENIAC—they could be read at full speed whereas a tape or card reader would have slowed it down very considerably.

A June 1945 paper produced by John von Neumann, also at Moore School, outlined the structure needed for a true electronic computer.[10] This was the EDVAC design. He was undoubtedly working from Turing's ideas and the favor was returned later that year when Turing produced a design for a computer called ACE for the National Physical Laboratory (NPL), quoting von Neumann's paper.[11] Turing also introduced the idea of running a sub-program that could be accessed a number of times in a main program. This was the concept of the subroutine, the last building block for a true computer.

The basic problem was still how to produce a suitable memory. Ideally, every store in the memory should be equally accessible so that the computer has complete freedom to read or write from any position at any time—it was called a Random Access Memory or RAM. At the time there were two contenders.

The first was the delay line. Initially this seems an odd idea, but the principle was to feed a series of pulses representing the data to be stored into the delay line, which was usually a tube of mercury. Some time later it would appear at the opposite end and then it was amplified and returned to the beginning. Thus it could store the data indefinitely. The problem was that it wasn't truly a RAM. It could only be accessed when the data appeared out of the end of the delay line so there could be a wait for data, thus slowing down the operation. Alan Turing's ACE design had elegant methods of minimizing the problem, but it still existed.

The other contender was an electrostatic store based on a cathode ray tube (CRT). Dr. F.C. (Freddie) Williams had spent the war working on radar and ended up at the Telecommunications Research Establishment (TRE) at Malvern in the UK. After visiting the US he became interested in the possibility of building a memory using a CRT. At the end of 1946 he was appointed a professor at Manchester University and set about the quest seriously. By November 1947, he and his co-workers, Tom Kilburn and Geoff Tootill, had succeeded in making a 2048-bit store based around a standard radar CRT.[12] Though it needed to be constantly refreshed before the dots decayed, this was interlaced with the access operations so any position on it could be read at any time.

To test the device properly it was decided to build a small computer around it. In this they were helped by the fact that Max Newman was also a professor at the university and interested in the possibility of building such a device. In addition, Alan Turing had left the NPL and joined the university. They thus had access to most of the background knowledge that existed in the UK on computing.

The machine, known as the 'Baby', was only tiny but, built around a Williams Tube of 32 stores of 32 bits, it had at last all the elements of a full stored program computer. The manual input was from a keyboard and the output was on a monitoring display screen. On the morning of June 21, 1948 it ran a 52-min factoring program and became the world's first stored program computer to operate.[13]

Having proved the principle, they steadily expanded the device, and in April 1949 'Baby' was running serious programs. By October 1949 it had 128 stores of 40 bit words in fast Williams storage tubes and 1024 words in slower drum memory. The input and output were now paper tape and a teleprinter. It was now known as the Mark I and had sprawled to something like nine racks of equipment and it was largely built out of war surplus parts supplied by TRE. So successful was the device that the government supported the Ferranti Company to make commercial versions.

Also out to build a computer was a team under Maurice Wilkes at Cambridge University. They used the EDVAC design as a starting point as they were interested more in what the computer could be used for rather than in their design.[14] They stuck to the mercury delay line as the store and by May 1949 their computer was operational, and they soon began offering a computing service on their EDSAC computer. Without Alan Turing it took until May 1950 to get the NPL's ACE computer operational. By then the concept of it supplying a national computing service had long been overtaken by events. Again, with its relationship to the EDVAC design, it used a mercury delay line store. The UK thus had three separate computer systems running.

In the US, the EDVAC was beaten to it by the company set up by Mauchly and Eckert to produce commercial computers after they left the Moore School. Their UNIVAC computer was delivered to the US Census Department in March 1951. EDVAC wasn't completed until late on in that year. Ferranti, however, delivered their first Mark I computer, based on the Manchester design, in February 1951 making it the first commercial computer.

Not far behind was another UK commercial company, but one that seemed most improbable to be involved in computers. That was J Lyons and Co, the teashop and cakes empire. In a far-sighted move they had given financial support to the Cambridge group in exchange for having one of their own people on the team and watching progress. When the EDSAC was complete they set about making their own machine based on that design. In the autumn of 1951 their LEO I was completed for use in-house.

For some years, Lyons had been concerned with rising administration costs and were searching for a way to control them. Aware of progress in computing they thought this might provide a solution. After the courageous decision to build their own, they started to run relatively unimportant tasks through LEO I while the team gradually improved its reliability. By the end of 1953, they were confident enough to set it the task they had always wanted it for—the payroll. The results were astonishing. Whereas it had taken an experienced clerk some 8 min to calculate an employee's pay it took the computer one and a half seconds.[15]

It didn't take long for other companies to become interested and Lyons hired out their computer for some customers until they set up LEO Computers to manufacture more of the machines for sale. They sold moderately well for the time, going on to an improved LEO II and LEO III. Almost single-handedly they had introduced the concept of the commercial computer when previously all the emphasis had been on scientific and computational work. Though they were very expensive, they were of interest to organisations such as insurance companies and government departments with large amounts of routine calculations.

In America IBM, with its huge investment in punched card systems, finally joined the battle in 1953 with their 701 computer which mostly went to government agencies.[16] Many other companies were also trying to get into the business. Once they got going the early British lead was quickly whittled away. With a larger potential market and more resources the American machines were soon better engineered and cheaper. The writing was on the wall for the British industry.

One of the next steps forward was the introduction of larger memory devices. The delay lines and CRTs were not readily expandable. A more promising area was the use of magnetics. One approach was to use tiny ferrite toroids, one for each bit, which could be switched between two states. These were threaded with a grid of wires so that the bits could be addressed at a particular position and read or written. By stacking a series of these alongside each other all the bits that made up a computer 'word' could be read or written at the same time. It was simple, elegant and reliable, and of course could be accessed randomly at high speed. It was much used for the next 20 years.

Also developed were various forms of magnetic drums and tapes. These basically used the same idea of flipping magnetic material between two states, but they were in the form of a drum or a continuous tape. Obviously the access was much slower, but they were useful for slower secondary storage. Now all the components for computers were in place and the industry could start to take off.

The one drawback was the use of electronic vacuum tubes. This made the machines large, and they consumed huge amounts of power. Once transistors became available and reliable it was an obvious move to use them instead. At first they were quite expensive so the logic was implemented in combinations of diodes and transistors which cut the numbers of transistors that needed to be used. The size of these second-generation machines dropped considerably and their computing power went up as much as their consumption went down.

In Britain, a number of companies tried to enter the business. In addition to Ferranti and LEO, there was International Computers and Tabulators, GEC, Marconi, Elliot Automation, English Electric and EMI. The market wasn't big enough for them all, particularly under American pressure, and a rapid series of amalgamations resulted in a single company, International Computers Ltd. (ICL) in 1968. The same process occurred to a lesser extent in the US.

Though the machines were being used by universities and government, a growing number was being used commercially. By 1959, in the UK there were over a hundred.[17] Half of these were being used for payroll and the next most popular use was in stock control, and costing. But this was just the beginning as the numbers were just about to increase dramatically. The next year there were over 300 and the next decade added some 5000 (Fig. 10.2). What separates computers from most other items is that the sales graph goes ever upwards and still shows no signs of levelling out into the classic 'S' curve experienced by almost every other product. This is driven by the extraordinary ability of the electronics industry to constantly innovate.

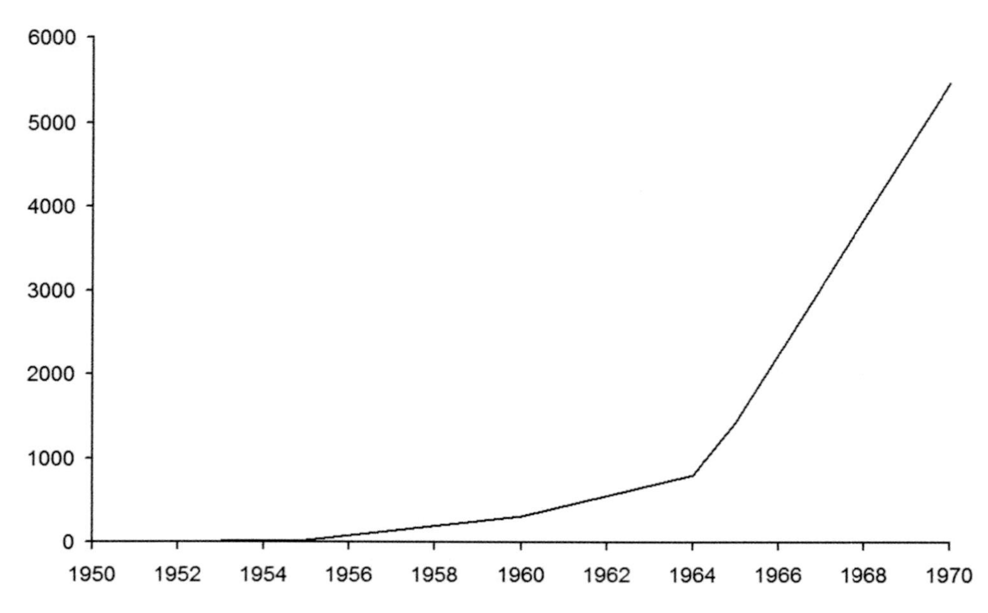

Fig. 10.2 Number of computers in use in the UK, 1950–1970. *Source:* Author[18]

To make these computers easier to use, higher-level languages were developed. These allowed the programming in a language that could be understood by people and the machine translated it into low-level commands that it understood. This made the programming much easier, and hence quicker, particularly when the many new programs were needed. Considerable advances were made in the development of languages and thus in computing science.

The first computers were run by the operator who received the results. When computers started to be used for business then the operation moved over to batches. Jobs were prepared previously and then fed into the computer, usually on cards. This meant that several jobs could be run in the course of 1 day. The computer lived in a special room and was only accessed by those in charge. Anyone wishing to run a program would have to approach them and get their agreement to run it.

The next stage was to be able to share a computer; a system known as timeshare. With this, more than one user could use the computer at the same time. The machines were now sufficiently fast to switch from one task to another so that each user thought they had it to themselves. It meant that a single machine could do most of the tasks in a business.

It was natural to go a stage further and place keyboards and screens in various departments around the company so that information could be entered directly. This was known as networking, as it required a network to connect all these terminals, as they were called, to the central computer. It was an irony of this system that vast numbers of people were necessary to input all the data, thus creating another form of drudgery which was what the computer was designed to remove.

Some forms of business had now become so complex that it was virtually impossible to run them without the use of computers. This was particularly true of supermarkets where the problems of stock control and ordering became more and more severe as the number of stores increased. It is interesting to note that the rise of computers and of supermarkets occurred almost at the same time during the 1960s. It is not much of an exaggeration to say that some forms of modern life would not be possible without the backing of computers.

Though computers were starting to have a significant effect by 1970 the real revolutions were still to come. At this stage, computers were still large machines that lived in a separate room in a business, and though terminals were appearing in departments they had not yet reached a stage of one on every desk. Computers were still very much under the control of the IT department. The future lay in providing computing power to the individual employee, and this was to be a complete change in the way that business is done.

NOTES

1. Dictionary of National Biography, Charles Babbage.
2. Babbage, available at: http://www.sciencemuseum.org.uk/onlinestuff/stories/babbage.aspx?page=3.
3. Sketch of The Analytical Engine Invented by Charles Babbage by L.F. Menabrea of Turin, Officer of the Military Engineers, from the Bibliothèque Universelle de Genève, October, 1842, No. 82. With notes upon the Memoir by the Translator Ada Augusta, Countess of Lovelace, available at: http://www.fourmilab.ch/babbage/sketch.html.
4. This comes from Dr. Doron Swade, who led the team who built the Difference Engine copy in the Science Museum and is a recognized expert on Babbage, at an IET London Local Network lecture on Babbage, Lovelace and Turing on February 18, 2014.
5. Turing, A.M. (1936–37) On computable numbers, with an application to the entscheidungsproblem. *Proceedings of the London Mathematical Society*, Series 2, 42: 230–265, with corrections from *Proceedings of the London Mathematical Society*, Series 2, 43 (1937): 544–546. Also available at: http://www.cs.ox.ac.uk/activities/ieg/e-library/sources/tp2-ie.pdf.
6. There seems to be no evidence to prove whether or not he was aware of Babbage and Lovelace's work at this point. His approach is rather different from Babbage's which suggests that he didn't know. It is interesting to note that modern computers are closer to Babbage's conception than Turing's with its single 'tape' and the ability to modify the information in the stream. Generally, like Babbage's arrangement, modern programs do not do this.
7. Dalakov, G. Konrad Zuse—the first relay computer, available at: http://history-computer.com/ModernComputer/Relays/Zuse.html.
8. Smith, M. (1998) *Station X: The Codebreakers of Bletchley Park*. London: Channel 4 Books, p. 148.
9. Richey, K.W. The ENIAC, University of Manchester—computer50, available at: http://ei.cs.vt.edu/~history/ENIAC.Richey.HTML.
10. von Neumann, J. (1945) First Draft of a Report on the EDVAC. Pennsylvania University. The paper only bears von Neumann's name but undoubtedly contains work from Mauchly and Eckert.
11. Turing, A. Proposed Electronic Calculator, National Physical Laboratory.
12. The University of Manchester, The Williams Tube, available at: http://www.computer50.org/kgill/williams/williams.html.

13. Lavington, S. Early British Computers 7, available at: http://ed-thelen.org/comp-hist/EarlyBritish-05-12.html#Ch-07.
14. Early Electronic Computers 1946–1951, University of Manchester—computer50, available at: http://www.computer50.org/mark1/contemporary.html.
15. Bird, P., Lyons, J. et al. Leo Computers, available at: http://www.kzwp.com/lyons/leo.htm.
16. Lexikon Services, History of Computing, available at: http://is2.lse.ac.uk/History/01HISTORYCD-Chrono1.htm.
17. Goldsmith, J.A. (1959) The State of the Art (a) Commercial Computers in Britain. *The Computer Journal* 2:3, 97–99.
18. Harrison B., Seeking a Role, The United Kingdom 1951–1970, p314/5.

11

Chips into Everything: Integrated Circuits

If the auto industry advanced as rapidly as the semiconductor industry, a Rolls Royce would get a half a million miles per gallon, and it would be cheaper to throw it away than to park it.

Gordon Moore

On May 6, 1952, Geoffrey W.A. Dummer, a British radar engineer who worked at the UK's Telecommunications Research Establishment in Malvern, presented a paper on 'Component Development in the United Kingdom' at the Electronic Components Symposium in Washington. In it he said: 'At this stage, I would like to take a peep into the future. With the advent of the transistor and the work in semiconductors generally, it seems now possible to envisage electronic equipment in a solid block with no connecting wires. The block may consist of layers of insulating, conducting, rectifying and amplifying materials, the electrical functions being connected directly by cutting out areas of the various layers.'[1]

As the transistor had not long been invented this was a far-sighted prediction of integrated circuits. However, he wasn't the first. In 1949, Werner Jacobi, working for Siemens and Halske in Germany, applied for a patent for a semiconductor amplifier.[2] In the same year, July 1949, Jack Morton, who headed Bell Labs transistor development, said in an internal report: 'Imagine a technique in which not only the connecting leads and passive elements... but also the active semiconductor elements are 'printed' in one continuous fabrication process. Savings in life, size and initial cost per element… seem great enough to warrant reduced maintenance costs on a large scale by replacement of whole assemblies. The block may consist of layers of insulating, conducting, rectifying and amplifying materials, the electrical functions being connected directly by cutting out areas of the various layers.'[3]

Dummer, as an expert on the reliability of electronics, understood that having everything in a single block was a way to overcome what Morton had called the 'tyranny of numbers'. This is where, as complexity increases, the amount of interconnection brings an escalating unreliability. When trying to build evermore complex computers, for example, just the number of wires and soldered joints would set a limit to its usability.

J.B. Williams, *The Electronics Revolution*, Springer Praxis Books,
DOI 10.1007/978-3-319-49088-5_11

Unlike the other two, Geoffrey Dummer did try to realize this dream. In 1956, work had produced resistors made of silicon, one of the necessary components. In 1957, his ideas were sufficiently advanced for him to commission the Plessey company to attempt to make an integrated circuit and a model of it was shown at a components symposium. However, no device resulted. The next year they changed their methods, and tried to use a thin film process instead of a semiconductor one. The resulting device only worked briefly before failing.[4]

Meanwhile, in America the military was very interested in miniaturizing electronics. The Navy initiated a project with the unlikely name of 'Tinkertoy' which used a set of ceramic substrates wired together at the edges with a miniature vacuum tube mounted on the top layer. This was later replaced by the Army's 'Micro-Module' which used transistors and, though smaller, still used the same concept. The drawback of these schemes was that though they shrank the electronics they still had the same amount of interconnections and so were thus subject to Morton's 'tyranny of numbers'.

A better attempt was the approach taken by Harwick Johnson working for the Radio Corporation of America (RCA) who filed the first patent for an integrated circuit in May 1953.[5] It described a series of resistors and capacitors together with one transistor which formed a phase shift oscillator all in a single slice of semiconductor. Curiously, he didn't attempt to make a more general claim for an integrated circuit, presumably thinking that it was already known. However, as the junction transistor had only recently been shown to work, there was no way of manufacturing the device.

Jack S. Kilby, at 6 ft 6 in. seemed like another American farm boy, but his father was an electrical engineer working for the Kansas Power Company and this drew his son into a similar field.[6] After university and wartime service on communication systems, he joined the Centralab division of Globe-Union and worked on miniaturizing electronic circuits. Capacitors, resistors and the connections were formed on ceramic substrates by printing processes. Initially vacuum tubes were also mounted to produce complete hybrid circuits as they were known.

In 1952, Kilby attended the Bell Labs Symposium on transistors and then set up a line to make germanium alloy devices for his employer. By 1957, they were using the transistors and the printing techniques to make tiny amplifiers for hearing aids in small quantities.[7] The real interest in miniaturization was in the military, but they were insisting on silicon transistors and Centralab didn't have the money to set up a silicon production line. With his interest now focused on further miniaturization he looked for another job, and found one at Texas Instruments, one of the early producers of transistors and famous for being the first company to make silicon devices. He was employed on a wide brief to look into miniaturization. It soon became apparent to him that a semiconductor company was only really good at making one thing—transistors. As he had just joined, he didn't have the holiday when the company largely shut down for the annual break so he sat and thought about the problem. His conclusion was that they should make all the components—resistors, capacitors as well as the transistors—all out of silicon. He worked out how this could be done, probably unaware of the previous efforts of other people.

His boss, Willis Adcock, was enthusiastic, but also skeptical, that such devices would work. This goaded the young man into making a crude circuit for a one-transistor oscillator using resistors and capacitors as well as the transistor all cut from bits of silicon or from parts of existing products. It was all wired up with the fine wires used for connecting transistors in their mountings and he was able to demonstrate how it worked on August 28, 1958. Though this wasn't truly integrated, it demonstrated that all the components could be made from silicon.

The next step was to make a circuit on one piece of semiconductor. Kilby couldn't find what he needed in silicon and was forced to use germanium. He took a production wafer with transistors and contacts already in place, and got the technicians to cut a slice out with a single transistor on it. With some clever manipulation and addition of some more contacts he was able to wire the whole thing up as an oscillator. It worked. Here was a true integrated circuit even if it was rather crude. A week later he had a working two-transistor digital circuit made in the same way.

Now the project was sucking in more people; by early the next year they had designed and succeeded in making a similar digital circuit, but this time it was a much better device and all fabricated on a single piece of germanium. They had achieved the first true functioning integrated circuit. These devices were used for the first public announcement of what Texas Instruments called the 'Solid Circuit' (integrated circuit) concept at the IRE show in March 1959, after a patent application had been filed in February. Where others had speculated about such devices, Jack Kilby had gone ahead and done it. He was lucky in his timing as the necessary diffusion techniques that were needed had only just become available.

Meanwhile, over at Fairchild Semiconductor, Robert Noyce had been thinking along similar lines but he had a different take on the problem.[8] They were making many transistors on a wafer, then breaking them up and packaging them only for the users to connect them back together in their circuits. His simple thought was, 'why can't we connect them together on the chip?'

When he seriously attacked the subject in January 1959 he had one great advantage. As described in Chap. 8, his colleague Jean Hoerni had come up with the Planar process. This had two features which assisted him: the connections were all on the one surface, and the chip was covered by a layer of silicon oxide which protected it but it was also a good electrical insulator. The scheme was to etch holes in the oxide where connections were needed and then evaporate a grid of metal connections on the top. On July 30, 1959 he filed for a patent.[9]

Noyce's connection system meant that the interconnections could be made in bulk in some more steps in the semiconductor manufacturing process rather than the time-consuming handwired connections that Kilby's method needed. Crucially, only the same process steps were needed regardless of the complexity of the interconnections. So a chip with more transistors on would not take more time to wire up. It seemed to point to a way of overcoming Morton's 'tyranny of numbers'.

There was still one problem left to be solved. Transistors all fabricated on a single piece of semiconductor were all connected together. Kilby's solution to this was to cut large slots in the chip so that air separated the individual sections. As Fairchild were to discover, this was unreliable and under environmental extremes the chips would crack.

Of course, as with many inventions, the time was right and many people were working on the same problem. Another one was Kurt Lehovec of Sprague Electric who tried to build an integrated circuit. His idea was to use reversed biased p-n junctions to isolate the sections of his chip. He quickly filed a patent application on April 22, 1959.[10] They were easy to fabricate as part of the process of making the transistors, and they didn't introduce the stresses in the chip that the slots method did. Hence they were reliable.

At Fairchild, Jay Last's team had been tasked with turning Robert Noyce's ideas into reality. First they tried the slot method and then abandoned it when the reliability problems became apparent. They then moved to the Lehovic approach but refined it, which led to Robert Noyce filing two further patent applications on variants of this method.[11] Fairchild now had all the pieces in place to produce fully satisfactory integrated circuits—at least in theory.

Despite all the technology, Texas Instruments (TI) beat them in the race to produce the first commercial integrated circuit. At the 1960 IRE convention, TI announced their first commercial digital IC, the SN502. However, this was a bit of a pyrrhic victory as it was made using air gap isolation and fine bonding wires and was only made in quantities of 6–10 a month.[12] They even came with a data sheet with its parameters filled in by hand as the unit was tested. It wasn't the way forward.

Meanwhile, Fairchild were struggling to make working devices. With 44 steps in the production process there was a lot to go wrong. The electrical isolation, while the right method in theory, required a very long diffusion time and the chips would often crack during the process. Slowly, however, solutions were found and by early 1961 good ICs were being made and the company felt confident enough to launch the first Micrologic product at that year's IRE convention. They went on to design a whole range of digital logic circuits but struggled to make enough even for the limited market.

It was only in 1963 that they finally got on top of the problems with yet another process advance. This was to use epitaxial deposition. Up to now the essential dopant impurities had been introduced by diffusing them into the chip at high temperatures. With epitaxy the process was the other way round. New layers were grown on the polished surface of the wafer but, crucially, with the same crystal orientation as though it was still part of the single crystal.

TI rapidly adopted the planar process and electrical isolation and started to produce their own range of logic circuits. Other companies such as Westinghouse, Signetics and Motorola joined the scramble to get into the business. Due to cross-licencing and, in some cases, out-and-out poaching of secrets, new processes developed in one place rapidly became industry standards.

It might have been thought that the industry was on its way but, surprisingly, it met a lot of resistance. As Jack Kilby reported: 'Gordon Moore, Noyce, a few others and myself provided the technical entertainment at professional meetings.'[13] The objections were concerns over low yields, that the resistors and capacitors were very poor compared to components made by other methods, while the transistor people didn't want their devices messed up by all this other stuff on the chip. These objections were basically true, but they missed the point.

A huge amount of work was put in across the companies in the field to improve the processes and steadily the yields increased. It was soon apparent that ordinary circuit designs should not just be converted to silicon. The emphasis in the design should be to use transistors as much as possible and aim to minimize the use of resistors and particularly capacitors. With careful arrangements the relatively poor quality of the components didn't matter. Thus, gradually the objections were overcome, but there were still those who wouldn't touch the circuits.

Despite development support from the US military, what saved the industry were two programs that chose to use integrated circuits because of their small size. The first was the guidance computers for the Minuteman intercontinental ballistic missile. In 1962, the few tens of thousands of ICs produced by the industry were all bought by the military and for the next 4 years they consumed the majority produced.[14] The requirements for minimum size and weight were such that they were prepared to pay the prices of $100 or so that the firms were charging for the first ICs.

The second stroke of luck was President Kennedy's announcement in 1961 that the country would aim to put a man on the moon within a decade. This lead to ICs being specified for the guidance computers for NASA's Apollo space programme. Here again the extreme requirements and extensive budget meant that the expensive devices could be justified. Without these specialized customers it would have been difficult for the industry to get started. Britain and other countries didn't have those sorts of customers and, as a result, the US gained a lead which they weren't to lose for 20 years.

Over the next few years, the companies introduced families of logic devices with improved designs. They also began to copy each others' products which gave users more than one source. Commercial users began to use ICs as the prices fell. The natural market was the computer manufacturers but IBM held off using them until the late 1960s when their reliability had been proved and the economics were better.

A curious absentee from the business was Bell Labs. After all the pioneering work on the transistor they missed ICs for a considerable time. Partly this was because AT&T, their parent company, had the same concerns as IBM and only started to use ICs on a similar timescale, but they were more fixated on the fundamental devices and didn't understand the commercial reasoning behind the IC. Besides, as they went into the 1960s they were beginning to lose their way.

So far, the concentration had been on digital devices which are largely made of transistors with few resistors and capacitors. Analogue devices such as amplifiers presented much bigger problems. It was here that an extremely loose cannon/genius entered the picture, working of course for Fairchild. His name was Robert Widlar. He was hired in 1963 on the whim of one person after other interviewers had turned him down because he had arrived at the interview drunk.[15] Despite his alcoholism he justified the faith put in him.

After assisting in improving the planar process he set to work to design better analogue devices. One of the basic building blocks was called an operational amplifier where a few external components defined its gain and characteristics. Though the theory was well understood they were not used extensively as they required a considerable number of components to construct them. Bob Widlar set out to make IC operational amplifiers and change all that.

His first design, the uA702, was a step along the way, but had numerous drawbacks. He didn't consider it good enough, but the company decided to produce it despite his protests. He then designed a much better device, the uA709, using very clever circuit techniques well suited to the IC process, which functioned well as an operational amplifier. They can be tricky things to use because the multiple stages and a large amount of feedback can cause them to oscillate when not desired. Keeping them stable can be difficult. The uA709

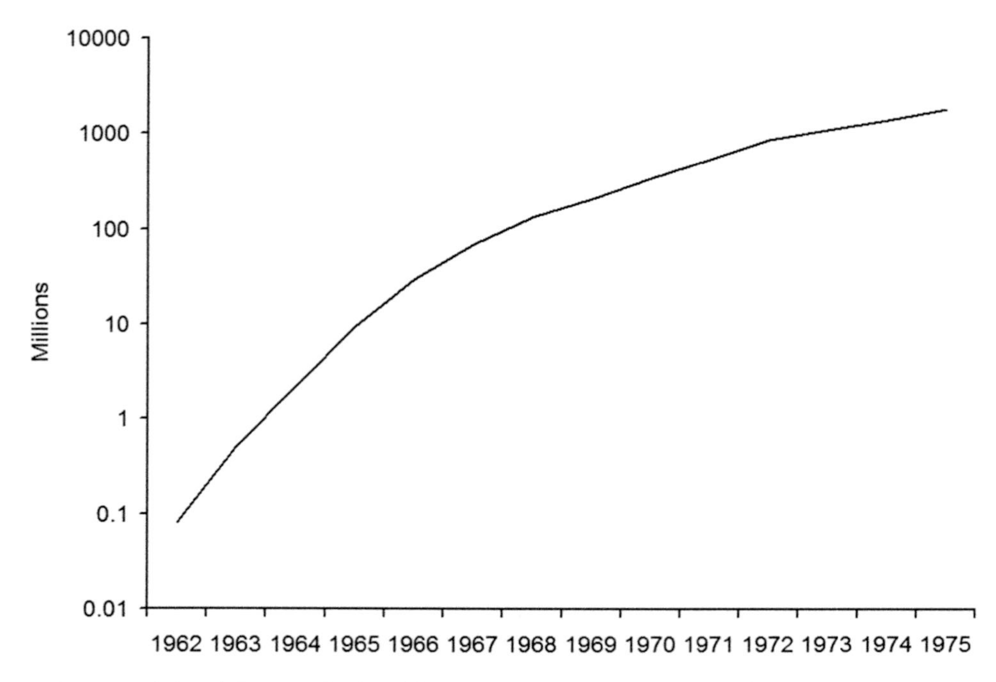

Fig. 11.1 Rise of the annual sales of integrated circuits—note the logarithmic scale. *Source:* Author[16]

fell into this category, and was also rather easy to destroy. Despite these problems, it was a great success and Fairchild couldn't make them fast enough.

At this point Widlar decided to leave Fairchild, claiming he wanted to be rich, and together with his close working colleague Dave Tarbert joined the new National Semiconductor company. Here he designed an even more revolutionary device, the LM101. This overcame the drawbacks of the uA709. The most innovative thing was the method of keeping it stable. The problem was caused by having multiple stages to achieve the necessary high gain, and he reasoned that if he could reduce these stages he could overcome it. The solution was a clever design with only two stages and only required a single capacitor to keep it stable under all conditions.

Back at Fairchild, British-born designer Dave Fullagar had been tasked with designing the next generation uA709.[17] He didn't like the way he was being directed and also felt that the LM101 could be bettered. He produced a design which met all the requirements, and by incorporating a 30 pF capacitor on the chip it was unconditionally stable. In 1968 it was sold as the uA741 and was as near to the ideal operational amplifier as made no difference for most requirements. It was such a success that it is still in use.

Despite all the difficulties the increase in sales of integrated circuits was phenomenal. In 1968 they reached more than 100 million, and by the early 1970s had passed a billion units (Fig. 11.1). This didn't tell the whole story of what was happening. Gordon Moore at Fairchild had noticed something else. In a paper he wrote in 1965 he spotted that the number of transistors on the most advanced chips was doubling each year.[18] He analyzed the underlying reasons for this and though at the time they had only reached around 50 he predicted that by 1975 it would reach 65,000 (Fig. 11.2).

More than a few people thought this was crazy, but those in the know didn't doubt that it was possible. It required the processes to keep improving, but that was steadily happening. One of the bigger concerns was whether it would be possible to find circuits that would be required in sufficient numbers to justify the production of such devices. Here again, they needn't have worried.

At first sight, the Planar process was also ideal to make field effect transistors (FETs). They required an oxide layer over the channel between the source and drain and then a metal layer. This was exactly what the process could provide, and it even required fewer process steps. The downside was that the resulting devices were unstable and unreliable.

It was to take until 1967 or 1968 before these problems were brought under control by methods that negated the effects of contamination and the replacement of the metal layer by silicon which could self-align over the channel. Then the potential of the devices could be exploited. Like bipolar transistors FETs come in complementary versions and this could be exploited to produce logic devices that only consumed power when they switched, leading to very low power designs. RCA exploited this and launched the 4000 series of logic devices.[19] Though slower than the bipolar families they consumed much less current and therefore were useful for battery-powered equipment such as watches.

The other area of interest was to use the Metal Oxide Silicon (MOS) device, with its gate unconnected, as a memory. The gate and its insulation layer formed a capacitor and it

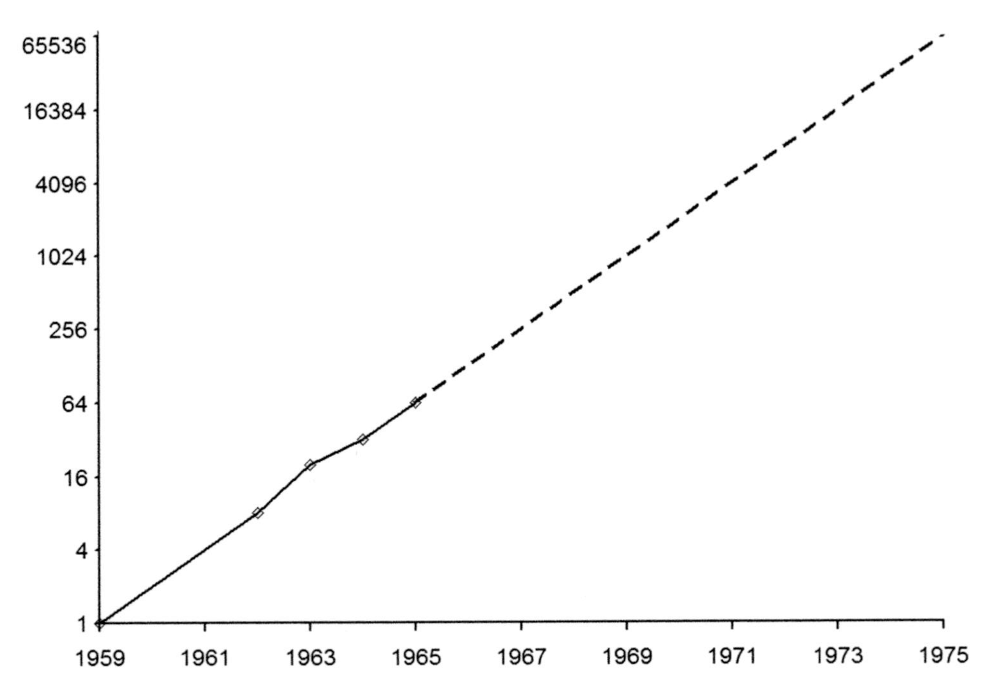

Fig. 11.2 Gordon Moore's projection of the number of transistors per chip. It would reach 2^{16} or around 65,000 in 1975. *Source:* Author[20]

had long been known at Fairchild that it could hold charge for some considerable time.[21] However, other opportunities used all the resources and it was not exploited.

In 1968, Gordon Moore and Robert Noyce finally left to found their own company, Intel.[22] Their objective was to make memory devices, as the computer industry was crying out for larger memories. Their first product in 1969 used small six-transistor circuits with two states to store each bit of information. There were 256 of these cells on a single chip which meant that they had fabricated more than 1000 transistors on one chip.

By the next year they had produced the 1103 which used the idea of storing the information on the gate of the MOS.[23] The charge slowly leaked away, so it needed to be refreshed from time to time. This dynamic random access memory, DRAM as it was called, had 1000 bits (actually 1024) using a three-transistor cell. It was sufficient to convince the computer manufacturers that the future lay in semiconductor memories rather than the ferrite cores that they had been using.

Over the next few years, as well as increasing the number of bits stored on each chip, Intel introduced a memory that could be written electrically and would then store the information more-or-less indefinitely. As a neat trick the package had a small window in the top and the information could be erased by placing it under an ultra violet lamp for a few minutes. Thus the Electrically Programmable Read-Only Memory (EPROM) was born.

Later, an even more useful device was introduced where the information, or parts of it, could be erased electrically. This was the Electrically Erasable PROM (EEPROM) which was to form the basis for memory sticks and portable devices too small for a magnetic hard drive to store their programs.

Here was the answer to the question of what to make with huge numbers of transistors on the chip that would have a wide application—memories of various types. They were soon to be joined by computer cores which led to yet another revolution which is investigated in Chap. 15.

In 1975, Gordon Moore returned to his predictions which were now known as Moore's Law.[24] He was nearly spot on with the largest current device having almost the 65,000 transistors he predicted. He looked at the factors that made up the increase and predicted that it would continue for a while and then drop from doubling every year to doing that every 2 years. He was right except in when this change took place. He was basing his ideas on Charge Coupled Device memories, but they turned out to be far too sensitive to radiation to be reliable, but later found a use in cameras. In fact, the change was already taking place in 1975.

The talk at the time was of reaching a million transistors on a chip, which seemed such an impossibly high number. However, the progress was steady, despite Moore's reduction in rate, and this milestone was quietly reached in 1986. By the end of the century it had increased by more than two more orders of magnitude and was heading for a billion. This was for devices that could be held in the palm of a hand. It was a truly extraordinary development.

Revisiting this subject later, Gordon Moore was to remark that what semiconductor companies were selling was 'real estate' and that cost stayed reasonably constant at about a billion dollars an acre.[25] Thus the more devices that could be packed on the cheaper they would get. He calculated the average price of a transistor and found that from 1969 to 2000 the price had decreased by a factor of a million. The effective price of a transistor on an IC had dropped from around 1$US to less than one ten-thousandth of a cent. The extraordinary reduction in the cost of a device is the secret behind the relentless march of the electronics industry.

Where the original breakthroughs were largely the result of advances in physics, what had driven later progress was chemistry. Gordon Moore, as an example, was a physical chemist, and it was people like him who ground out the problems in the processing and steadily improved what was basically cookery. So successful was his law that, by the 1990s the Semiconductor Industry Association was setting a 'Road Map' which defined feature sizes and other parameters for years in the future.[26] The law had changed from being a prediction to a map that was guiding the industry.

The question had always been—how long can this go on? There are fundamental physical limits to how small the features can be which are determined by the size of atoms, but well into the twenty-first century they still haven't been reached. There are those who predict that by the time they are met, some way, perhaps using a different technology, will allow the progress to continue (Fig. 11.3).

It is this extraordinary progress, unlike any other business, which has made electronics unique. If you counted all the devices on all the chips, by the century's end the industry had shipped in the region of a million million million (10^{18}) transistors and they were everywhere. It wasn't so much that an electronic unit was cheaper—they were usually much the same price—but it could do so much more. Computers to calculate the weather

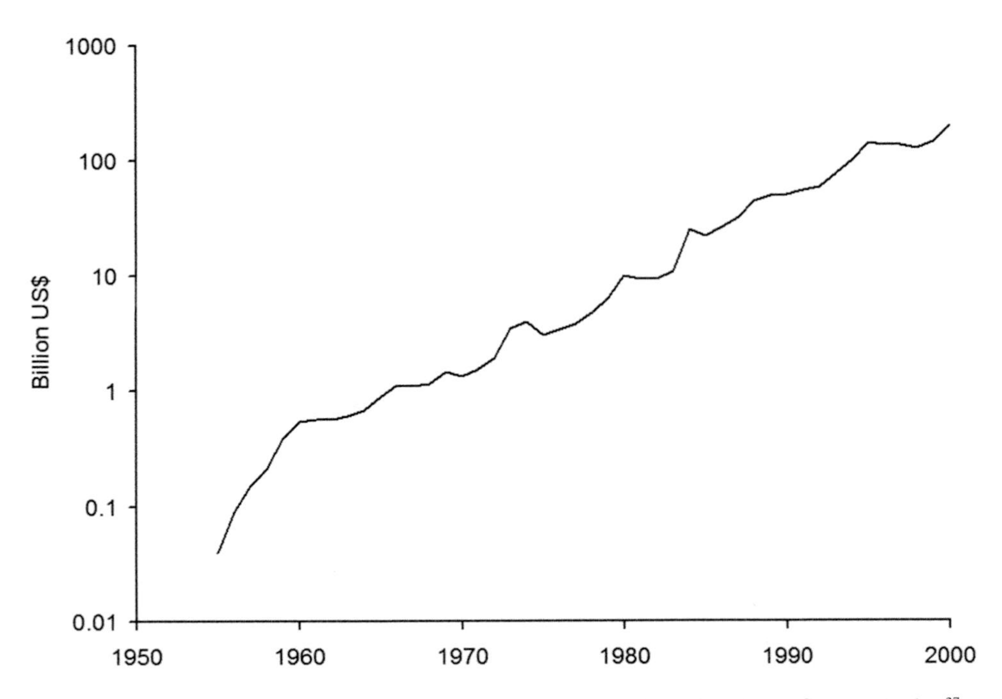

Fig. 11.3 Semiconductor sales by value—showing no signs of slowing down. *Source:* Author[27]

were not so very different in size to some of the early machines, but their power was many orders of magnitude better.

Television sets have changed little in price since their introduction—that is, the price in pounds or dollars—despite the considerable inflation and the increase in features. Electronics have crept in everywhere—cars, cameras, telephones, and kitchen appliances—in all aspects of our lives. The convergence of computers and semiconductors has changed the way we work, while those blending with telecommunications have profoundly altered the way we communicate.

The various aspects of these advances are examined in further chapters, but so much of the progress in the last part of the last century depended on this extraordinary development in semiconductors that this needed to be tackled first. While most developments follow the famous 'S' curve, starting slowly before taking off and then levelling out, the number of semiconductors just continues to rise and rise with no sign of slowing down. In this respect, they are unique.

NOTES

1. Dummer, G.W.A. (2005) Integrated electronics development in the United Kingdom and Western Europe. *Proc IEEE* 52:12, 1412–1425.
2. German patent number 833366.
3. Laws, D.A. and Riordan, M. (2012) Making micrologic: The development of the Planar IC at Fairchild Semiconductor, 1957–1963. *IEEE Annals of the History of Computing* 34:1, 20–36.
4. Dummer.
5. US patent 2,816,228; Ahrons, R.W. (2012) Industrial research in microcircuitry at RCA: The early years, 1953–1963. IEEE Annals of the History of Computing 34:1, 60–73.
6. Riordan, M. and Hoddeson, L. *Crystal Fire: The Invention of the Transistor and the Birth of the Information Age*. New York; WW Norton & Co., p. 256.
7. Kilby, J,S, (1976) Invention of the integrated circuit. *IEEE Transactions on Electron Devices*, 23:7, 648–654.
8. Laws and Riordan.
9. US patent 2,981,877.
10. US patent 3,029,366.
11. US patents 3,150,299 and 3,117,260.
12. Phipps, C. (2012) The early history of ICs at Texas Instruments: A personal view. *IEEE Annals of the History of Computing* 34:1, 37–47.
13. Kilby, J.S. (2000) Turning potential into realities: The invention of the integrated circuit. Nobel Lecture, December 8, 2000.
14. Slomovic, A., The government's role in the microelectronics industry, Rand Corporation, 16 December 1988, available at: www.dtic.mil/dtic/tr/fulltext/u2/a228267.pdf.
15. Bob Widlar, available at: en.wikipedia.org/wiki/Bob_Widlar.
16. Derived by dividing value of sales by average price - Data for sales value from: Slomovic, A. (1988) Anteing up: The government's role in the microelectronics industry, Langlois, R.N. and Steinmueller, W.E. (1999) Strategy and circumstance: The response of American firms to Japanese competition in semiconductors, 1980–1995, *Strategic Management Journal*, 21:1163–1173; and average price data from: Morris, P.R. *A History of the World Semiconductor*

Industry, p. 50; Atherton, W.A. (1984) *From Compass to Computer*, p. 257. It correlates quite well with the curve in Kilby, J.S. Invention of the Integrated Circuit.

17. Rako, P. (2007) Voices: Dave Fullagar, analog-IC designer and entrepreneur, EDN Network 20 November 2007, available at: http://www.edn.com/electronics-news/4,326,905/Voices-Dave-Fullagar-analog-IC-designer-and-entrepreneur.
18. Moore, G.E. (1965) Cramming more components onto integrated circuits. *Electronics*, 38: 8, 114–117.
19. Ahrons, R.W. (2012) Industrial research in microcircuitry at RCA: The early years, 1953–1963. *IEEE Annals of the History of Computing* 34: 1 60–73.
20. Moore, G.E. (1965) Cramming more components onto integrated circuits. Electronics, 38:8, 114–117.
21. Sah, C.T. (1988) Evolution of the MOS transistor: From conception to VLSI. *Proceedings of the IEEE*, 76: 10, 1280–1326.
22. Intel Timeline. A history of innovation, available at: http://www.intel.com/content/www/us/en/history/historic-timeline.html.
23. Byström, M. CPU Zone, RAM history, available at: http://www.cpu-zone.com/RAM_History.htm.
24. Moore, G.E. (1975) Progress in digital integrated electronics. International Electron Devices Meeting, IEEE, *Technical Digest* 1975, 11–13.
25. Moore, G.E. (2006) Moore's Law at 40. in D.C. Brock (Ed.) *Understanding Moore's Law: Four Decades of Innovation*. London: Chemical Heritage Foundation, Chapter 7. This book is no longer available in a print edition, but is available at: http://www.chemheritage.org/community/store/books-and-catalogs/understanding-moores-law.aspx.
26. Brock, D.C. (2006) Reflections on Moore's Law. In D.C. Brock (Ed.) *Understanding Moore's Law: Four Decades of Innovation*. London: Chemical Heritage Foundation. This book is no longer available in a print edition, but is available at: http://www.chemheritage.org/community/store/books-and-catalogs/understanding-moores-law.aspx.
27. Data from: Slomovic, A, The government's role in the microelectronics industry; 1976–2000, Semiconductor Industry Association; Global Billings Report History (3-month moving average) 1976–September 2010, available at: http://www.semiconductors.org/index.php?src=directory&view=IndustryStatistics&submenu=Statistics&srctype=billing_reports&query=category.eq.Billing%20Reports&pos=60,10,63.

12

From Signboards to Screens: Displays

When you have a few billion people connected with screens, not voice... screens are important. You can transmit a thousand times more information.

Yuri Milner

In 1900 signs, whether for giving information or as advertisements, had to be either paint on boards or print on paper—the traditional poster. Variable information presented even greater problems and here the choice was to use pointers over some form of fixed scale; the most obvious versions being the clock face and the electrical voltmeter and its variants. More random information needed manual intervention, such as the station master changing the painted wooden board to show the destination of the next train.

The advent of the electric incandescent lamp gave other possibilities, such as shining the light through a mask to show a word or outline—the exit signs in cinemas and theatres, for example. The lamps could be turned on and off and grouped to show more complex images. In practice, they were often merely flashed on and off to attract attention or switched from one to the next which could show a sequence of numbers, such as which floor the lift had reached.

In wasn't until the 1920s that the gas discharge 'neon' tube, a derivative of the Geissler tubes mentioned in Chap. 2, was added to the mix. These were very popular in the US but less so in other countries. They were bright and confident and came to symbolize America. They could be produced in a number of colors and bent into various shapes. By switching various sections on and off in sequence, fixed messages—usually advertising—could be displayed. However, they still could only display a predetermined message.

A further step was the arrival of television with its use of the cathode ray tube (CRT) as the means of displaying the image. With the electronics at the time it couldn't do a lot more than show the image collected by a camera. This could, of course, easily be switched from one image or caption to another rapidly, but without a means of storing the pictures or a way of generating them synthetically the uses were limited to displaying real-time scenes.

One device derived from the CRT was the 'magic eye' which was popular as a tuning indicator for radios.[1] Though it was possible to tune them by ear, the advent of automatic gain controls meant that the output signal varied much less in amplitude, but the price of

J.B. Williams, *The Electronics Revolution*, Springer Praxis Books,
DOI 10.1007/978-3-319-49088-5_12

this was that it made exact tuning more difficult. It was a round device which showed a glowing part ring which extended further around the circle as the amplitude increased, hence showing the point of best tune.

There the matter remained until after the Second World War. With the advances in electronics, and particularly television, the search was on for better ways to display information. Television CRTs became bigger, with the 9-in. or 12-in. sizes from before the war steadily increasing during the 1950s from 14-in., to 17-, 19-, and 21-in.[2] The difficulty was that as the screen increased so did the length of the tube, so the sets became bulkier and bulkier with huge projections out at the back.

Attempts were made to overcome the size by increasing the deflection angle from the usual 90–110°, but even this didn't shorten the tube very much. Though the CRT was now reasonably satisfactory there was still a hankering for something much flatter—the ideal was a panel sufficiently slim to hang on the wall. A number of attempts at flat tubes with clever optics were made, but they never progressed into production.[3]

The other thing that was lacking was color. As described in Chap. 6, this problem was solved with the elegant 'Shadow Mask' tube where three beams pick up separate dots of red, green and blue phosphors. By adjusting the amounts of each that are displayed a color picture can be built up. Though it took until the 1960s before the sets were cheap enough to be made in large numbers, they made a very practical television set.

By the 1950s there were many attempts to find better ways of displaying written or number information. A typical problem was arrival and departure boards at railway stations and airports. An ingenious solution was the split flap or 'Solari' board. This consisted of a large number of flaps connected to a central cylinder. Each pair of flaps could open on the front face like the pages of a book. On these were printed the required information such as the name of a destination, half on the upper and half on the lower flap.

By flipping the flaps over up to 40 different messages could be displayed.[4] This was usually achieved electromagnetically. These systems were much used in stations and passengers became adept at listening for the clack clack as the flaps flipped over to announce the next train. They began to come into use in the late 1950s and remained a popular system for the rest of the century. Their drawbacks were that they were slow to operate and setting up new multiple messages could be a complex and time-consuming operation.

Though the split flap arrangement could be used for clocks where the numbers came up in sequence, they were not suitable where random numbers needed to be displayed such as in electronic instruments. The obvious way to do this was to use small incandescent lamps, and many ingenious arrangements were produced. Some used a set of lamps each shining through a number shaped mask which showed on a frosted screen at the front. Another arrangement had a set of masks on a disc moved by a moving coil meter again shining on to a frosted glass front.

A less complex arrangement was to use individual filaments and for simplicity this used the '7 segment' format. The basic idea went back a long way. In 1908 a patent contained an 8 segment arrangement for showing numbers.[5] The difference was that it had an extra bar for the 4, and this pointed up the one thing that people didn't like about the 7 segment system and that was the way the 4 was displayed with a vertical bar instead of the angled one (Fig. 12.1).

Fig. 12.1 An untilted 7 segment 4 compared with normal shaped 4 that the 'Nixie' tube could show. *Source:* Author

If the display bars were tilted slightly the other numbers looked quite reasonable, but the 4 was the stumbling block. People simply weren't used to it, and it generally wasn't acceptable in the 1950s and into the 1960s. Of course an extra bar could have been added, but curiously this simple adaptation was never done, though later displays with additional diagonal bars appeared so that more letters could be shown.

What was needed was something better and, ideally, could display normally shaped numbers. To achieve this, the American calculator maker Burroughs Corporation reached back to an old technology. It had long been known that a gas discharge tube under the right conditions would glow around the cathode or negative electrode. If this was suitably shaped then it could display a bright number in its characteristic red/orange colour. By stacking a set of numbers 0–9 behind each other and having a thin grid for the anode, then by powering the appropriate cathode it could show that number.

The story goes that the draughtsman drawing up the device numbered his drawing NIX 1 standing for Numerical Indicator Experimental number 1.[6] From this it earned the nickname Nixie. It may not be right, but it seems a plausible explanation for the unusual name. Despite the numbers not quite being in the same plane, and the distraction of the fine anode and other cathodes, it made a very satisfactory display. They could easily be stacked side by side to show a multi-digit number.

When they were introduced in the late 1950s the high voltages that they needed was not a problem when they were driven from vacuum tubes. As transistors took over this became more difficult, but suitable driver integrated circuits that could switch the voltages eventually arrived. The advantage of Nixies was that they could be manufactured in a similar way to the well-established thermionic vacuum tubes, and had reached around $1.50 in large quantities and so were fairly cheap compared with other devices. The result was that they were the dominant numerical indicators, reaching sales of some 5 million units in 1972 after which they started to decline as other technologies took over.

Adapting techniques developed in the semiconductor industry allowed flat versions of these to be made in the 7 segment format. These were very good-looking, bright displays but the high voltages still put them at a considerable disadvantage to some of the newer technologies. If the 7 segment format was acceptable, then devices that were simpler to implement might as well be used.

Another technology was produced soon after, and that was yet another variant on well-known -vacuum tube- and CRT technology. Vacuum fluorescent displays, as they were

Fig. 12.2 A vacuum fluorescent display showing all the bars and legends, which are a *light green* color. *Source:* https://en.wikipedia.org/wiki/Vacuum_fluorescent_display#/media/File:VFD2.jpg

known, were like miniature CRTs (Fig. 12.2). They contained very small cathodes which were heated to emit electrons, a grid to control the current flow and a set of anodes which were coated in phosphor. With a suitable voltage on an anode the phosphor would glow as the electrons hit it.[7]

By having 7 anodes which formed the segments, a 7 segment display could be produced. Normally they had a distinctive pale white/green colour and were mostly produced in Japan, though due to problems with driving them they were never very popular elsewhere. In the 1970s and beyond, Japanese-made calculators, video recorders, microwave ovens and so on usually had vacuum fluorescent displays.

One of the advantages was that multiple digits or more complex arrangements could all be put inside a single glass envelope. The various sections could be selected by having separate grids. By connecting sets of anodes together the number of connections to the device could be minimized. The appropriate sections could be turned on for a very short time, then those in the next section and so on. This system is known as multiplexing and cuts the amount of circuitry and interconnection drastically as the size and complexity of the display goes up.

With all the work on semiconductors and solid state physics it was only to be expected that sooner or later a display of this type would be made. As early as 1907, H.J. Round, who did so much work for Marconi on vacuum tubes, was investigating the behaviour of carborundum crystals and how they showed a rectifying effect. He also noticed that under some conditions they produced light.[8] At the time no one knew what to make of this effect.

It was rediscovered in the 1920s by the Russian Oleg Lossev, and though he investigated the phenomena fairly thoroughly, without a theoretical understanding of what was happening little progress was made. That had to wait until after the intensive work on semiconductor devices at Bell Labs, and elsewhere, that led to the invention of the transistor. As a result, in the 1950s it began to be understood that it was the p-n junction in the semiconductor that was responsible for generating the light.[9] (See Chap. 8 for more on the transistor and the physics behind it.)

Unfortunately, the semiconductor materials most studied, germanium and silicon, were very poor generators of light and even that wasn't in the visible band. Fortunately, there are more materials that are semiconductors, in particular what are known as group III/V alloys. Mixtures of group III elements such as aluminum, gallium and indium, together with group V elements such as phosphorus, arsenic and antimony, formed the stable lattices that the semiconductors needed and could be doped to produce p-n junctions.

The work had originally been carried out to see if any of these materials produced superior transistors to germanium or silicon, but it was noticed along the way that they could generate light. However, the first materials investigated either were very poor generators of light or produced their radiation in the infrared region. As the 1950s progressed, the search for an efficient semiconductor light source was not really yielding results.

However, Dr. Nick Holonyak, working for General Electric in America, was investigating an even more involved material, gallium arsenide phosphide (GaAsP), in which he could vary the amounts of arsenic and phosphorus. In 1962, he was able to show that this could produce red light. These LEDs were not very efficient devices, but General Electric soon put them into production. At the high prices necessary at the time, the market was not large.

Spurred on by Holonyak's work, others started to investigate better methods for producing LEDs. Monsanto, a chemical company, was interested in producing the basic materials and tried to do a deal with Hewlett Packard (HP) in which they would make the material and HP the LEDs. Unfortunately, the two companies didn't really trust each other and the result was that they both ended up making the material and the LEDs. It took until 1968 for Monsanto to get the devices into production; unlike GE, they decided to sell the LEDs at affordable prices and they began to take off as indicators.[10]

By the following year, HP was producing 7 segment type displays still using GaAsP which produced a red output.[11] These used 28 separate LEDs, 27 for the digit and one for the decimal point. They produced quite nice looking digits (though the 4 still was not quite in the conventional form) at the cost of rather more LEDs than were strictly necessary. Later devices reduced the number to 14, i.e., two for each bar of the digit. The problem with these early displays was that they were very small and dim. Often a lens was placed over the top to make them seem bigger. Nevertheless, manufacturers of calculator were soon making them in large numbers (Fig. 12.3).

During the 1970s, two things drove LEDs forward. First, Fairchild started to produce them using their planar process, driving the price right down which of course produced an enormous expansion in their use.[12] The next was that George Craford, who had been a student of Holonyak's, joined Monsanto and by doping the GaAsP devices was not only able to increase the brightness but also to produce orange and yellow and even

Fig. 12.3 An early LED calculator display with its very small digits which show *red*. The lenses over each one can just be seen. *Source:* https://en.wikipedia.org/wiki/Calculator#/media/File:LED_DISP.JPG

yellow-green LEDs. A good green was produced by going back to gallium phosphide but using the newer doping methods.

Steadily over the next 20 years the brightness improved around ten times in each decade as new materials were found. In the range of colors there was one thing that was missing—blue. It wasn't until the mid-1990s that this was finally solved with the production of gallium nitride devices. (Nitrogen is also a group V element.) This led on to bright green and yellow devices. With the addition of a suitable phosphor to 'down convert' the blue and yellow the holy grail of a white LED was reached.

Though by the end of the century LEDs were used everywhere, as indicators, digital displays, signboards and so on, the one area that they were never to conquer was large display panels such as for television. Each lit spot or pixel has to be a separate device and there is no way of making them in the mass as for some other technologies. Some large panels are produced for signboards for buses or trains, but they are made from a large matrix of individual devices. These are necessarily expensive but their reliability makes them suitable for that duty. LED panels have neither the resolution nor low-enough cost suitable for use in a television screen.[13]

Though these display technologies were satisfactory for many uses they consumed significant amounts of power and were thus not really suitable for portable applications. For example, a digital watch was produced with LEDs but it was necessary to press a button to see the time otherwise the battery would have been flat in no time. What was needed was a truly low-power technology.

In 1888, an Austrian botanist, Friedrich Reinitzer, discovered a strange material which melted at 145 °C, becoming cloudy white and viscous. However, if the temperature was raised to 179 °C it became clear and liquid. He had discovered a 'mesophase' between solid and liquid. He wrote to Otto Lehmann, Professor of Physics at the Technical University Karlsruhe, Germany, who started to investigate the phenomenon. He found that the mesophase had some characteristics of both a liquid and a crystal, so he called it 'fliessende krystalle' or 'liquid crystal' in English.[14]

It might have been expected that this discovery would lead to a flurry of activity, but it was largely ignored although the German company Merck did supply some materials from 1907. It wasn't until 1962 that Richard Williams at RCA discovered that the visual appearance of the liquid crystal was altered by applying an electrical field to a thin cell of the material sandwiched between two glass plates. Now there was some interest as this had the possibility of making a low-power display because it required only a very small current to produce the effect.

George Heilmeier and his colleagues at RCA took up the challenge and found a material that changed from clear to milky white when an electrical field was put across it. They called this phenomenon Dynamic Scattering Mode (DSM). The trouble was that the effect only occurred between the temperatures of 83 and 100 °C. Clearly the next step was to find a suitable material that exhibited the effect at room temperatures. On 28 May 1968, RCA held a press conference and announced the new flat low-power display technology. RCA failed to exploit this commercially, but it was the starting point for the development of liquid crystal displays.

One of the quickest off the mark was the Sharp company in Japan. They were heavily involved in the calculator business which had become very competitive. Looking for an edge in the market they decided that a liquid crystal display coupled with a CMOS calculator chip was the way forward and that, by using this, a small portable device could be made. There was only one problem—they didn't have a display.

They set to work to find a suitable material and after a considerable amount of effort produced one that would work from 0 to 40 °C. The next question was how to make connections on the two pieces of glass that were transparent. Fortunately, the answer to this was already available in the form of thin layers of tin oxide or, even better, indium tin oxide (ITO). This could be masked to produce any desired pattern of segments and bars to make a complex display all in one go.

The resulting calculator, the Elsi Mate EL-805, was launched in May 1973 and proved very popular. It ran from a single AA battery, was small enough to fit into a pocket and was a fraction of the size and weight of existing machines. The potential of LCDs starting to be acknowledged, but at this stage few realized that the DSM was a dead end. It was slow which, while adequate for a calculator, meant that it couldn't be used in many applications particularly for the flat panel television. Besides, it still drew more current than was ideal.

The focus then moved to Switzerland where the electrical company Brown Boveri and chemical concern Roche were cooperating to research liquid crystals. In December 1970 Dr. Wolfgang Helfrich and Dr. Martin Schadt at Roche filed a patent for a new electro-optical effect which was called the twisted nematic field effect (TN).[15] This used a material which was arranged to twist by 90° across the cell by making micro-grooves on the glass surfaces at 90° on the two plates.[16] Applying an electric field destroyed the twist.

With a polarisers on each face, the display could go from transparent to dark as it was energized, or vice versa with one polariser the other way round. With a suitable mirror behind, a reflective display could be produced with a much better contrast than the DSM devices. At first there was opposition to the increased complexity of the twist and polarisers, but it was soon realized that the current consumption was very much lower at a few microwatts per square centimeter, and the contrast better. Around the world, research switched to TN displays, and in particular the search was on to find suitable materials.

Dr. George Gray at Hull University in the UK had been undertaking blue skies research on liquid crystals for many years before their electro-optical properties had been discovered. His university began to take his work seriously after the announcement by RCA in 1968.

In 1967, John Stonehouse, the UK's newly-appointed Minister of State for Technology, visited the Radar Research Establishment at Malvern.[17] As part of their remit they had a CVD section which dealt with components, valves and devices. He put it to them that the UK was paying more in royalties for shadow mask CRTs than the cost of the Concorde development and what they should do was to invent a flat panel TV display.

RRE had no idea how to make such a thing, but they set up a working party to investigate. One area that looked possible was liquid crystals and they produced a report to this effect at the end of 1969. One of the results was that, in 1970, they gave Gray a contract to develop liquid crystal materials that worked at room temperature.

By 1972, Gray and his small team had found a group of substances that looked hopeful and sent them to RRE for tests. Unlike the materials that others had been using they were stable, unaffected by light or moisture. They still didn't quite meet the desired temperature range, but that was the next stage. If two substances are mixed the combination will have a lower melting point which is lowest at a particular ratio, which is known as the eutectic point.

By mixing two or more of these substances the working range could be lowered and widened. It took some clever computer calculations to work through the myriad combinations and find the eutectic points where up to four substances were used. By 1973, Gray's team had produced mixtures with good working ranges around room temperature and there was a huge demand for samples. Obviously they couldn't satisfy this and the British Drug House (BDH) company was brought in to make the materials in quantity.

By the following year they had improved the mixture further so that it was stable and would work over the range of −9 to 59 °C, which was almost exactly the −10 to 60 °C required by manufacturers of calculators and watches. However, at that point Glaxo, the owner of BDH, in a curious decision, put the company up for sale. The buyer was the German company Merck who had been in the liquid crystal business for many years. It was a good decision as they dominated the supply of the LC materials from then on, under licence from the Ministry of Defence in the UK.

Despite the interest, things started slowly. Then Sharp in Japan changed to the TN mode for their pocket calculators, and sales went from 10 million units in 1973 to 50 million by 1981. Seiko also turned to LCDs and produced a digital watch featuring a TN display. Soon they were using the technology in clocks and any and all products that required low power, especially portable items. One of the great advantages was that the pattern of what was displayed was produced by a simple mask of the ITO layers on the glass. Thus multiple digits and almost any desired symbol could be produced.

Obviously, as the complexity increased, the need to minimize the number of connections and the driving circuitry meant that the displays should be multiplexed. However, the threshold between on and off was not very sharp so the displays could only be multiplexed with a small number of ways. This meant that it was not possible to produce large panel displays.

The research on materials didn't stop with the first successful material as the advantages of the LCD could only be fully exploited with even better materials. Particularly in the Far East, there was a need for more complex display that could show Chinese characters. By 1982, panels of 480 columns by 128 lines of dots had been achieved but these were reaching the limit.

In 1982, researchers at RRE in the UK found that increasing the angle of twist in the displays to 270° gave a much sharper threshold. Soon the team at Brown Boveri in Switzerland had developed this further, but in 1985 it was the Japanese companies of Sharp, Hitachi and Seiko who turned this super twist nematic mode (STN) into commercial displays. These were reaching in the order of 640 by 200 dot displays. There was one drawback; they were colored either green or blue and not the desirable black and white. Subsequently, a clever film was applied to the face which was able to remove this effect to produce a white display.

However, while useful for many applications, the STNs were still not enough for a full television display, particularly as the target had moved with the introduction of color. What was needed was the application of an idea that had been around for a while which was to have a transistor switching each dot or pixel, the so-called active matrix (AM) mode. To do this, it was essential that these devices should be formed on the same glass substrate as the display. The question was how to achieve this. Various attempts used cadmium selenide, tellurium and crystalline silicon to make the transistors, but none of these made satisfactory transistors that could be fabricated on a large panel.

G.L. Le Comber and W.B. Spear and colleagues at Dundee University in Scotland had been working on amorphous silicon for solar cell applications. Around 1979 they managed to fabricate satisfactory field effect transistor (FET) devices. In 1982, Spear was on a lecture tour in Japan and described their work, mentioning the FETs in passing. This produced a flurry of activity which soon yielded small—around 3-in.—AM displays from a number of companies.

Sharp were late into this but they had one advantage, the size of the glass panels that they were using in production. It was big enough to make a 14-in. display, which was the minimum acceptable size for a television. After some difficulties, and having designed a special integrated circuit, Sharp produced a television panel. By using triads of three dots with colored filters over them a full-color display ws produced, which they announced in June 1988. It took the company a further 3 years to make the first wall-hanging televisions.

In the race to produce flat panel televisions, LCDs didn't have it all their own way—there was a potential competitor. In 1964, Donald Bitzer and Gene Slottow at the University of Illinois wanted a more satisfactory display for their PLUTO computer-assisted instruction system.[18] It led them to make another variant of the gas discharge display where each discharge is in a tiny cell within a glass plate. When energized with high voltage it glowed with a plasma and hence this became the name for this type of display.

From then on, these displays had a chequered career before, in the 1990s, a variant was produced where the same three phosphors as used in color CRTs were put in front of sub pixels and a full color flat screen display produced. They could be produced in large sizes; in fact, there was a minimum practical size of 32 in., and for a while they had the market for very large flat panels to themselves.[19] However, there were drawbacks. They were expensive to manufacture and complex to drive with the need to switch high voltages. In addition, they consumed quite a lot of power.

However, LCD panels were coming along; they made the laptop computer practicable, and spread out to mobile phones and all manner of small visual screens. By the turn of the twenty-first century they were running neck and neck with CRTs in the television market, and a few years later were to take over completely. As the technology improved, LCDs steadily increased in size, squeezing plasma televisions into a smaller and smaller section of the market.

Without all this development of display technologies many devices such as mobile phones would not be possible. Many others, such as microwave ovens, video and DVD recorders, would not be easy to use. Displays have become so universal to give users information that it is now difficult to imagine a world without them, even though they are a much more recent development than might be imagined.

NOTES

1. Magic Eye tube, available at: https://en.wikipedia.org/wiki/Magic_eye_tube.
2. *The Guardian*, September 2, 1953.
3. For example, Denis Gabor's tube, *The Guardian*, April 6, 1956.

4. The Harlem Line, Solari departure boards, available at: http://www.iridetheharlemline.com/2010/05/14/fridays-from-the-historical-archives-solari-departure-boards-photos-history/.
5. US Patent 974,943.
6. Sobel, A. (1973) Electronic numbers. *Scientific American*, 228: 6, 64–73.
7. Furr, R. Electronic calculator displays, available at: http://www.vintagecalculators.com/html/calculator_display_technology.html.
8. Round, H.J. (1907) A note on carborundum. Letter to the Editor, *Electrical World*, Volume XIX.
9. Dupuis, R.D. and Krames, M.R. (2008) History, development, and applications of high-brightness visible light-emitting diodes. *Journal of Lightwave Technology*, 26: 9, 1154.
10. Invention History of Light Emitting Diode (LED), Circuits Today, available at: http://www.circuitstoday.com/invention-history-of-light-emitting-diode-led.
11. Borden, H.C. and Pighini, G.P. Solid-state displays. *Hewlett-Packard Journal*, February 1969, also available at: http://www.hpl.hp.com/hpjournal/pdfs/IssuePDFs/1969-02.pdf.
12. Fairchild, FLV102 data sheet, available from: http://www.lamptech.co.uk/Spec%20Sheets/LED%20Fairchild%20FLV102.htm.
13. An LED TV isn't really that. It is actually an LCD with LED backlighting instead of a cold cathode tube.
14. Kawamoto, H. (2002) The history of liquid-crystal displays. *Proceedings of the IEEE*, 90: 4, 460–500. This section relies heavily on this paper which was written by someone who at various times either worked for or was involved with nearly all the main players in this history.
15. Wild, P.J. First-hand: Liquid crystal display evolution. Swiss Contributions, IEEE Global History Network, available at: http://www.ieeeghn.org/wiki/index.php/First-Hand:Liquid_Crystal_Display_Evolution_-_Swiss_Contributions.
16. The grooves were made by stroking the surface with a tissue or cotton bud. It seems very crude but it remained the standard method as nothing better was found.
17. John Stonehouse was later to disappear in strange circumstances, apparently drowned on a Miami beach. In reality, he had run off with his secretary and was later found living in Australia under an assumed name. He was extradited to Britain and spent some years in prison.
18. Jay, K. The history of the plasma display, available at: http://voices.yahoo.com/the-history-plasma-display-361716.html?cat=15.
19. Plasma display, available at: https://en.wikipedia.org/wiki/Plasma_display.

13

Distributing Time: Clocks and Watches

I am a sundial, and I make a botch
Of what is done much better with a watch.

Hilaire Belloc

Great Britain, November 1840: As their tracks moved steadily westwards, the directors of the Great Western Railway came to a decision that all the clocks on their line would be set to London time and not to the local time at a particular station.[1] With their fast trains and tracks running more-or-less east–west scheduling was a particular problem. At the time it was for operational convenience, but its consequences rippled down the years and are still with us.

Before this date the clocks in a particular town or city were set to their local time, which was ahead or behind London time depending on their longitude. In Bristol, midday was 10 min later than in London; they even had a clock with 2 min hands to show this. Any local clocks were set to local time but, as they didn't keep time very well, it was necessary to check them against some other means, such as the sun or the stars. Normally, it would be the sundial, suitably adjusted for the date by looking up the correction in an almanac.[2] A stationmaster would adjust his clocks using tables supplied by the railway company to convert local time to London Time.[3]

During the 1840s, other railways followed GWR's lead. By 1847, the Railways Clearing House was recommending that all railways followed suit to make it easier for travellers changing from trains of one railway to another. By the end of the decade, almost all had changed to a standardized time, despite the difficulties of maintaining it.

London Time or, more specifically, Greenwich Mean Time, was established by astronomical observations. Clocks were made as accurate as possible and corrected when the observations were made. The mean time referred to the average of the variable length of the day throughout the year. The problem was how to distribute this time around the country.

At first, men with pocket chronometers attended the clocks and set them against the Greenwich Mean Time. However, in 1852 the Greenwich Observatory installed a Shepherd electric pendulum master clock which was capable of producing a synchronized electrical signal. This was connected to the railway's telegraph network, and signals on a particular hour were sent to stations around the country so that their clocks could be checked.

J.B. Williams, *The Electronics Revolution*, Springer Praxis Books,
DOI 10.1007/978-3-319-49088-5_13

Within 2 years, it was connected to the general telegraph network and available right across the country.

Though this had started as an initiative by the railways, by 1855 it was estimated that 98% of public clocks in Britain were set to Greenwich Mean Time.[4] Though there were some who tried to stick to their old local time the march of standardized time was unstoppable. Finally, in 1880 the Statutes (Definition of Time) Act was passed and GMT became the standard for the whole country. Thus the needs of the railways, aided and abetted by the telegraph system, had standardized time across the land, and subsequently across the world.

The first attempts to apply electricity to clocks occurred in 1840 when Alexander Bain, who we met in Chap. 4, applied for a patent.[5] His first ideas were for a contact to be made and broken by the swing of the pendulum of an ordinary clockwork clock.[6] The electrical impulse so obtained was used to click the seconds of a slave clock, which of course could be some distance away. It seemed a simple and elegant way to distribute time.

Over the next few years he explored other options, such as using a contact and electromagnetic impulse generator to maintain a pendulum and remove the normal escapement. He applied for a further patent to cover these ideas.[7] Despite a dispute with Charles Wheatstone, who behaved in a most ungentlemanly manner towards him, his patents stood. They were the basis for numerous attempts to improve clocks over the next 50 years or so.

There was a fundamental flaw in all these attempts to produce accurate clocks –the contacts and/or impulses disturbed the natural timing of the pendulum and degraded its timekeeping potential. Thus it was only in the twentieth century with Frank Hope-Jones' Synchronome, and finally the Shortt clock with its 'free' pendulum coupled to a slave one that did the work, that a considerable improvement in accuracy was produced. In the meantime, electrical distribution of time at acceptable accuracy was achieved.

The coming of wireless allowed time signals to be transmitted across the world with considerable benefits for navigation, as timekeeping is important in the determination of longitude (see Chap. 28). However, for ordinary people it was the start of radio broadcasting that brought the transmission of the six pips for the Greenwich time signal so that everyone could correct their watches and clocks.[8] As most only kept time to the order of a minute a day this allowed them to be simply checked every day.

Next to enter the fray was the telephone network which produced the 'speaking clock'. This was introduced in 1936 under the letter code of 'TIM' which was convenient for users to remember. It was a rather convoluted machine with its array of motors, glass disks, photocells and vacuum tubes and was the size of a small room. Nevertheless, because it was regularly checked against the Greenwich time signals, it was accurate to much better than a second and helped to spread standard time.

One of the drawbacks of mechanical clocks and watches at the time was that they needed winding either daily or sometimes weekly. If this was forgotten they stopped. Various attempts had been made in the nineteenth century to produce clocks that were wound electrically. Though this sounds simple, in practice it was quite complicated and so only used for specialized clocks.

In 1918, the American Henry Ellis Warren came up with a better idea.[9] A synchronous motor run from an alternating current supply will run at a fixed speed determined by the frequency. It wasn't difficult to then have a chain of gears to turn the hands at the correct speeds to produce a clock. There was only one problem; the timing depended on the frequency being accurately held.

Warren also had a solution to this problem and built a special clock which had 2 min hands, one driven by an accurate pendulum clock and the other by a synchronous motor connected to the ac supply. The clock was then installed at the power station; all the operators had to do was to keep the two hands together and they knew that the frequency they were generating was correct.[10]

In Britain, it was not until the Grid was set up in the mid- to late-1920s that this could be introduced. However, it turned out to be vital to keep tight control of the frequency when connecting numbers of power stations together on the Grid. In 1931, mains-run clocks started to appear in Britain and became popular over the next few years.[11] They were very reliable as long as there wasn't a power cut or the electricity generator supplying the load was having difficulty and reduced the frequency. Even then, the supply organization would try to correct the clocks within 24 h.

Despite the success of the free pendulum developments, the search was still on for even better methods of measuring time. One approach that seemed to hold out hope was to use a tuning fork. These had been used to tune musical instruments since the 1700s, but research into their shape and materials in the late nineteenth century produced highly accurate frequencies. Just after the First World War, with the availability of vacuum tubes, oscillators were built using the tuning fork as the feedback element determining the frequency.[12]

In the 1920s, the tuning fork system was developed to a high degree of accuracy, but a better device was to be developed. In 1880, the Curie brothers, Jacques and Pierre, had discovered a strange phenomenon in some crystals; when compressed, they produce an electrical voltage and when placed in an electrical field, they deform. This is called the piezoelectric effect. One crystal used to exhibit this was the highly stable quartz.

This remained a curiosity until during the First World War when piezoelectric crystals were used under water for sending and receiving sound signals for communicating with submarines and depth sounding. After the war, Walter G. Cady investigated the properties of quartz resonators in vacuum tube oscillators with a view to producing stable frequencies. The publication of his results inspired many other researchers to look at this field and many improved circuits were soon produced.

The one disadvantage of the quartz crystal was that it produced a high frequency, and means had to be invented to scale this down in order to drive a synchronous motor suitable for a clock. However, it was soon discovered that if the crystal was cut in a particular way it had a near zero temperature coefficient – i.e., the frequency was unaffected by changes in the ambient temperature around the crystal. It opened the possibility of very accurate clocks.

In 1927, Warren A. Marrison at the Bell Labs used this approach to build an accurate clock. Rapidly, standards laboratories around the world produced their own variants and improvements and soon this became the standard way of producing an accurate clock. The accuracies achieved left the pendulum clocks behind and produced a new level of timekeeping. As an offshoot it was also possible to both produce and measure frequency very accurately.

The coming of the Second World War again brought a need for reliable radio communication. Of importance in this was to produce signals at a known frequency, and traditional methods had not been very satisfactory for field equipment. It was here that the crystal-controlled oscillator came into its own. By using the same type of circuit which had been developed for precision clocks, radios that accurately maintained their frequency could be made.

However, the manufacture of the crystals was not so straightforward. Before the war crystals were produced by hand in a small number of companies mainly for amateur radio enthusiasts. The situation had improved before America entered the war in 1941, with its industry producing 100,000 units in that year, but by the end of the war this had rocketed to 2 million units a month.[13]

After the war there was a move to bring some of these improvements to watches, but it was not easy. The first attempt to produce an electric watch that didn't need winding was made in 1957 by the Hamilton Watch Company, Their 500 model[14] used a battery instead of a spring as a power source, but still retained the balance wheel as the timing element. As it used very fine contacts it still needed periodic servicing and was not the answer.

Max Hetzel, a Swiss electrical engineer who worked for the Bulova compay, thought he could do better. His design, which became the Accutron, used a tuning fork vibrating at 360 times a second as the timing element. Although still powered by a battery, the Accutron was considerably more accurate than even a well-adjusted mechanical watch and when it was launched in 1960 it made a marked impression on the traditional watch makers. In Switzerland, American and Japanese teams began to study how to make an electronic watch.[15]

A large Japanese watch manufacturer, Seiko, saw the danger and put together a team to investigate how to make accurate timepieces that responded to the Accutron's challenge. Their first response was to produce a portable electronic chronometer for use at the 1964 Tokyo Olympic Games where they were the official timekeeper. Their approach was to use a quartz crystal rather than a tuning fork as the timing element. They then moved on to the greater challenge of a watch.

On the other side of the world, Gerard Bauer, the president of the Swiss Chamber of Watchmaking, also saw the threat. He managed to get the competing companies to set up the CEH (Centre Electronique Horloger) alongside the existing research institute in Neuchâtel. The first problem was that Switzerland had no electronics industry, so they recruited Swiss-born engineers in America with the right skills and set them to work. Much of the existing industry was very skeptical, but they ploughed on.

As part of the background to the watch industry, every year the Neuchâtel Observatory held a competition where the watch makers could show off their best timepieces. In 1967, CEH submitted some prototype units and they filled the top ten places with Seiko's prototypes close behind. They were all way ahead of the performance of the best mechanical watch. A marker had been put down.

Two years later, Seiko were the first to sell a commercial product with their Astron SQ. It used a hybrid circuit with 76 transistors and was difficult to make, so only 200 were produced. The Swiss were a year behind but their design used an integrated circuit chip to divide down the frequency of the quartz crystal and was a better solution.[16] Sixteen companies launched watches based on the same electronic module. Both the Swiss and Japanese designs used a micro motor to drive the conventional gear train and hands.

By 1967, Hamilton had realized that their 500 model needed replacing by an all-electronic model but they were also aware that the Swiss and Seiko were likely to produce commercial models soon and they would be well behind. They decided that they needed something different, and chose to attempt an all-digital watch.

They teamed up with a small Texas electronics company, Electro-Data, and designed a quartz watch with a red Light Emitting Diode (LED) display. By 1970, they had six prototypes, but it took another 2 years to bring it to production with its single Integrated Circuit

(IC) chip. They called it the Pulsar—another suitably space-age name—and it sold for $2100, which wasn't exactly mainstream.

Hamilton had done the right thing in playing into the strengths of the American electronics industry. However, the big semiconductor players like Texas Instruments, Fairchild and National Semiconductor, also saw the opportunity. They were the experts in electronics and it didn't take them long to enter to watch market. Others quickly joined them and by the mid-1970s there were some 40 American companies mass-producing electronic watches. The result was a price war and tumbling prices.

There was one problem with the use of LEDs—the battery consumption. Though the chip could be made low power, the display used so much that the battery rapidly flattened. Thus these watches had a button to press to light the display, which only stayed on for a second or so. This rather defeated the whole point of a watch, which is to tell the time one handed. The only advantage was that it could be used in the dark—if the user could find the button.

Another option was the Liquid Crystal Display (LCD). The first attempts at this used the Dynamic Scattering mode (DSM) (see Chap. 12). This suffered from poor lifetime and, though low power, still had too high a battery consumption. It was inevitable that Seiko, as one of the biggest watch makers, would look at the digital market. They rejected DSM and went straight to the Twisted Nematic mode (TM) devices which had truly low-power consumption sufficient for the tiny battery to last a year or more.[17]

In 1973, Seiko launched their digital watch with the snappy name of 06LC. It had black characters on a near white background and was likely to last a great deal longer than 2 years before its battery needed replacing. The display only worked by reflected light, and so that it could be used in the dark a tiny lamp was included which could be lit by pressing a button. It was a great success in the marketplace.

The formula was quickly copied and by around 1977 LEDs had dropped out of favor. As components improved so did reliability and LCD watches began to dominate the market. The center of production moved from America and Switzerland to Japan, and then to Hong Kong as the prices tumbled. Now even a cheap quartz watch kept much better time than the best of the earlier mechanical devices.

For Switzerland this was a disaster as the watch industry was an important part of their economy. Between 1970 and 1980 the number of firms fell from 1600 to 600 and the number of people employed dropped from 90,000 to 47,000.[18] and. But then in the early 1980s they hit back. They designed the Swatch, which was an analogue quartz watch but, significantly, made largely of plastic on an automated production line. The number of parts had been reduced from 91 to 51 and it was intended as a second watch – hence the name.

By these means the price was kept low and it soon made an impact on the market after it was launched in 1983. To distinguish the product they built it up as a separate trendy brand to be casual and fun. The approach was very successful as in less than 2 years more than 2.5 million Swatches were sold. It brought the Swiss watch industry back into the game.

The Swatch led a trend back from digital watches to the conventional analogue dials, but still with a quartz 'movement'. There were worries that children would lose the art of 'telling the time' by which people meant interpreting the conventional clock or watch face. Though there are some advantages in a digital watch, apart from the presentation, in that it is simple to add other functions from automatic calendars to stopwatches and alarms, the trend was still back towards analogue presentation.

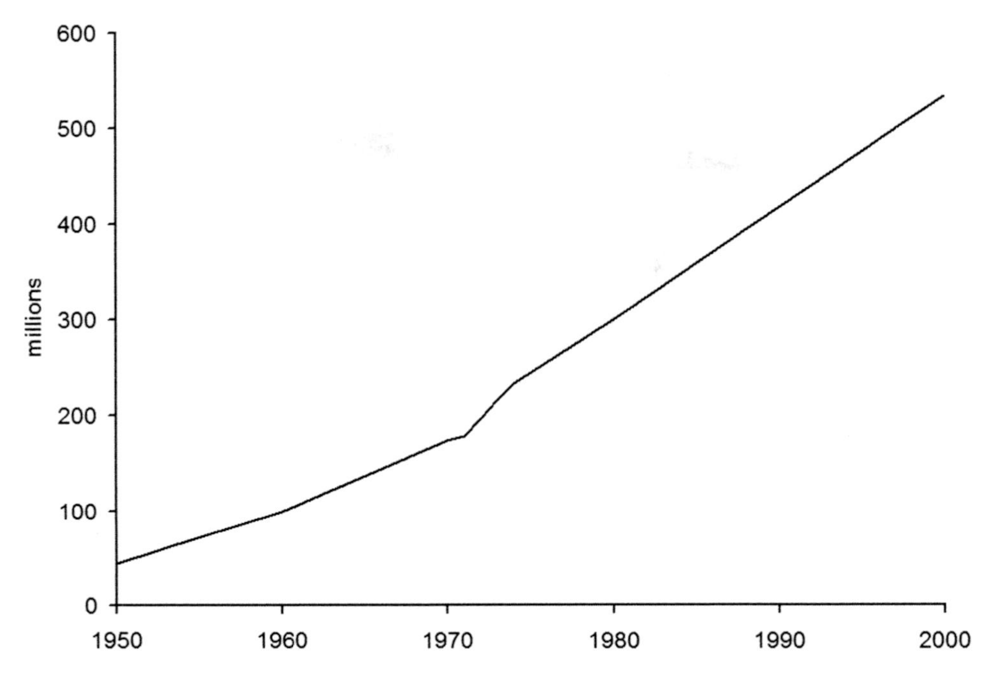

Fig. 13.1 Estimated annual worldwide sales of watches with some acceleration once the quartz watches arrive in the mid 1970s. At the end of the century there was one for almost every ten people on the planet. *Source:* Author[19]

Although clocks were so cheap that they appeared everywhere, far from killing the watch market, sales increased worldwide to more than 500 million by the end of the century. What was even more surprising was that mechanical watches didn't disappear. The Swiss watch makers found that there was still a market for luxury mechanical watches, some of which even had clear backs to display the 'beauty' of the movement. By 1997, three million mechanical watches were being made which, though ten times as many analogue quartz watches were made, represented half the value (Fig. 13.1).

As the world became more globalized, it also became much more time sensitive. Where 'within 5 min' had been regarded as quite adequate, now a minute or even tighter was regarded as the norm. The hunt was on for clocks that were always correct.

One way to achieve this was to use a radio signal, derived from a standard clock such as at Greenwich, to periodically automatically correct a clock. In the mid 1970s the low frequency transmitters at Rugby were rearranged to transmit accurate coded time signals on a very low carrier frequency of 60 kHz.[20] Around the world other similar systems were set up. By the early 1980s, commercial clocks with built-in receivers were starting to appear. These received the radio transmission and periodically adjusted the clock to keep it in step with the accurate transmitted time signal kept linked to standard Greenwich time.

Of course, the standards laboratories were still trying to find ever more stable ways of measuring time. Astronomical measurements and work with the accurate crystal clocks had shown that the rotation of the Earth was not entirely constant and was subject to a 'wobble' in addition to a slow decline due to the friction of the tides.[21] There was clearly a need for a more reliable standard than the very time-consuming astronomical methods.

Interest soon focused on the properties of atoms, and on caesium in particular. This was based on the work of Isidor Isaac Rabi, who won the Nobel Prize for Physics in 1944 for his method of recording the magnetic properties of atomic nuclei. With the ending of the Second World War work began on trying to build a clock based on these principles. Though this began in America it was Louis Essen and fellow physicist John Parry at the National Physical Laboratory (NPL) in Britain who succeeded in making their Caesium Mk 1 in 1955.[22]

Caesium Mk1 was so accurate that it would only gain or lose 1 s in 300 years, but that was only the start. Over the next few years this accuracy was further improved. By 1967, the atomic clocks were so good that the basic unit of the second was redefined in terms of caesium atom vibrations (despite a rearguard action by the astronomers who didn't want to give this up and hand it over to the physicists). They had to accept, though, that the definition of the second as 1/86,400 of a mean solar day had to be abandoned because the Earth's motion was not reliable enough.

This extreme accuracy of time measurement appears to be rather estoteric, but it does have an area of importance outside the rarified world of the standards laboratories. Because light has a fixed speed, the time that it (or a radio wave) takes to travel between points allows distances to be determined very accurately. This has become an important method of detemining position on the Earth and hence is the basis to modern navigation systems (see Chap. 28).

Timekeeping, which had begun as the preserve of the astronomers and was taken over by the mechanical clockmakers, in the twentieth century became absorbed by physics and electronics, transformed from a position where accuracies of a few minutes were regarded as sufficient to absolute precision. Without that accuracy, the modern world could not function.

NOTES

1. GMT, Railway time, available at: http://wwp.greenwichmeantime.com/info/railway.htm.
2. Because the axis of the Earth is tilted and it travels in an elliptical orbit, midday varies by up to ±16 min throughout the year, but the corrections can be calculated and published in the form of tables.
3. Railway time, available at: http://en.wikipedia.org/wiki/Railway_time.
4. Harrington, R. Trains, technology and time-travellers: How the Victorians re-invented time, available at: http://web.archive.org/web/20080828054933/http://www.greycat.org/papers/timetrav.html.
5. British patent 8783, year 1840.
6. See Hope-Jones, F. *Electric Clocks and How to Make Them*, pp. 6–10. ebook available at: http://www.electricclockarchive.org/Documents.aspx.
7. British patent 9745, year 1843.
8. Royal Museum Greenwich, The 'six pips' and digital radios, available at: http://www.rmg.co.uk/explore/astronomy-and-time/time-facts/six-pips-and-digital-radios.
9. Antiquarian Horological Society, History, available at: http://www.ahsoc.org/ehg/electricalindex.html.

10. Holcomb, H.S. and Webb, R. The Warren Telechron Master Clock Type A, available at: http://www.clockguy.com/SiteRelated/SiteGraphics/RefGraphics/WarrenTelechron/Telechron.TypeA.MasterClock.pdf.
11. Smith, B., Brief history of Smiths English Clocks Ltd., available at: http://www.electric-clocks.co.uk/SMITHS/sm-hist.htm.
12. Marrison, W.A. (1948) The evolution of the quartz crystal clock. *The Bell System Technical Journal*, Vol. XXVII, 510–588. The following section leans heavily on this article.
13. Thompson, R.J. (2011) Crystal Clear: The Struggle for Reliable Communication Technology in World War II. New York: Wiley-IEEE Press, p. 2.
14. Smithsonian Institute, The Hamilton Electric 500, available at: http://invention.smithsonian.org/centerpieces/quartz/coolwatches/hamilton.html.
15. Stephens, C. and Dennis, M. (2000) Engineering time: Inventing the electronic wristwatch. *British Journal for the History of Science*, 33, 477–497, available at: www.ieee-uffc.org/main/history/step.pdf.
16. Smithsonian Institute, Beta 21, available at: http://invention.smithsonian.org/centerpieces/quartz/coolwatches/beta21.html.
17. Kawamoto, H. (2002) The history of liquid-crystal displays. *Proceedings of the IEEE*, 90: 4, 460–500.
18. Young, A. (1999) Markets in time: The rise, fall, and revival of Swiss watchmaking. FEE, available at: http://www.fee.org/the_freeman/detail/markets-in-time-the-rise-fall-and-revival-of-swiss-watchmaking.
19. Data from: Adukia, G. et al., World Watch Industry, available at: http://www.scribd.com/doc/31816052/Watch-Industry; Anindita, et al., The watch industries in Switzerland, Japan and the USA 1970s, available at: http://www.slideshare.net/abhikaps/the-swiss-watch-industries; Economist Intelligence Unit, A Study on the United Kingdom Market for Wristwatches, Table 5; Young, A. Markets in time: The rise, fall, and revival of Swiss watchmaking, FEE 1 Jan 1999, available at: http://www.fee.org/the_freeman/detail/markets-in-time-the-rise-fall-and-revival-of-swiss-watchmaking. There is further information at the Federation of the Swiss Watch Industry, The Swiss and World Watchmaking Industry in 2004, available at: http://www.fhs.ch/statistics/watchmaking_2004.pdf. This suggests even higher figures for the end of the century, but this would mean that 1 in 5 of the world's population buys a new watch each year which seems impossibly high.
20. Hancock, M. The official History of Rugby Radio Station, available at: http://www.subbrit.org.uk/sb-sites/sites/r/rugby_radio/indexr69.shtml.
21. Forman, P. Atomichron: The atomic clock from concept to commercial product. IEEE Electronics, Ferroelectrics and Frequency Control Society, available at: http://www.ieee-uffc.org/main/history-atomichron.asp.
22. NPL, The world's first caesium atomic frequency standard, available at: http://www.npl.co.uk/educate-explore/what-is-time/the-worlds-first-caesium-atomic-frequency-standard.

14

From Desktop to Pocket: Calculators

Our object in developing the HP-35 was to give you a high precision portable electronic slide rule. We thought you'd like to have something only fictional heroes like James Bond, Walter Mitty or Dick Tracy are supposed to own.

Hewlett-Packard, HP-35 User Manual 1972

The need to have a more efficient way to calculate numbers had been recognized for centuries. Many attempts were made by the likes of Pascal, Leibniz and Babbage to produce machines, but none were entirely satisfactory or found wide use. Frenchman Charles Xavier Thomas de Colmar first made public his Arithmometer in 1820, but it wasn't ready and it was only in 1851 that he started to make it available commercially.[1]

The Arithmometer was made of brass in a wooden case in the usual amnner of nineteenth-century instrument makers. It had sliders to enter numbers, dials to show the result, and a handle to turn to perform the requested operation. It was capable of adding, subtracting, multiplying and dividing numbers. Adding and subtracting were straightforward, but multiplying was by repeated addition, and division by repeated subtraction which became quite complicated.

The construction left something to be desired and though the Arithmometer found a body of users in insurance actuaries, government departments, colleges and observatories, sales never reached more than about 100 per year. In those 40 years up to 1890 the total only reached about 2500 machines.[2] However, it had paved the way and around that date a number of competing machines appeared. Some were improved variants of the same principles but others were based on rather different mechanisms (Fig. 14.1).

The Comptometer, invented by the American Dorr Eugene Felt, was patented in 1887, and manufactured by the Felt and Tarrant Manufacturing Company of Chicago.[3] Keys were used to enter the numbers and, as they were pressed, the digit was automatically added to the accumulator. As the keys were positioned in columns the whole number could, in principle, be entered at once by a skilled operator hence making the machine very fast for additions.

J.B. Williams, *The Electronics Revolution*, Springer Praxis Books,
DOI 10.1007/978-3-319-49088-5_14

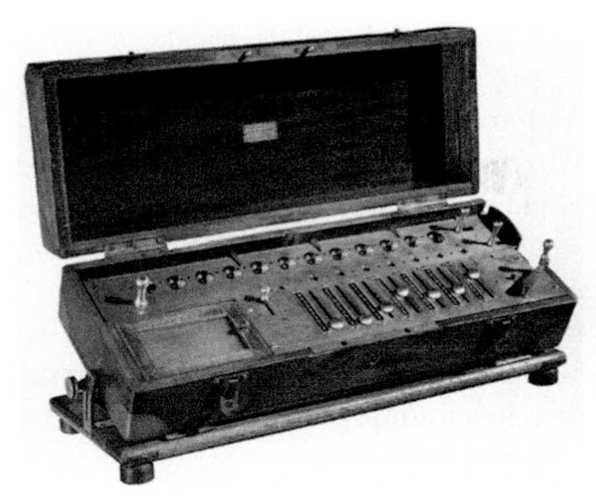

Fig. 14.1 An arithmometer (*left*) and a comptometer. *Source:* http://en.wikipedia.org/wiki/File:Arithmometer_Veuve_Payen.png; http://en.wikipedia.org/wiki/File:Comptometer_model_ST_Super_Totalizer.png

It could do the other arithmetic functions but they were less convenient. To subtract, the first number was entered and then the second had to be decreased by one and then entered in nines complement. This was quite simple as the keys were also marked with a second smaller number which was the nines complement. This was then added and the overflow thrown away, producing the subtraction.

Multiplication was by repeated addition and division by a variant of repeated subtraction which was a bit of a nightmare. Needless to say, operators had to be trained to use these machines, but they became quite popular and the generic name for a mechanical adding machine became 'Comptometer'. By 1903, cumulative sales had reached 6300, well ahead of the Arithmometers, and by 1926 had climbed to 126,150.[4] They and other adding machines proved reasonably reliable and increasingly found their place in offices.

Comptometer-type machines were ideal for commerce, but for more complicated calculations where addition didn't predominate then some of the other designs were a little easier to use. However, none of them could undertake a really simple division. Thus scientific or engineering work requiring large amounts of computation were extremely tedious and hence prone to error.

A popular method of multiplying and dividing was by the use of logarithm tables. The logarithm of each number is looked up and then added to multiply and subtract to divide. The resulting number is then looked up in an antilogarithm table which returns it to a normal number. It works quite well but isn't particularly quick and is limited to the four digits of the available tables. (Five-figure tables are too unwieldy for normal use.)

An extension of this idea is the slide rule. A logarithmic scale is printed on a ruler and a second, movable scale is placed alongside it. By placing the 1 of the second scale against the required number of the first scale the result of the multiplication can be read off opposite the second number. It is quick and simple though not particularly accurate. In practice, it was the mainstay of scientific and engineering calculation for more than two-thirds of the century.

It was surprising that the electrical industry didn't contribute to calculators sooner than it did, as for over 50 years its only contribution was to add motors to some of the machines. While it removed some of the handle turning, this didn't greatly improve the complexity of the operations, particularly division. As a result, only a proportion of the mechanical calculators had electric motors.

In the mid-1950s, the Kashio brothers in Japan decided to make an electrical calculator. They used 342 relays and the result was a machine built into a desk to hide the size of the works. It was launched in 1957 by their Casio Computer Co as the 14-A because it displayed 14 digits. One of their difficulties was how to display the result and here they used a matrix of numbers lit by the appropriate lamp behind it.[5] Because of its complexity there was only one display which showed the numbers as they were entered and then the result.

A further simplification was that there was only a single set of 10 keys to enter each number, unlike the 10 keys for each digit used in the Comptometers. In this it was showing the way forward. Here the numbers were entered a digit at a time as is regarded as normal today. Crucially, it could add, subtract, multiply, and divide simply at the press of a button without all the complexity necessary with the mechanical calculators.

The baton then passed to Britain where the Sumlock Comptometer Ltd. produced their ANITA (A New Inspiration to Arithmetic) calculator. The parent company, Bell Punch, had migrated from ticket punches to tote machines and mechanical calculators. In 1956, they started to look at the feasibility of making an electronic calculator, and by 1958 they had succeeded in building a prototype. It took a further 3 years to get the machine into production.[6]

ANITA took advantage of the arrival of Nixie tubes (see Chap. 12) to display the numbers. Basically, it used various types of vacuum tubes to undertake the four function calculations. It retained the 'full keyboard' with a set of 10 keys for each digit to make it familiar to the mechanical Comptometer users, but inside was a completely electronic machine. Again it achieved multiplication and division more simply than mechanical devices. It was a considerable achievement to pack all of this into a desktop-sized case and sell it at a competitive price. By 1964, 10,000 ANITAs were being sold per year.

The appearance of ANITA motivated a number of companies that had been considering the possibilities of electronic calculators. Germanium transistors had now reached a point where reliability and price were such that they could be considered for a calculator where some hundreds would be required.

There is some debate as to who produced the first transistorized calculator, but in 1964 the American Friden EC130, the Japanese Sharp CS-10A, and the Italian IME 84 all appeared.[7] The Friden used a cathode ray tube as a display while the other two used Nixies. These were all desktop machines and the Friden and IME used the single 10-key system introduced by Casio. However, Sharp, who had been building ANITAs under licence, retained the 'full' keyboard that it used.

The machines were expensive, priced in the region of $1500 to $2000, but their fast operation and ease of multiplying and dividing with a single key press made them attractive. They were also silent, which was an advantage where a number were used together as the noise from Comptometers could be considerable. Last, but not least, the absence of mechanical movement made them more reliable and requiring considerably less maintenance.

With the rapidly-developing semiconductor industry in the 1960s (see Chaps. 8 and 11) the opportunities for transistorized calculators opened up. Casio, for example, switched from relays to transistors in 1965 with the introduction of their 001.[8] In the next few years, some 50 companies joined in, eager to exploit the possibilities but all making relatively large desktop calculators suitable for the office.

As integrated circuits became available at practical prices these were incorporated into calculators, which held out the possibility of smaller units which wouldn't take up so much desk space. The calculator and the semiconductor business now became mutually dependent. Semiconductors were essential for calculators and advancements in the technology held out the possibilities of driving down prices to create bigger markets. On the other hand, calculators had become one of the biggest markets for semiconductors. Despite this, the number of calculators sold around 1968 was around 2.6 million and they were still predominately mechanical.[9]

Jack Kilby at Texas Instruments, who had been one of the inventors of the integrated circuit (see Chap. 11), was looking ahead in 1965.[10] At the rate that integrated circuit complexity was increasing it was foreseeable that, before long, all the circuitry required for a calculator could be fabricated on a single chip. The effect of this, if it could be made, was to reduce a calculator to something that could be held in the hand. It also raised the possibility of the price coming right down, opening up a mass market.

Kilby put a team to work and by 1967 they were in a position to patent their work.[11] The objective was not to make calculators but to use the design to show the capabilities of integrated circuits and particularly Texas Instruments technology. Though the semiconductor part was very advanced, they decided to use a small printer as the output device. Though this seemed a reasonable decision at the time, it was something of a blind alley.

Other people soon had similar ideas. The Japanese company Sharp produced their QT-8 in 1970. Rather than waiting for the technology to be capable of making a single chip, their design used four chips, not made by Texas Instruments but by another American company, Rockwell. It used a vacuum fluorescent display, one of the new types beginning to appear, and was small by desktop standards at 245 × 132 × 70 mm. More remarkable was the price. When it appeared in the UK in 1971 it sold for £199.[12]

While the first offering ran from the mains, it was followed by a version with rechargable batteries, producing a portable machine. This was soon followed by the Pocketronic from Canon which would have required a very large pocket! It used the Texas Instrument development, still in several chips, and a small printer to display the results. The Sanyo ICC-0081 Mini Calculator used chips from General Instrument and amber gas discharge displays. Again, it wouldn't go anywhere near a pocket. The pattern was thus Japanese calculator manufacturers using American integrated circuits.

Each of the chips in these sets had something like 900 transistors, but such was the pace of development of the semiconductor industry it was inevitable that soon someone was going to be able to put the whole of the calculator circuits onto a single integrated circuit. The winner of this race was a young American company called Mostek. It had been founded by ex-Texas Instruments people to make Metal Oxide Semiconductor integrated circuits, as its name implies.

Mostek had been very successful in making memory devices, and they were approached by the Nippon Calculating Machine Corporation, otherwise known as Busicom, to design and make a single calculator chip. At the same time, they approached Intel, another young company, to make a chip set for a more complex calculator range.

The device that Mostek produced was the MK 6010 and its clever design only required around 2100 transistors.[13] By January 1971, they were able to make the part and it was included in the Busicom Handy LE 120A. This was a true hand-size calculator using dry batteries and featuring another advance—LED displays. Its only drawback was that it was quite expensive.[14]

Over in Scotland, UK, a team who had originally worked for the computer maker Elliott Automation were made redundant when the company was absorbed by English

Electric. They went to work for the American semiconductor maker General Instrument in Glenrothes before setting up on their own as Pico Electronics Ltd.[15] There they designed a single calculator chip which was made by General Instrument as the GI 250. By later in 1971, this was in used in the Royal Digital III from Monroe, an American calculator manufacturer in the Litton Industries group, and soon in a number of other calculators. What was significant about this design was that it was actually a small computer in its architecture and all fabricated together in the single piece of silicon.

Also following this approach was Intel with their four chip set which was made in this form partly because they didn't have the packages with larger numbers of connections that the others were using, and partly that it gave greater flexibility to Busicom in the range of calculators that they wanted to use it for. Unfortunately, Busicom had misjudged this part of the market and it wasn't very successful. However, these chips were to have another incarnation which turned out to be even more important (see Chap. 15).

In September 1971, Texas Instruments announced their single calculator chip. Again, it was a computer-type architecture and more sophisticated than some of the others containing some 5000 transistors.[16] Variants were soon taken up by budding calculator makers. Among these was Bowmar, an LED manufacturer who found that the Japanese calculator companies just wouldn't buy their displays, and so decided to enter the market itself.

In Britain, electronics entrepreneur Clive Sinclair also looked at the Texas Instruments chip. He decided that it consumed too much power as it required sizable batteries and made the so-called pocket calculators rather large. Now Sinclair was very inventive and would often do things that other engineers wouldn't. In some cases he got away with it and sometimes he didn't.

He and his team found that by pulsing the power to the chip it would still work and would hold the information on the internal capacitances between the times when it was powered. By a clever circuit which would run the chip relatively fast when calculating and more slowly when it was not, he reduced the power consumption from 350 mW to 20 mW. The result was that he could reduce the battery size from AA to small deaf-aid mercury cells and still have a reasonable battery life.[17] This allowed him to make the Executive calculator which was smaller and lighter than anything else on the market. Crucially, it was only 3/8-in. thick and really could be put into a pocket. It wasn't particularly cheap, but Sinclair reasoned that it was aimed at the 'executive toy' market which wasn't particularly price sensitive. As more and more single calculator chips became available he was to go on to make a whole range of calculators.

The small calculator market really took off in 1972. Casio, for example, launched their 'Mini' which sold a million units in 10 months.[18] However, the American companies weren't going to leave the market to the Japanese. The single calculator chips had changed the rules of the game. Where previously the advantage had been with the Japanese who had cheap labor to assemble the many components, it now lay with the chip makers as the new designs required few components, but the cost was dominated by the chip and display.

Hence Texas Instruments and Rockwell both produced their own calculators.[19] They were only one of many as more and more companies thought that it was simply a matter of putting together a few components to get into the expanding calculator business. For the makers of desktop calculators it was an essential defensive move, but the extremely competitive market meant that prices would halve in each of the three half-year periods up to November 1973 and hence there would be a huge shakeout.

One way to avoid the worst of this was to offer more functions. The first to seriously do this was the instrument maker Hewlett Packard. As they also made LEDs it was logical for them to use those for the display of what they called a scientific calculator. Not only could it add, subtract, multiply and divide, but it had a whole range of logarithms, trigonometric functions and so on. They claimed that the slide rule was now obsolete, but at the time one could buy many slide rules for the price of one of their calculators.[20]

However, the HP 35 was very successful, selling nearly 200,000 units in Western Europe alone in its first year.[21] Texas Instruments responded with their SR10 and prices started to fall rapidly for scientific calculators as well.[22] Over the next couple of years, most scientists and engineers had to invest in one of these devices which did indeed make the slide rule obsolete. The calculators could do everything the slide rule could, and a lot more, and with much greater accuracy. There was no competition.

The Japanese calculator makers were under pressure as the average price of their exports fell drastically from early 1971 to late 1973.[23] Finding that their American integrated circuits suppliers had become their competitors, they turned to local semiconductor companies for their large-scale integrated circuits (LSIs).[24] This gave a considerable fillip to the Japanese semiconductor industry. As the competition became more intense, some of the Japanese calculator companies began to fight back.

Though fairly satisfactory, the combination of the LSIs and LEDs being used had one drawback – they used too much power, requiring a sizable battery pack or even rechargeable batteries which hence defined the minimum size of the calculator. As Clive Sinclair had realized, what was needed was a much lower power consumption, but his method was not the way forward.

The solution was available but, curiously, both halves were tried before they were finally put together. What was required was the use of complementary MOS (CMOS) for the LSI which only drew power when it switched, and liquid crystal displays (LCDs). With the Americans heavily committed to LEDs it was likely that this move would come from Japan and in 1973 Sharp announced their EL-805.[25]

They had gone further than necessary by placing the chips as well as the display all on the piece of glass. It turned out to be an expensive solution. The further problems were that they had to use two chips as the CMOS is more difficult to make than the PMOS that had been used up to that point, and the LCDs were still not fully developed. Nevertheless, this pointed the way forward. One warning that the American industry didn't heed was that the LSI chips were made by Toshiba – in Japan.

Over the next couple of years, competition became extremely fierce. In Britain, Sinclair reduced the price of his four function Cambridge calculator to under £20 by the middle of 1974. By November of the following year, prices were a quarter that for the cheapest models; it was estimated that 50 million units had been sold that year, which was remarkable as the industry hadn't existed 5 years before. Inevitably, some of the manufacturers were either going bust or getting out of the business.

Having reached really low powers, the next step was to add a solar cell so that the calculator never needed to have its battery changed. This was achieved around 1978 by Tokyo Electronic Application Laboratory (TEAL) in their Photon model.[26] On the other hand, Casio produced their Mini Card LC-78 in the same year, and though it didn't have a solar cell and was relatively expensive, it was only the size of a credit card. Calculators were now approaching their ultimate forms.

Fig. 14.2 A scientific calculator with solar cells and more than 200 functions from the 1990s. *Source:* Author

As the sophistication of the integrated circuits increased it became possible to pack in more functions (Fig. 14.2). After the scientific calculator the programable one appeared. At first sight, this appears to be converging on a computer, but there is a fundamental difference. A calculator is for calculations, whereas a computer is capable of a great deal more. Thus the programable calculator allows the user to set up a sequence of calculations but no more than that.

Before the widespread use of personal computers, the programable calculator was useful for scientists, engineers, and those who needed to do repetitive complex calculations. Even in the age of computers it is sometimes useful to have a small hand-held device, already set up to do some awkward calculation, which only requires the input of some data and then will give the answer directly.

Once more complex displays were available the graphing calculator appeared which was capable of displaying a mathematical function in the form of a graph. Again, it was not the thing for everyone, but for a small minority with particular requirements it was very useful. It merely went to show how far the basic concept of the calculator could be pushed when large amounts of calculating power could be produced cheaply.

Driven by the phenomenal march of the integrated circuit industry, the growth of the calculator market was spectacular. This rise can be shown by the cumulative sales of Casio, who stayed in the market when many others dropped out. From a base of around 100,000 units

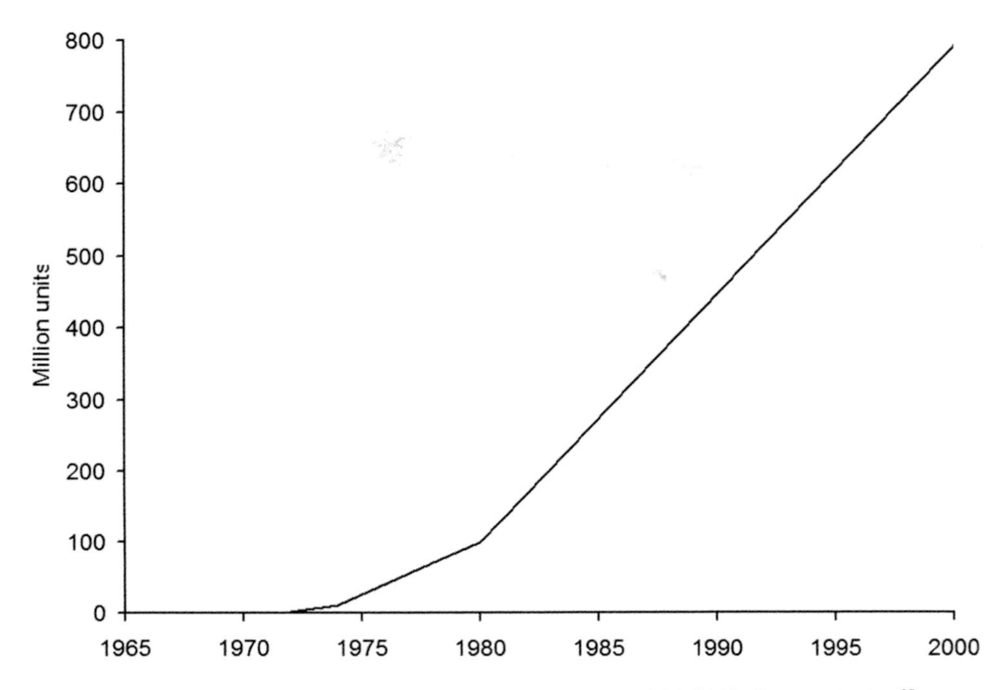

Fig. 14.3 Cumulative growth of Casio calculator sales, 1965–2000. *Source:* Author[27]

in 1969 sales rocketed to a billion a little after the turn of the century (Fig. 14.3). And this was the output of just one company. Over the years, so many calculators have been made that a considerable proportion of all the people in the world have at least one.

Though many calculators are simple four function devices which are quite adequate for most people to add up their shopping or check their power bills, there are considerable numbers of more sophisticated devices, often with more functions than their owners will ever use. Calculating power is so cheap that it is easier for the manufacturer to pack on more and more functions so that a particular group of users will find that the calculator contains the special functions that they need, even if these are only a very small subset of those available.

Small portable 'pocket' calculators have become so much part of everyday life that it is difficult now to believe that they didn't exist before 1970. The speed with which they became universal was astonishing. The slide rule became obsolete almost overnight, and calculations that took ages with pencil and paper could be completed in a fraction of the time. Mental arithmetic, once a prized skill, virtually disappeared as everyone reached for a calculator the moment they needed to do a sum.

The pocket calculator business also had another effect. At its beginning it drove the large-scale integrated circuit makers to greater endeavors. Later, it was to have an even more profound effect as it brought together the semiconductor and computer industries with far-reaching effects. This is the subject of the next chapter.

NOTES

1. Johnston, S. (1997) Making the arithmometer count. *Bulletin of the Scientific Instrument Society*, 52, 12–21, also available at: http://www.mhs.ox.ac.uk/staff/saj/arithmometer/.
2. Calculator, available at: http://en.wikipedia.org/wiki/Calculator.
3. Tout, N. Mechanical calculators: Comptometer, vintage calculator web museum, available at: http://www.vintagecalculators.com/html/comptometer.html.
4. Early Office Museum, Key-driven calculating machines, available at: http://www.officemuseum.com/calculating_machines_key_driven.htm.
5. Casio Calculator Collectors, Casio 14_A, available at: http://www.casio-calculator.com/Museum/Pages/Numbers/14-A/14-A.html.
6. Tout, N. The Development of ANITA, available at: http://www.anita-calculators.info/html/development_of_anita_1.html and succeeding pages.
7. Bensene, R. Friden EC-130 Electronic Calculator, The Old Calculator Web Museum, available at: http://www.oldcalculatormuseum.com/friden130.html; Tout, N. Sharp Compet CS10A, available at: http://www.vintagecalculators.com/html/sharp_compet_cs10a.html ; Tout, N. IME 84rc, available at: http://www.vintagecalculators.com/html/ime_84rc.html.
8. History of CASIO's electronic calculator business, available at: http://www.casio-intl.com/asia-mea/en/calc/history/.
9. Electronics, June 1968, quoted in Tout, N. The story of the race to develop the pocket Eelectronic calculator , available at: http://www.vintagecalculators.com//html/the_pocket_calculator_race.html.
10. Kim, I. (1990) Functions at your fingerprints. *Mechanical Engineering Magazine*, 112: 1, also available in: http://www.xnumber.com/xnumber/kilby3.htm.
11. The original 1967 filing was abandoned but eventually became US patent No 3819921 in 1972.
12. Tout, N. The Sharp QT-8D micro Complet, available at: http://www.vintagecalculators.com/html/sharp_qt-8d.html.
13. Other integrated circuits, Mostek MK6010, available at: http://leoninstruments.blogspot.co.uk/2013_01_01_archive.html.
14. Tout, N. Busicom LE-120A HANDY-LE, available at: http://www.vintagecalculators.com/html/busicom_le-120a___le-120s.html.
15. McGonigal, J. Microprocessor history: Foundations in Glenrothes, Scotland, available at: http://www.spingal.plus.com/micro/.
16. Woerner, J. Texas Instruments IC - Integrated circuits, available at: http://www.datamath.org/IC_List.htm#First commercial available single chip calculator.
17. Pocket calculators add up to a big market. *New Scientist*, July 20, 1972.
18. History of Casio's electronic calculator business, available at: http://www.casio-intl.com/asia-mea/en/calc/history/.
19. Tout, N. The story of the race to develop the pocket electronic calculator, available at: http://www.vintagecalculators.com//html/the_pocket_calculator_race.html.
20. Swartzlander, E.E. (2013) Stars: Electronic calculators: Desktop to pocket. *Proceedings of the IEEE*, 101: 12, 2558–2562.
21. Evans, P. (1976) Calculators in perspective: A review of the industry that surprised everybody. *The Production Engineer*, July/August 1976.
22. Valentine, N. The history of the calculator—part 2, The calculator site, available at: http://www.thecalculatorsite.com/articles/units/history-of-the-calculator-2.php#microchip.
23. *Electronics Today International*, August 1974, quoted in Tout, N. The story of the race to develop the pocket electronic calculator, available at: http://www.vintagecalculators.com//html/the_pocket_calculator_race.html.

24. Kim, Y. (2012) Interfirm cooperation in Japan's integrated circuit industry, 1960s–1970s. *Business History Review* 86, 773–792.
25. Chiba, T. et al. (2012) History of developing and commercializing families of solid-state calculators. Third IEEE History of Electro-technology Conference (HISTELCON), pp. 1–5.
26. Tout, N. Teal photon, available at: http://www.vintagecalculators.com/html/teal_photon.html.
27. Casio, available at: http://www.casio-intl.com/asia-mea/en/calc/history/.

15

Shrinking Computers: Microprocessors

One of the things I fault the media for is when you talk about microprocessors, you think about notebook and desktop computers. You don't think of automobiles, or digital cameras or cell phones that make use of computation.

Ted Hoff

In April 1970 Frederico Faggin, a young Italian working in America, followed a number of his bosses, who had left Fairchild Semiconductor, to the young Intel that they had formed. He had been promised a challenging chip design project though he didn't know any of the detail. Little did he realize quite what he had let himself in for. The worst was that the customer was due in a few days' time and no work had been done on the project for some 6 months. When Masatoshi Shima, Busicom's engineer, learned of this, he was furious and called Faggin all sorts of names despite him trying to explain that he had only just arrived.[1]

As if this wasn't enough, the task that Faggin had been set was to design four chips in far too short a timescale. The scheme had been invented by Ted Hoff, the manager of the Application Research department, who had taken the customer's more complex arrangement and reduced it to what was basically a computer. It consisted of a central processing unit (CPU), a memory to store the program (Read only memory—ROM), a store for intermediate results (Random access memory—RAM) and some input and output (I/O) connections (Fig. 15.1). Each of these sections was to be in a separate chip.

Once Shima had calmed down, he was persuaded to remain in America and help Faggin with the design work on the chips. Surprisingly, Busicom agreed to this. Then began a furious effort to try to make up for lost time by a young company which had few design tools to assist the process and no spare resources of people. The work was almost entirely done by Faggin and Shima with the help of a draughtsman and eventually two draughtswomen.

The RAM and ROM didn't present too many problems in a company that had been set up to make memories, and the I/O chip was reasonably simple. However, the CPU needed around 2300 transistors and was at the limits of the technology at the time. It was thus a staggering achievement to have all the chips completed within 6 months and the first ones

J.B. Williams, *The Electronics Revolution*, Springer Praxis Books,
DOI 10.1007/978-3-319-49088-5_15

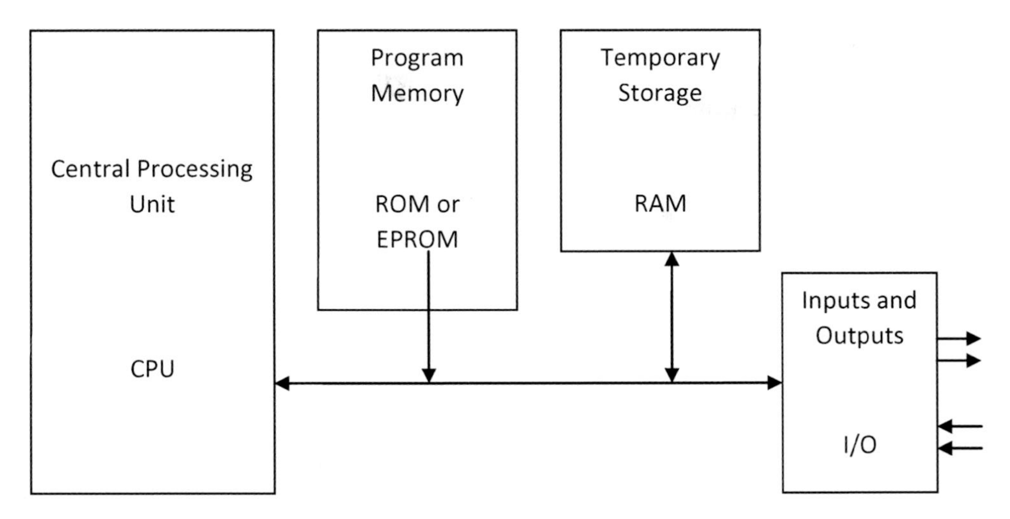

Fig. 15.1 The main sections of a computer. A microprocessor is just the CPU while a micro-controller contains all the elements. *Source:* Author

coming off the production line. However, when the first CPU chips appeared just before the end of the year, they didn't work. After much sweat, Faggin realized that a mistake had been made in the production which was nothing to do with him.

Three weeks later he received more chips, made correctly this time, and they worked. As he said 'the first microprocessor was born'. Though they contained some minor errors, these were easy to fix. Soon sets of chips were on their way to Japan and being fitted into Busicom calculators. Intel could claim to be the inventor of the microprocessor though this was usually ascribed to Ted Hoff who had come up with the idea, not to Frederico Faggin and Masatoshi Shima who had sweated to make it work.

This is the usual story that is told of the invention of the microprocessor. As you might have guessed, it is not that simple. It ignores the work taking place at Texas Instruments, and at Pico Electronics working with General Instrument in Scotland as described in Chap. 14. There are also a number of other contenders for the title. Again, it was a case of simultaneous invention and who got there first is a moot point. It is notable that Intel didn't try to file a patent, as they thought it obvious at the time. The importance of the device, known as the 4004, was in the effect that it had on the future.

Like so often with these things, it nearly didn't happen. Intel had a contract with Busicom that gave the customer the rights to the chip. When Faggin discovered this he campaigned for the company to buy back the rights, but the management took the view that they were a memory company and the project was not important.

However, when Shima—now back in Japan—tipped off Faggin that Busicom was in financial trouble he saw his opportunity to press his management. His case was also bolstered by the fact that he had built a wafer-sort tester for the 4004 using one of those chips inside it. He had discovered how useful it was and felt sure that other people would find it so as well. In the event, Intel regained the rights to market the chip before Busicom went out of business.

One thing the tester project had shown was the importance of another Intel invention, the Electrically Programmable Read Only Memory (EPROM). The program in the ROM normally used for the calculator was permanently made into the device at the time of manufacture. Thus it could not subsequently be changed. With the EPROM the program was written to it electrically after manufacture. Better still, the device had a window in the top and when placed under a UV lamp for around 20 min the program would be erased and could then be written again.

This meant that changes could easily be made during the development process before the final program was transferred to the ROM. Perhaps more important was that where, like with the tester, only a small number of items were being made there was a practical solution. For these cases the cost of having a ROM made was far too expensive. The result was that using the microprocessor for one item, or a few, became practical and not just for the likes of calculators where the volumes were huge.

In November 1971 Intel launched the microprocessor and the other chips as the MCS-4 family. There were still those in the company who were concerned that if they called it a computer they might be regarded as competing with the computer companies who were their customers for their memory chips.[2] However, Faggin and those involved with marketing the device were convinced and continued to push it.

The 4004 was a 4 bit device which means that it handled 4 bits with their possible 16 states at a time. This was quite convenient for a calculator where with suitable manipulation it could handle the ten states of a decimal digit. These could be manipulated one at a time with the intermediate results stored in the RAM. The I/O chip took in the key presses and drove the display.

However, for coding letters 4 bits is not enough. There was already a 7 bit code known as ASCII (American Standard Computer Information Interchange) which was used for Teletype machines, so it was natural to use that in computing.[3] In practice, this meant that the computer or microprocessor needed to handle 8 bits at a time.

In fact, Intel had taken on another project to produce an 8 bit microprocessor around the same time as the Busicom one. This was for CTC (Computer Terminals Corporation of San Antonio, Texas) for use in a computer terminal. At the time, large computers often had many users, each one with a terminal which consisted of a keyboard and a Cathode Ray Tube (CRT) screen. It had a communication link to the main computer, but was really a simple computer in itself. Clearly, to handle letters it needed to be 8 bit.

For various reasons progress was slow, and it wasn't until Faggin had finished the 4004 and had a proven methodology that the layout of the 8 bit device started in earnest. Eventually, CTC decided the device would be too expensive and so didn't use it. Intel reclaimed the rights to it and it was launched in April 1972 as the 8008 (twice the 4004).[4]

Intel's marketing department soon realized that it was one thing to have the chips but the customers needed large amounts of information and also systems to help them develop the software. Progress was inevitably slow while potential users learned about the devices and what they were capable of, and experimented with them. At first few chips were sold.[5] There was a further problem in that they were mounted in 16 and 18 pin packages which meant that the data and addresses for the memory chip had to be multiplexed, i.e., come out one at a time.

The other chips for the 4004 family dealt with this, but the 8008 required some 20 standard logic integrated circuits to provide the interface between the processor, memory, and I/O as no support chips were designed. To use standard memory chips a considerable amount of circuitry was needed and this made the devices inconvenient to use. However, the devices pointed the way forward and the potential market began to be educated as to the possibilities of microprocessors.

Clearly, there was a need for a more advanced 8 bit device. It needed to use a larger package of 40 pins so that standard memories could be used with a minimal number of external logic devices. In addition, the instruction set could be improved. All of this was understood by Faggin but it took him some time to convince his management who were still fixated on memories.

The result was the launching of the 8080 in early 1974 with a hefty price tag. Here was a simple processor that could be used for all manner of tasks and was powerful enough to tackle them. The market was given a further boost when Motorola launched their competing, but perhaps superior 6800, a few months later. It only had 4000 transistors to the 8080s 6000, showing the difference between the efficiency of the designs. The potential users were now ready and the chips started to be used seriously.

Many other companies began to pile in with their own offerings, but it was the families of derivatives of these two initial devices that were to dominate the market for some 20 years. The 8080 led to the 8088 and 8086 and all the other devices that were the work horses of the IBM type personal computers, while the 68,000, a 6800 derivative, formed the family for many of Apple Computer's offerings. The rise of the personal computer is examined in Chap. 17.

As with so many things there is a glamorous side and the great unsung mass. It is the same with microprocessors. The ever more powerful chips used in personal computers gain all the publicity and provide the froth, earning a disproportionate amount of the revenue while selling a tiny fraction of all the devices. The rest are buried away in products where the user often doesn't know they are there.

The original objective had been to produce a computer on a chip. Some of the early devices mentioned in Chap. 14 were just that, though they were optimised for use in calculators. Though in theory the program ROM could be altered for another task, this doesn't seem to have been done. It was only in 1974, after Intel's devices with external memories had shown the way, that Texas Instruments made revised versions of their chip available as the TMS1000 series. They were 4 bit devices with limited capabilities, but they were cheap. For those companies who had products in sufficient volumes to justify the high costs of having their own ROM versions manufactured they could be attractive.

However, for the average user that just wasn't practicable. What was needed was a single chip device with a flexible means of programing it. In 1976, Intel came up with a solution with their 8048 family. First, they produced a version of the device (8748) where the program memory was an EPROM which could be electrically programed and erased with a UV lamp. This was quite courageous as the EPROM technology was new and combining it with a microprocessor was a considerable step into the unknown.

Intel's idea was that users could develop their software using these devices and then transfer it to ROM for production. The customers had other ideas and for many small run projects, where development cost was more important than the product price, they simply used the EPROM version for production. Better still, members of this microprocessor family could be switched to use an external program memory. In fact, Intel sold devices with faulty ROMs for this purpose.

At first sight, using an external program memory seems like a retrograde step, but the chip contained the CPU, the RAM, I/O and a timer which was usually necessary for setting time periods. To have just the program memory outside the chip wasn't a particular problem, particularly as standard EPROMs started to fall in price, and bigger memories than those supplied internally could be used. It meant that for many products little more than the two chips was necessary. The way was now open for anyone and everyone to produce microprocessor-based products.

Next, Motorola produced their 6801 family which was more powerful than the 6800 and also included the RAM, I/O, and timers. Some versions included the program memory but again it could use an external memory. Finding these a bit expensive for some purposes they also produced a simpler and cheaper 6805 family. Of course, Intel and Motorola weren't the only players. By 1975, a dozen companies were making various sorts of microprocessor while by 1980 this had doubled.[6]

The question, of course, was what they could be used for. Once their capabilities had been demonstrated effectively, then the impetus moved to the user. At first, it was difficult for older engineers to consider putting a whole computer—the 4004 in a small chip had similar computing power to the 30 tons of the early ENIAC machine—into their product. However, many young designers had no such inhibitions and went ahead with the exciting possibilities opened up by microprocessors.

In 1976, the American car industry was under pressure from new government regulations to reduce emissions. They turned to electronics, and particularly microprocessors, to give improved engine management and so meet the regulations. This was the high-volume breakthrough that the microprocessor makers needed. Once the car manufacturers had introduced microprocessors into their vehicles they were to use them more and more over the coming years to solve all sorts of problems, from ABS to air bags to trip computers, entertainment and navigation systems. By the end of the century, a sophisticated car could easily have tens of microprocessors hidden away.

In Japan, the consumer electronics manufacturers saw the opportunities. Microprocessors appeared in many products such as televisions where additional features could be implemented relatively cheaply. They also made possible products that would not have been practical without them such as video cassette recorders (VCRs), camcorders and microwave ovens.

Many users could see their possibilities in a variety of smaller-volume fields. All the early lectures on microprocessors used the 'blood analyzer' as their example of a possible product, while the typical learning project was often a traffic light system. It was not long before these became real products. Microprocessors crept in everywhere that required some sequencing and control such as lifts, robots, machine tools and even milking machines.

In the high street, ATM machines and point-of-sale tills with their barcode readers rely heavily on the computing power of the microprocessor. In offices, as well as the personal computer, there were many more buried in printers, copiers, fax machines, private telephone exchanges and so on. Microprocessors were starting to shape the modern world without most people realizing that they were there (Fig. 15.2).

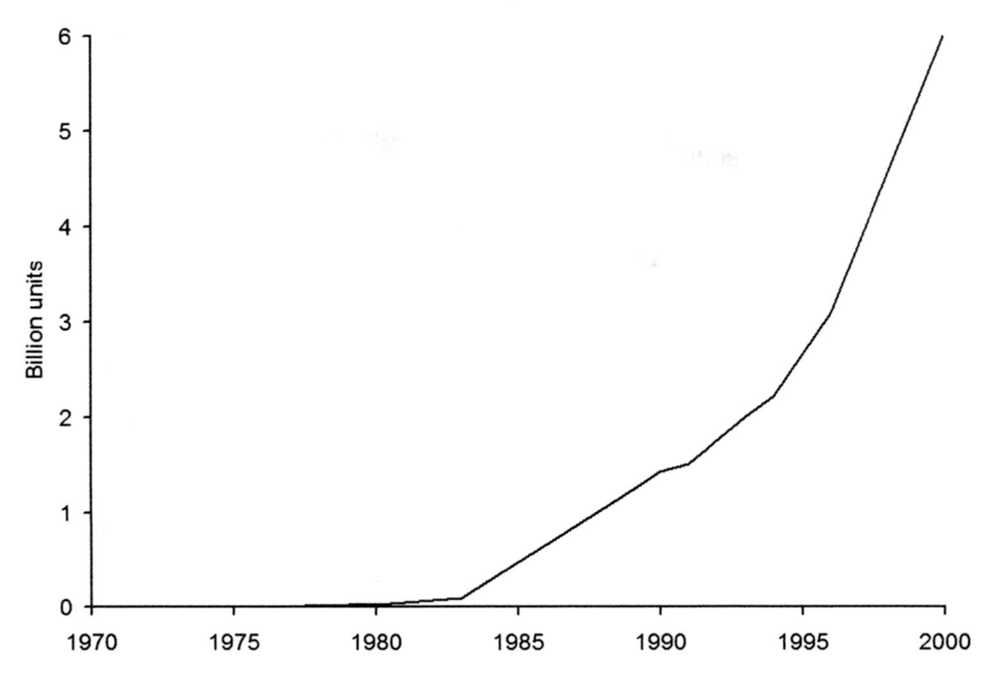

Fig. 15.2 Annual production of microcontrollers, 1970–2000—true computers on a chip. *Source:* Author[7]

Some technical improvements in the program storage ROMs helped to increase their usage. First was the obvious idea of the One Time Programmable ROM (OTP). This really was just an EPROM with no window in the device, which made it cheaper to produce. The user could electrically program it but couldn't change the result. It was suitable for reasonable production runs where the program was fully developed and didn't need to be changed.

Better still was the electrically erasable programmable ROM or flash EEPROM. This could be erased and rewritten electrically simply by using suitable signals externally which meant that it didn't need to be removed from its circuit to be reprogramed. Steadily, microprocessors with their memories on board—often known as microcontrollers—moved to this format. With a suitably large program memory built in it contained everything that was needed but was flexible if changes needed to be made. It was the ultimate form.

Though the families of microprocessors available were very successful, there were those who thought that they could be improved. The instruction sets, though satisfactory, had grown rather haphazardly and took at lot of logic to implement. The result was that the designs used large numbers of transistors, ran hot, and were expensive to produce.

The alternative approach was called Reduced Instruction Set Computer (RISC). It was based on the counterintuitive idea that simpler instructions would run faster, even if more of them were needed. Considerable work was done on the concept, both in academia at Stanford and Berkley universities and at IBM and other companies, but though products were produced and they worked, they didn't make much impact in the marketplace.[8]

In Cambridge in England, the computer company Acorn was looking for a microprocessor to produce the next generation of computers after its very successful BBC computer (see Chap. 17). They looked at all the current offerings but none seemed to have the speed and performance they were looking for. They began to consider making one of their own.

As part of the investigation two engineers, Sophie Wilson and Steve Furber, visited the Western Design Centre in Phoenix, Arizona, where Bill Mensch and a small team was designing a successor to their favorite device, the 6502, in a bungalow in the suburbs.[9] Wilson and Furber had expected to find a large team backed by extensive and sophisticated resources. Their visit showed them that designing a microprocessor didn't need that and they ought to be quite capable of doing it themselves.

This was quite an optimistic view as IBM had spent months using their large computers to simulate the instruction set of their RISC processor. Not knowing this, Wilson plunged right in and did the whole thing in her head. Furber's task was to turn these ideas into logic. Then the chip layout designers took over.

Eighteen months later, in April 1985, the first Acorn RISC Microcomputer appeared. They plugged it in and it worked perfectly, but the one thing that puzzled them was that it appeared to be consuming no power. The fault was soon traced to a missing track on the test board they were using which meant that no power was being supplied to the chip. Amazingly, it was deriving its power from the surrounding logic circuits, and Wilson and Furber began to appreciate how little power it was consuming. Only then did they realize what they had achieved. They had a powerful 32 bit microprocessor which could perform to compete with the best of them, but it only had 25,000 transistors compared with Intel's latest offering which used 135,000.[10] Not only that, the ARM was faster and consumed much less power. They had a winner here if only they knew what to do with it.

The obvious thing was to build a desktop computer and the result was the Acorn Archimedes. This was a very effective design, but by the early 1990s the battle had been lost and the IBM PC format based on the Intel processor range had conquered all. However, the chip was found to be so small that other circuitry could be incorporated on the same piece of silicon producing systems on a chip.

Across the Atlantic, the American computer company Apple was interested in this low-power microprocessor to build portable devices where battery lifetime was important. They were concerned that Acorn as a computer maker was a competitor, but the solution was to set up a separate company as a partnership to exploit the processor—ARM. A third partner was VLSI who had actually been manufacturing the chips.

ARM decided that they would not make chips, but carry on designing them and licence them to any manufacturer who was interested, taking a royalty on their sales. The clever bit was that the licencees became part of the ARM club which had an entrance fee, hence giving it income until the royalties came through. Steadily, chip manufacturing companies joined this club, starting with two in 1992. By 2000 there were 31 licencees and the numbers were rising fast.[11]

The ARM processors were continually updated, but they were still recognizably based on Sophie Wilson's brilliant initial conception. The devices were finding their way into all manner of portable devices where the low power consumption and high performance were vital. This began with Apple's Newton device which, though not so successful in the marketplace, pointed the way. By the turn of the millennium most of the Personal Digital

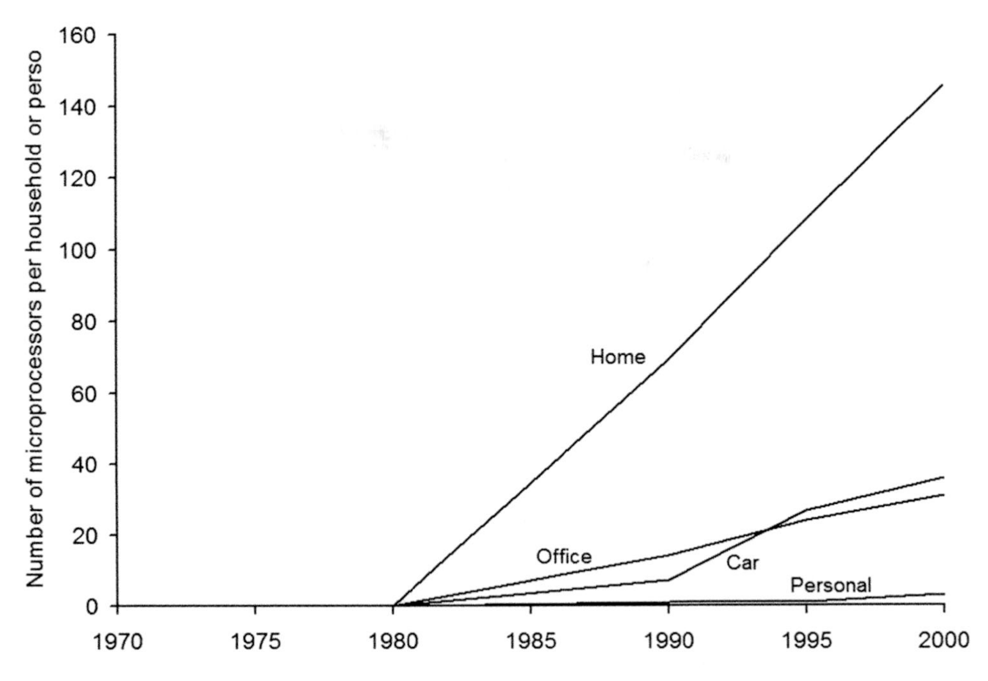

Fig. 15.3 The penetration of microprocessors into everyday life, 1970–2000. This is perhaps a bit optimistic outside America. *Source:* Author[12]

Assistants (PDAs) and mobile phones were moving to ARM-based chips from various manufacturers.

By that point the ARM chips were selling at the rate of some hundreds of millions a year which was some four times Intel's total for their personal computer chips. The total number of ARM chips that had been sold was in the order of a billion, and rising fast.[13] It only took a few more years and one had been made for every person on the planet. So what had started as a failed attempt to compete with the Intel stranglehold on the personal computer ended up ruling a different and much bigger world, and one that may well take over devices such as tablets—nearly all ARM processor based—steadily progress.

Acorn soon found they couldn't compete, and faded away. It is easy to see the company as a failure, but its legacy was immense. Not only ARM but over 100 companies were spawned from it either directly or indirectly by people who had worked there. Its influence was rather like Shockley Semiconductor who spun out a whole industry. The Cambridge phenomenon was largely created by Acorn.

By the turn of the millennium most people were carrying around mobile phones or other personal items hiding at least one microprocessor and sometimes more. Cars were full of them, controlling everything from the engine to the brakes. In the office or factory, everything from phones to printers to robots was dependent on them. Perhaps the biggest impact was in the home where they had crept into most appliances, games, televisions, cameras, and a multiplicity of other devices (Fig. 15.3).

The modern world would be utterly different without this recourse to huge amounts of processing power. The remaining chapters mostly investigate how they were used and created so many of the things we now take for granted.

NOTES

1. Faggin, F. (2009) The making of the first microprocessor. *IEEE Solid State Circuits Magazine*, Winter 2009.
2. Stachniak, Z. (2013) This is not a computer: Negotiating the microprocessor. *IEEE Annals of the History of Computing*, 35:4, 48–54.
3. ASCII, available at: http://en.wikipedia.org/wiki/ASCII.
4. Faggin, F., Hoff, M.E., Mazor, S. and Shima, M. (1996) The history of the 4004. IEEE Micro, 16:6, 10–20.
5. Betker, M.R., Fernando, J.S. and Whalen, S.P. (1997) The history of the microprocessor. *Bell Labs Technical Journal*, 2:4, 29–56.
6. Daniels, G.R. (1996) A participant's perspective. *IEEE Micro*, 16:6, 21–31.
7. Approximate growth of microcontrollers, although this should not be taken too literally as there are considerable discrepancies in the few available sources of data. Data calculated from: Malone, M.S. *The Microprocessor: A Biography*. New York: Springer, pp. 26, 150, 209, 210; Elsevier, *Profile of the Worldwide Semiconductor Industry – Market Prospects to 1994*, p. 20; Slater, M. (1996) The microprocessor today. *IEEE Micro*, 16:6, 32–44; Turley, J. The two percent solution. *Embedded*, available at: http://www.embedded.com/print/4024488.
8. Birkby, R. A brief history of the microprocessor, Lexikon's history of computing, available at: http://www.computermuseum.li/Testpage/MicroprocessorHistory.htm.
9. Bidmead, C. ARM creators Sophie Wilson and Steve Furber, *The Register*, May 3, 2012, available at: http://www.theregister.co.uk/2012/05/03/unsung_heroes_of_tech_arm_creators_sophie_wilson_and_steve_furber/.
10. European Patent Office, European Inventor Award, An unsung heroine of the 21st century, Sophie Wilson, inventor of the ARM processor, available at: http://www.epo.org/learning-events/european-inventor/finalists/2013/wilson/feature.html.
11. ARM company milestones, available at: http://arm.com/about/company-profile/milestones.php.
12. Data from graph in: Daniels, G.R. (2006) A participant's perspective, *IEEE Micro*, 16:6, 21–31.
13. The next 100 billion chips, available at: http://www.next100billionchips.com/?page_id=21.

16

Instant Cooking: Microwave Ovens

If children are not taught the rudiments of cooking, yet another generation will raise their children on 'ping' meals.

Delia Smith

Percy Lebaron Spencer was yet another American farm boy. He was born in 1894 in Howland in Maine and had a very hard upbringing as his father died when he was still an infant. His mother could not cope and he was taken in by a poor uncle and aunt and as a result had little formal schooling. However, he became interested in electricity and educated himself in the subject. Joining the Navy before the First World War he went to Radio School and learned about the upcoming field of electronics. When discharged at the end of the war he was thoroughly conversant with the subject.[1]

In 1925, Spencer joined the infant Raytheon company and soon became their chief designer of vacuum tubes. His moment came when, in 1941, one of the first cavity magnetrons, which were to become so important for radar, was taken to America as described in Chap. 5. He was one of the people who got sight of the new device and his company gained the opportunity to manufacture them. Soon he had instigated a much quicker method of manufacture and drastically increased the production rate.[2] As a result, Raytheon made the bulk of the devices used in the war.

One day after the war he was in the laboratory where magnetrons were being tested, and he noticed that the peanut bar in his shirt pocket was starting to melt. He was not the first person to notice this phenomenon. Most people just moved away, concerned at what was happening, but Spencer was curious and wanted to know why it was happening. He got some popcorn and held it near the output of the magnetron and it popped all over the lab. He then put an egg in an old kettle with a hole drilled in its side to let in the microwaves from the magnetron. It exploded in the face of a too curious observer.[3]

With the war over, the demand for magnetrons for radar had dropped away and a possible new market was of great interest to the company. With management encouragement he pursued the idea and soon they produced a microwave oven, marketing it in 1947 as the Radarange. It was quite a brute and stood more than 5 ft. tall, weighed 750 pounds and cost

J.B. Williams, *The Electronics Revolution*, Springer Praxis Books,
DOI 10.1007/978-3-319-49088-5_16

about $3000, but the cooking compartment was only the size of a modern microwave. Radarange wasn't a great success but was used in commercial kitchens and for airlines and railways.

The way this story is often told would suggest that Percy Spencer invented the microwave cooker but, as usual, it is a bit more complicated than that and the process was probably slower and more deliberate. Since the 1930s, it had been known that, when subjected to radio waves, various substances would become warm. This was known as dielectric heating and was caused by the molecules of the water or some other substances in the material being vibrated, which produced the heat.

Probably Percy Spencer was not the first person to realize that this worked better with the higher frequency microwaves, but he was the one to appreciate the possibilities and, crucially, do something about it. What was different was that his company was interested in using the method to heat food whereas other organizations were more concerned about heating various commercial products such as tires, or curing glues as part of production lines.[4]

The issue now became—what frequency to use? As it increased into the microwave region the heat increased, but the amount the waves penetrated decreased. Contrary to popular belief, microwaves do not cook from the inside. They penetrate from the outside, heating that layer and to a depth which depends to some degree on the nature of the material. Percy Spencer favored a higher frequency and settled on 2450 MHz, whereas General Electric, also interested in the area, favored around 900 MHz.

In the end, the regulatory authorities decided on both these bands. Spencer's argument was that, at 2450 MHz, the wavelength was around 12 cm and so more suitable for heating small pieces of food such as frankfurter sausages for hotdogs. In the end, his view prevailed and 2450 MHz became the standard frequency for microwave ovens. This was quite convenient as it was not far from the 10 cm wavelength used in the original development of the cavity magnetron for radars and the industry knew how to make these.

Microwave cooking is very efficient because all the available power is absorbed into the food as it bounces off the metal walls of the oven chamber without heating them. This has considerable advantages as there is no need to preheat the oven, and the air inside it, before this warmth transfers to the food. There are, however, a couple of problems.

As said above, the waves penetrate from the outside of the food. If the meal has a reasonable consistency as far as absorbing the waves—a reasonable but not too high water content—then penetration is adequate to heat right through. On the other hand, if there is a layer of high absorbing material on the outside the waves are not able to penetrate to the middle which is insufficiently heated. The system works best with a reasonably homogenous material, though leaving the meal to stand for a while to allow the heat to soak through works well enough in most cases.

A second problem results from the inevitable design of the oven. The waves of energy cross the chamber and bounce back. They end up with a fixed pattern known as a standing wave. This means there are points where the radiation is strong and others where it is weak. This, of course, is exactly what is not wanted for an efficient oven.

The pattern is exploited in a half serious physics experiment to estimate the speed of light (or in this case, electromagnetic waves). The normal turntable is removed and some cheese on toast is placed in the chamber. (Buttered bread or marshmallows can also be used.) The oven is switched on for just long enough to start to melt the cheese or other material which is then removed and a careful measurement made of the distance between the melted spots. This is half the wavelength of the microwaves and should be just over 6 cm.

The second half of the experiment is to read the rating plate on the back of the oven, which should show a frequency of 2450 MHz. By doubling the measurement and then converting it to meters and multiplying it by the frequency achieves a number close to 300 million meters/second which is the speed of light. What it also demonstrates is the uneven heating in the chamber.

Originally a 'stirrer', usually placed in the roof of the oven, was used to break up this pattern but later the simpler and more elegant solution was to place the food on a turntable which slowly rotated, ensuring that the hot and cold spots in the wave pattern were moved around to different parts of the food, thus ending up with a reasonably even heating. Most microwave ovens use this solution.

Despite the first Radarange microwave ovens not being a great success, Raytheon continued development. The first models had used water cooling of the magnetron, which was far from desirable. The problem was that the magnetrons were only about 60% efficient. This meant that if it was producing 1600 W into the oven a further 1000 W needed to be removed from around the magnetron. Clearly a more efficient device was needed.

In 1955, the American cooker manufacturer Tappan did a deal with Raytheon which allowed them to produce a built-in microwave cooker.[5] It was quite large but still considerably smaller than the early Radarange units. This was achieved by reducing the power to 800 W and now the magnetron could be air cooled with a fan. Hotpoint produced a unit the following year and the two ovens generated a lot of interest but neither achieved significant sales.

In the early 1960s, some Japanese companies started to get interested in microwave ovens. In 1962, Toshiba and Sharp both produced units using a Toshiba magnetron, which appears to have been largely a copy of American designs. The ovens were aimed at the commercial market where a relatively high price and size could be tolerated to achieve rapid heating of many meals, such as on a railway train.

Around this time there was a brief fashion for 'cook it yourself' restaurants. The user selected their precooked frozen meal from a vending machine, waited while the food was heated in a microwave oven and ate it.[6] It seemed like a good idea but somehow it never really caught on.

The problem with trying to produce household microwave ovens was that the magnetron itself was expensive to make. Here we turn to Japan. With no military market, her primary interest was in the domestic area and so work started on overcoming this problem. One of the companies involved was the New Japan Radio Company (NJR). In 1961, it became a joint venture between their parent company, Japan Radio Co, and Raytheon.[7] Its task was to produce electronic components and also to investigate the magnetron problem.

By 1964, Keisha Ogura at NJR had succeeded in producing a small, efficient magnetron which, crucially, could be produced cheaply.[8] Raytheon took this device and further refined it, which meant that they at last had the means to really attack the home market. Their next step was to take over Amana Refrigeration, a company with experience in the domestic market. This was to lead 2 years later to the launching of the Amana RR-1 which was a small 'countertop' microwave oven. This set the ball rolling and other companies soon followed suit.

Amana realized that potential users would have no idea how to use a microwave oven and that a huge effort was needed to educate both the retailers and the public. This had to start even before the product was made available. Gradually the campaign brought some success and sales started to rise. For those with busy lifestyles, the speed and flexibility of the microwave oven were a considerable boon.

However, there were problems lurking. There were considerable fears that (a) something that could heat a piece of meat that quickly might also heat the user, and (b) they weren't safe. These worries seemed to be endorsed in 1970, when government tests showed that the ovens leaked microwaves and new standards had to be introduced. This required changes in the designs but did lead to safer units. The fears began to dissipate.

Microwave ovens were ideal for thawing frozen food quickly, but often this food came either wrapped in foil or containing far too much water. It took some time to persuade the producers to modify their packaging and produce meals suitable for microwaving but, by the early 1970s, they saw the opportunities and rapidly converted their offerings. The ovens started to take off in America with sales of over a million units in 1975.

One of the essential features on a microwave oven is a timer. With cooking times measured in a few minutes, or sometimes in seconds, a suitable built-in electrical timer is a must. Though the early units used basic electromechanical devices, it was the coming of the microprocessor in the 1970s which allowed the timer to be implemented simply and in an attractive way. From there it was fairly easy for the manufacturers to add other features, such as defrost time calculators, to enhance their product without adding significant cost.

Though Americans had taken to microwave ovens slowly, the Japanese had been more enthusiastic. In a country where domestic space is at a premium, and there is a culture of rapid acceptance of new products, it is hardly surprising that the local manufacturers should attack the home market. The competition was fierce and the take-up was rapid. By 1976, when 4% of American homes had a microwave, in Japan the comparable figure was 17%.

Sales in the US rose steadily throughout the rest of the 1970s but really took off in the 1980s, rising from 10% of homes possessing one at the start of the decade to 80% at the end. This was a remarkable increase and at a rate not seen before for a domestic appliance (Fig. 16.1). All the elements were already in place. Frozen food had been around for many years and so had the freezers to store it in. The microwave oven to thaw it fast was just what the busy housewife needed.

Britain, as usual, lagged behind other countries. In this case, 1% of homes had microwaves in 1980 just 7 years behind the US which was much less than for all previous domestic appliances. Prices started at around £299 and gradually drifted downwards, but about £250 was common for quite a time.[9] Once established, however, sales took off at a similar rate to America, or even a touch faster. In 1989, the rate of sales slowed despite prices falling again and eventually units could be had as cheaply as £80.[10] The slow-down followed a spate of food scares.[11] These began with government minister Edwina Currie saying that eggs were contaminated with salmonella and spiralled from there to listeria in cheese and so on.[12]

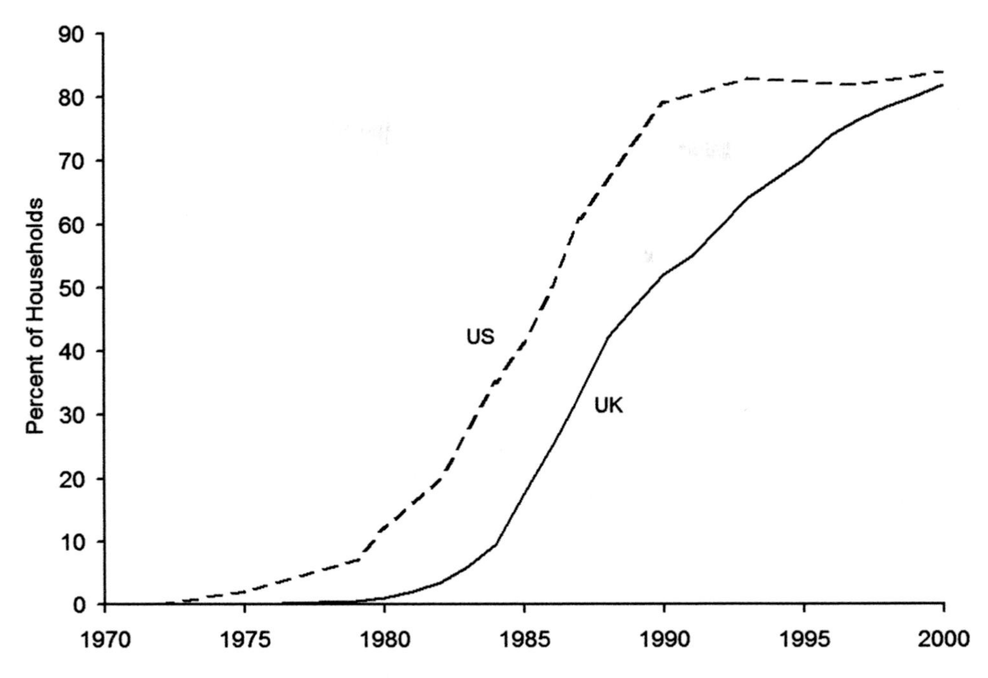

Fig. 16.1 The rapid rise of microwave oven sales in the US and UK, 1970–2000. *Source:* Author[13]

The issue as far as microwave ovens were concerned was that there was some doubt whether they were getting the food hot enough right through to kill any contaminating organisms.[14] This fear was sparked by the rising levels of food poisoning in the late 1980s which coincided with the rise of microwave ovens, and was compounded when an official study found that some microwave ovens failed to heat food thoroughly.[15] The industry reacted, trying to minimize the damage and explain that their products were safe. They stressed that common sense should be used and, just like a conventional oven, if food didn't seem hot enough it should be cooked some more.[16]

Despite the later slowdown, the rise of microwave ovens in the UK was extraordinarily rapid. It was much faster than for other household appliances and this requires some explanation. There is a theory that household electrical goods divide into two categories: time-saving appliances such as washing machines and refrigerators, and time-using items such as radio and television which are basically entertainment items. What was noticed was that, surprisingly, the time-saving appliances took off more slowly than time-using ones,[17] but microwave ovens were accepted at a rate similar to radios and televisions and not other appliances. Various explanations have been put forward for their take-up and the first is the increase in the number of people living alone, for whatever reason.[18] However, the number of single-person households only increased from about 20% to just over 30% from 1975 to 2000.[19] This is nothing like the numbers needed to explain the rise, though it may be a contributing factor.

The next suggestion is the increasing number of women going out to work who wanted a quick meal when they got home. Over the same period, the percentage of women in work in the UK rose from around 55 to 65%.[20] Again, this in no way explains the rapid rise of microwave ovens, but may be a contributing factor.

Microwaves are often used to rapidly thaw frozen food, but that had been around long before they became common. Sales of ready meals rose steadily from around 1970 until 1995 when they expanded sharply.[21] Again, this in no way correlates with the rise of microwave ovens, which took place before this date. Thus none of the usual explanations hold water when examined carefully.

In 1979, in the UK, retailer Marks and Spencer introduced its Chicken Kiev and this was different in that it was chilled and not frozen. There followed a steady rise in chilled meals throughout the 1980s. Clearly, they are used for many other purposes, but may well have been the incentive for many families to buy a microwave, seeing this as a quick way to provide a meal when everyone had been out at work.

There is some evidence for this relationship, as the UK consumes far more ready meals than France and Italy and has a much higher penetration of microwave ovens. For example, at the turn of the century when some 80% of UK households had a microwave oven, the equivalent figure for Italy was only around 27%.[22] Though the time pressures may not be as great as people think, it is the perceived lack of time that drives the behavior to accept more ready meals and so use microwave ovens.[23]

There have been numerous attempts to explain rates of adoption of innovations from Everett Rogers onwards,[24] trying to divide down the various motivations, and the advantages that the product has. The microwave oven is an oddity. Virtually everyone already had a conventional oven to heat food. Its advantage was purely in speed and at first sight that seems an insufficient reason for most potential users to buy one, let alone at this sort of take-up rate.

One has to fall back on individual motivations and that would suggest that it became the fashion to have a microwave. Certainly word-of-mouth and the showing off of one to friends, family and neighbors were the ways to spread the message once the initial early adoption phase was over. The decline in the rate of take-up after the food scares of 1989 strongly supports this. Some people had a residual doubt of the ovens' efficiency and safety and this restrained their probability of purchase.

In view of the fact that microwave ovens were an extra in the sense that there was already a way to cook food, what did people want them for? Obviously there was the element of keeping up with one's neighbors, but they also needed to be useful. Originally, the manufacturers had seen their prime use in the rapid defrosting of frozen food. To this end they added many elaborate features which were supposed to make this easier, such as the oven calculating the time needed when given the weight and type of material. Probably most users never used these.

While heating frozen food was one use, it was also of value with chilled food kept in the refrigerator. This could be either commercially produced precooked food, or simply the home-produced leftovers. Who could blame someone with a full-time job, still expected to prepare the food, for reaching for the ready meal and popping it into the microwave.[25] For preprepared home meals, cooks rapidly discovered which foods microwaved well and which didn't. Some, it was found, remained in better condition when heated this way rather than in a conventional oven.

Steadily, users realized what they were and weren't good for, and that you mustn't attempt to dry the wet cat in a microwave oven! Many individual ways to make life easier were found, even if it was only in reheating the forgotten cup of coffee that had gone cold, and it became an essential adjunct. This child of radar had become an indispensable piece of equipment in the modern kitchen and contributed to a change in living styles.

NOTES

1. Percy Spencer, available at: http://www.nndb.com/people/766/000165271/.
2. July 19, 1894, Percy Spencer, Inventor of the microwave oven born, available at: http://www.massmoments.org/moment.cfm?mid=210.
3. Murray, D. (1958) Percy Spencer and his itch to know. *Readers Digest*, August 1, 1958.
4. Osepchuk, J.M. (1984) A history of microwave heating applications. *IEEE Transactions on Microwave Theory and Techniques*, 32: 9, 1200–1224.
5. Smith, A.F. (2003) Microwave oven. In *Encyclopedia of Food and Culture*, available at: http://www.encyclopedia.com/topic/microwave_oven.aspx.
6. E.g., *The Times*, May 10, 1963, p. 16.
7. New Japan Radio Co, Company History, available at: http://www.njr.com/corporate/ history.html.
8. Southwest Museum of Engineering, Communications and Computation, Microwave ovens, available at: http://www.smecc.org/microwave_oven.htm.
9. *The Times*, January 15, 1977; p. 5; e.g., *The Guardian*, February 3, 1977; *The Times*, March 18, 1978, p. 12.
10. *The Times*, August 25, 1988.
11. *The Times*, June 22, 1990, p. 27.
12. BBC, On this day 3rd December 1988, Egg industry fury over salmonella claim, available at: http://news.bbc.co.uk/onthisday/hi/dates/stories/december/3/newsid_2519000/2519451.stm; Miller, D. and Reilly, J. Food and the media: Explaining health scares, available at: http://www.dmiller.info/food-scares-in-the-media.
13. UK data from: General Household Survey, Table 6, Consumer durables, central heating and cars: 1972 to 2001; Beynon, H., Cam, S., Fairbrother, P. and Nichols, T. The rise and transformation of the UK domestic appliances industry, Cardiff University, School of Social Sciences, Working Paper 42. There are some discrepancies which have been ironed out. US data from: Chart by Nicholas Felton, New York Times, February 10, 2008, available at: http://www.nytimes.com/2008/02/10/opinion/10cox. html?ex=1360299600&en=9ef4be7de32e4b53&ei=5090&partner=rssuserland&emc=rss&pagewanted=all&_r=1&_ga=1.125098697.291417130.1408289727&.
14. Eley, A. (1992) Food safety in the home. *Current Microbiological Perspectives, Nutrition & Food Science*, 5, 8–13.
15. Prentice, T. (1989) Government sets up microwave study as food poisoning rises. *The Times*, August 23, 1989.
16. E.g., Sharp advert, *The Times*, December 7, 1989, p. 26.
17. Bowden, S., and Offer, A. (1994) Household appliances and the use of time. *Economic History Review*, XLVII, 4.

18. Winterman, D. The rise of the ready meal, *BBC News Magazine*, available at: http://www.bbc.co.uk/news/magazine-21443166.
19. General Household Survey 2000, Household size 1971 to 2000, GHS 2000_tcm77–171,197, Table 3.1.
20. ONS, Labour Market Statistics, LFS: Employment rate: UK: Female: Aged 16–64 (%).
21. Department for Environment, Food & Rural Affairs, Family food datasets, available at: https://www.gov.uk/government/statistical-data-sets/family-food-datasets.
22. Time-starved consumers fill up on ready meals, *The Retail Bulletin*, February 19, 2003, available at: http://www.theretailbulletin.com/news/timestarved_consumers_fill_up_on_ready_meals_19-02-03/.
23. Celnik, D., Gillespie, L. and Lean, M.E.J. (2012) Time-scarcity, ready-meals, ill-health and the obesity epidemic. *Trends in Food Science & Technology*, 27, 4–11.
24. Rogers, E,M, (2003*) Diffusion of Innovations*, 5th ed. Cambridge: Free Press.
25. Madill-Marshall, J.J., Heslop, L. and Duxbury, L. (1995), Coping with household stress in the 1990s: Who uses "convenience foods" and do they help? *Advances in Consumer Research*, 22.

17

Essentials or Toys: Home Computers

There is no reason for any individual to have a computer in his home.

Ken Olson, president/founder of Digital Equipment Corp., 1977

With the development of microprocessors as described in Chap. 15, it was only a matter of time before they were used for the most obvious task—to form the basis of computers. The problem at the time was that they were very expensive. The first company to overcome this was a small American manufacturer called Micro Instrumentation Telemetry Systems (MITS). They managed to do a deal with Intel to obtain the 8080 microprocessor at a much reduced price and produced their Altair 8800 system in 1975.

The system they built was a sizable metal box with switches and lamps on the front. Inside, there were sockets to allow additional boards to extend its performance. The basic kit of the system sold for around $400, with the additional cards costing $100 or so.[1] The one card system was also available, ready built for $650. The basic device was rather limited, but a quite usable system could be put together with some extra cards. The era of the home computer had begun.

Of course, the buyers of such a device were all hobbyists or those with an interest in the possibilities of computing. There was no suitable keyboard available, so the user had to insert a program by clicking switches—quite a laborious task. There were lamps to indicate what was happening, but it was the input and output, together with the program storage that was very limited. However, some 2000 were sold.

This set others thinking what they could do, and with the arrival of two lower-cost microprocessors, the 6502 and Z80, they had the means. Two Steves, Jobs and Wozniac, produced their Apple I kit in 1976, but it was in 1977 that things started to happen.[2] They produced their Apple II which was a real computer and could work in color. They soon had competition from the Radio Shack TRS-80 and the Commodore PET. What these had in common was that they came in more satisfactory cases and were not kits. They were quite expensive when fitted with a useful amount of memory and input and output arrangements, but they started to sell in significant numbers.

Though some of these found their way to Britain, one group of people in particular were also looking at the possibilities. It began with Clive Sinclair, the calculator and

J.B. Williams, *The Electronics Revolution*, Springer Praxis Books,
DOI 10.1007/978-3-319-49088-5_17

electronics entrepreneur (see Chap. 14). After the collapse of the calculator market into a price war, he had had to retrench. He had diversified into a digital watch and was heavily involved in trying to produce a pocket television. So when the possibility of a computer kit came along he largely left the development to his employee, Chris Curry, who had been with him for many years.

The result was the MK14 which was very cheap, in kit form, and came with a tiny memory of only 256 bytes of RAM, but it had a hexadecimal keypad for entering programs, and an 8-digit display to show what was happening.[3] It was moderately successful but what should follow it was a matter of contention and the two men parted company over which route to take.

Chris Curry left Sinclair and, backed by the Austrian Herman Hauser and Andy Hopper, set up Acorn computers also in Cambridge. They had chosen the name of the company because it was part of their philosophy to produce products that were expandable and took account of future possibilities. By the next year they had produced their System One which, though still a kit, was a more sophisticated product based around the 6502 microprocessor, although the price had doubled. Chris Curry then had a clear idea of where he wanted to go, and that was to have a computer, ready assembled, in a nice case. It should include a full-size keyboard, a useful amount of memory, an interface so that the results could be displayed on a television screen, and a connection to use a tape recorder to store the programs.[4] This was called the Atom.

Clive Sinclair had also decided to take the computer business seriously, but his concept was different. He wanted the price to be low, and he thought that this would open up a market for home computers. In January 1980 he brought out his first offering, the ZX80, based on the other popular microprocessor family the Z80. He claimed that it was the world's smallest and cheapest computer and it sold for £99.95 assembled or £79 in unassembled kit form.[5]

The competition between the two companies now began in earnest, but not only were they using different microprocessor families, they had entirely different concepts of the sort of products they wanted to produce. Acorn wanted to make usable, expandable products that everyone could use, while for Sinclair price was everything. One of the ways the low price was achieved was by using a very unsatisfactory membrane keypad which made the Sinclair useless for serious work, whereas Acorn used a proper keyboard.

The British government had become interested in computers and particularly what should be taught in schools to further this subject. In 1980, it announced the Microelectronics Education Programme (MEP) for England, Wales and Northern Ireland.[6] This consisted of two parts: using computers to assist in the teaching of traditional subjects, and also providing education about computers and their software. It was to lead to a considerable buzz and interest in the subject.

In 1981, the BBC decided that it should have a computer program. As part of their Computer Literacy Project they wanted to be able to offer a computer which featured on the program and approached various manufacturers. After a considerable amount of discussion, they eventually settled on a new design from Acorn which was provisionally called the Proton. This became the BBC Microcomputer and the contract was for 12,000 units.[7] Due to delays in getting the computer into manufacture the program wasn't screened until January 1982. Even then it was well into the new year before significant numbers of users saw their order.

There were two models, A and B, the latter having more memory and other features. They were relatively expensive but they were neatly packaged with good keyboards and built with Acorn's expandable approach. They used what had become a standard system of having a television as the display. This was achieved with a UHF modulator so that the computer output could be plugged into the TV's aerial socket. However, the model B had an output which could connect to a display monitor.

Program storage was on tape cassettes but Model B could optionally be fitted with a controller so that it could use a much more satisfactory floppy disk drive. These used a thin flexible disk of magnetic material which was spun inside its cover by the drive. Data were stored in sectors on the disk which could be written or read at high speed as the sector came under the head. This could also be moved in and out so that multiple tracks could be picked up.

The model B also had additional outputs, for example for driving a printer. However, the great feature of both models was the Basic language that came installed. This meant that the user could straight away write a piece of code, and by typing 'Run' it would execute immediately as the code was 'interpreted' by the Basic chip in the machine. This was an enormous boon to all the budding teenage programmers.

In 1981, Kenneth Baker became the UK's Minister of Information Technology, and one of his first acts was to declare 1982 as the Year of Information Technology.[8] Part of the government's Microelectronics Program was a scheme to encourage computers in schools with a grant of 50% of the cost as long as the rest could be raised. The schools took to the BBC micros with great enthusiasm. Of course, what the young people were using at school also became a desirable thing to have at home.

At the beginning of 1982 Acorn were struggling to meet the demand, but steadily they ramped up production. By Christmas, they had shipped 67,000 machines, a far cry from the original 12,000 units projected by the BBC. This was only the beginning and more than a million were made before the computer was withdrawn from production. Though not all of these were sold in the UK the machine had an enormous impact on a whole generation, teaching vast numbers how to program and introducing many more to computing.

Meanwhile, Sinclair was still pursuing his vision of really cheap computers for the home. In March 1981, he released his ZX81 at a phenomenally low price for a fully assembled unit.[9] One of the ways he achieved this economy was having a very limited 1 K of RAM memory. An add-on 16 K pack that plugged into the back could be had for a further small sum. It still featured a poor membrane keypad, like its predecessor, together with the use of a television as the display and a cassette recorder for storing programs. While he sold 50,000 ZX80s in its lifetime, the ZX81 totalled around a million worldwide.[10]

Having got the low end of the market covered, Sinclair then moved on to the ZX Spectrum. This came with 16 K or 48 K of RAM. Though it retained the TV and cassette recorder interfaces of its predecessors it now had color graphics. It also had a rubber keypad, which was an improvement on the membranes previously used but had a dead fish feel. Later, this was improved slightly. By the beginning of 1984, Sinclair had sold a million of these as well.[11] In the end, he sold around 5 million units.[12] Though some of these went abroad, an extraordinary number were absorbed by the UK.

What were they used for? In the schools, enterprising people wrote small programs which helped children learn other subjects from maths to spelling. It was in the subject of computing that they had the greatest impact. With their built-in Basic language these computers could be used easily to develop simple programs. The thrill of getting something to work became quite addictive for many youngsters.

The real favorite was games. Sometimes, these came already programmed, ready to be read in from the tape, but the many magazines that sprang up often printed the instructions which would then have to be laboriously typed in by hand. The more adventurous either adapted these or created their own from scratch. The severe limitations in RAM, particularly in the ZX81, led to some very innovative programming to enable something useful to run with so little code.

In the US, things developed very differently. Though the Apple II, the PET and its successor the VIC20 were used in schools, some of the other machines were simply too expensive for either schools or the home. Despite an interest in the subject there was no one like Clive Sinclair to hit the bottom of the market and really get it going. Nevertheless, the affluence of the country, and its general approach to new technology, still meant that computers took off.

The surprising difference between the countries was highlighted by the American *BYTE* magazine, the bible for all those interested in computing.[13] It found a greater interest in Britain, which could be seen in the greater numbers attending computer exhibitions, which was extraordinary considering the population was around one quarter that of the US. This was put down to the government initiatives, the BBC and its computer and, of course, Clive Sinclair. One theory was that as Britain hadn't been doing so well economically, people were concerned they might be put out of a job by computers and it was essential to learn about them.

BYTE was also impressed by the many other low-cost machines that were available, and the many features that had been packed into them, often in surprisingly small memories. American machines had made very little impression in the UK, as they were simply too expensive, particularly as the suppliers tried to sell them at higher prices than in their home market. This technique, which had often worked in the past, wouldn't any more as there was a thriving indigenous industry.

However, in August 1981 there occurred an event that changed the whole industry. IBM entered the personal computer market with their 5150 model. While this was not the company's first attempt to produce a desktop computer, everything about this device was different from IBM's usual approach to products. It had been developed in 12 months at the company's Boca Raton plant in Florida by a team dedicated to this task.[14]

It was built as far as possible from tried and tested parts largely sourced from outside suppliers who could react much more quickly than the sluggish IBM itself. The first crucial decision was the processor around which it was to be built. For performance they wanted to use one of the 16 bit devices that were just becoming available, but were put off by the cost of the resulting hardware. The solution was to use a new device, the 8088 from Intel which, while it was a 16 bit device inside, externally only had an 8 bit bus resulting in reduced hardware.

The second important decision was to find an external source for the crucial disk operating system software. While the basic system came with a cassette tape interface a disk drive could be added as an extra. The company they went to for the operating system was an embryonic Microsoft, who didn't have a suitable program but bought one in, modifying it to IBM's requirements. It was to be their lucky break.

The basic offering was a large box with the computer in it together with a series of slots where expansion boards could be plugged in, an idea that dated back to the original Altair 8800. It also came with a good keyboard. It had color graphics when connected to a television. Available as options were a monochrome display monitor, a printer, and disk drives.

At first sight the 5050 didn't appear to have any advantages over the offerings from other makers, and was very expensive. However, it was a 16 bit machine and it did have the IBM logo on it and, as everyone knows, you never got sacked for buying IBM. Their innovation was to have an 'open architecture' which meant that they published all the technical information about the machine and encouraged other firms to make additional hardware and software for it. This also meant that it was easy for someone to make a copy or 'clone' of the computer. At first, IBM encouraged this in their attempts to make the machine become the standard. Later it was to run away from them. There was a great rush for these machines, and by the end of 1982 some 250,000 had been sold just in the US.[15] In the following year it was sold abroad.

The other manufacturers weren't standing still. While many were moving up market to try to produce business machines, Commodore came out with their 64 model which was generally available from September 1982. It had the usual television and cassette recorder interfaces, but good graphics and sound, with the 64 denoting that it had 64 K of RAM memory. It was substantially built and was priced like the cheaper models but with generally more power.[16] It was good at what it had primarily been designed for—as a games machine.

It was very successful for a few years, selling more than 2 million units in 1983 and dominating the market in many countries as well as the US.[17] In Britain, it was the only foreign computer to make a significant impression in a market led by Acorn with their BBC machines, and Sinclair. However, by the end of the decade its heyday was over and sales fell away sharply with Commodore going bankrupt in 1994.[18] Despite that, with some 15 million produced it was probably the highest selling computer from a single manufacturer.[19]

For business use the IBM PC continued to progress. Not only had IBM made the information on its hardware available but Microsoft, who supplied the Disk Operating System (DOS), had made a very smart move when their arrangement with IBM had left them free to sell it to other manufacturers. All another company had to do was copy the hardware and make their own version of the Basic operating system (BIOS) to avoid IBM copyright and then they could make a compatible computer that could run the DOS and hence any programs that ran on that.

It wasn't long before 'clones', as they were called, appeared from Columbia Data Products and Compaq. They were the forerunners of a whole industry. As the numbers increased, the IBM computer architecture using the Intel processor and running Microsoft's Disk Operating System steadily became the industry standard. Though many manufacturers tried to build business machines to their own different designs they were all doomed to failure as the IBM standard marched on.

There was only one company that seemed able to withstand this and that was Apple. They used a different microprocessor family that was not compatible and had entirely their own software. At the start of 1984, they came up with their Macintosh computer which featured a graphical interface using a 'mouse' which was so much more user friendly than the previous command line systems.[20] IBM and Microsoft, though they announced their

Windows system at much the same time, took until late 1985 to make it available. Through such innovative products, Apple developed a loyal following and were able to maintain their place holding a few percent of the market.

Over the next few years, IBM continued to upgrade their offerings with the addition of hard drives, color monitors and faster microprocessors; and in particular the introduction of the 'Windows' graphical interface. As more software such as word processors and spreadsheets became available, the capabilities of the machines increased and ever more companies found them useful. The whole thing got into a virtuous circle.

Gradually the users, familiar with these systems at work, wanted the same things at home. As prices fell, driven by far eastern clone makers, these became affordable for the home. If the user wanted to do serious work and not just play games, then they became the first choice. At this point, particularly in the US, in the late 1980s home computer ownership really began to take off.

In the UK, after the initial surge with the BBC Micro and Sinclair machines things calmed down with a slower take-up of IBM-type computers due to their relatively high prices. However, in the 1990s with the coming of email and the Internet the take-up started to rise again in all developed countries. By the end of the century, something like half of homes in these countries contained a computer.

What was surprising was the similarity of the take-up of computers in the home in the US and UK (Fig. 17.1). Most appliances entered homes in America often a decade or more ahead of Britain, but here it was quite different. For a period in the late 1980s the UK was ahead of the US. Some of the reasons for this have been explored, but there still remains a question about why this should be the case. While the delay in accepting new things had been decreasing as the century went on, the enthusiasm in Britain for computers is still a puzzle. It can only be put down to a critical mass of hype, together with the low-cost offerings.

What were all these devices now being used for? This question was being asked as early as 1980, before things had really got going. Market research at the time suggested three areas: education, entertainment, and personal finance.[21] What this meant was the production of numerous simple programs of the type 'early learning fun' which aimed to teach everything from spelling to math. The entertainments were games varying from chess to 'shoot 'em up' types such as Space Invaders.

While people claimed they used computers for their personal finances many found that the tedium of entering every item soon caused them to lose interest. It was often an excuse to buy a machine which was really wanted as the thing to have and spent most of its time—when it was turned on—for playing games of one sort or another.

As the computers improved, particularly in their modes of storage, then their role started to change. The first big step was the abandonment of unsatisfactory tape cassettes and their replacement with floppy discs. This meant that programs could be loaded quickly, but particularly work files could be saved and reloaded easily. Hence they became useful for real work. The coming of hard drives completed this process.

There was now some point in entering financial data because the program could do all the math for you and keep the results which could be added to subsequently. This meant that doing simple accounts for clubs and societies was very useful. Also, the advantages of a word processor program in a computer over a typewriter began to be appreciated. This really opened up with the arrival of low-cost printers.

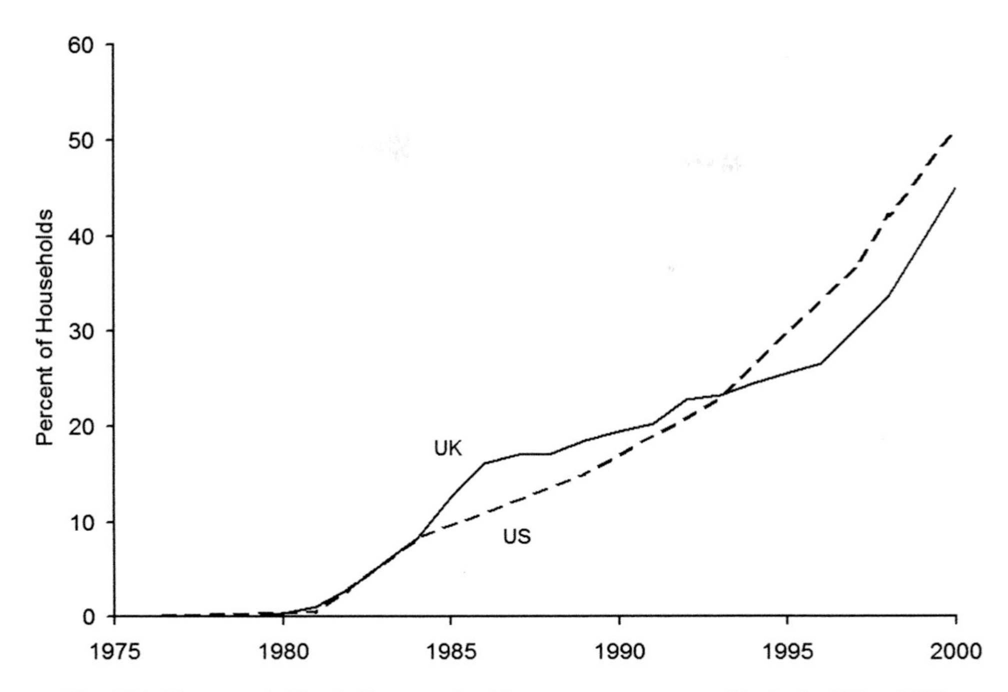

Fig. 17.1 The remarkably similar growth of home computer ownership in the US and UK, 1975–2000. For a period in the late 1980s the UK was ahead before crossing over in the 1990s. *Source:* Author[22]

With the rise of email in the 1990s, followed by the growth of the Internet, the uses of home computers started to multiply. Though the purchase was often justified on a particular application, users soon found many other uses. It was this wide area of applications that began to make them indispensable to many homes. The need to stay in contact with far flung families, or to find information, became driving forces in themselves. The potential of these was beginning to be understood by the millennium.

It was noticeable that the uptake of computers was greater in families with a married couple and children with the lead person in the 40–54 age group.[23] It was also the case that this helped the children with their school work. It was found that teenagers who had access to home computers did noticeably better at their exams than those that didn't.[24] This was despite most having access at school.

So the microprocessor, only invented in 1971, was now used to produce small computers. By 1980 they were starting to appear in homes and by the end of the century around half homes had them and they were being used for many different tasks, many of which wouldn't have been dreamed of a few years before. Its potential was just beginning to be understood.

NOTES

1. MITS Altair 8800, available at: http://oldcomputers.net/altair.html.
2. Knight, D. Personal computer history: The first 25 years, available at: http://lowendmac.com/2014/personal-computer-history-the-first-25-years/.
3. Stobie, I. Chris Curry, available at: http://chrisacorns.computinghistory.org.uk/docs/Mags/PC/PC_Oct82_CCurry.pdf.
4. Hornby, T. Acorn and the BBC micro: From education to obscurity, available at: http://lowendmac.com/2007/acorn-and-the-bbc-micro-from-education-to-obscurity/.
5. Hetherington, T. Clive Sinclair, Allgame, available at: http://www.allgame.com/person.php?id=3575.
6. The Microelectronics Education Programme Strategy, National Archive of Educational Computing, available at: http://www.naec.org.uk/organisations/the-microelectronics-education-programme/the-microelectronics-education-programme-strategy.
7. Smith, T. BBC micro turns 30, *The Register*, available at: http://www.theregister.co.uk/2011/11/30/bbc_micro_model_b_30th_anniversary/?page=4.
8. Editorial, The new industrial revolution, *The Times*, January 11, 1982.
9. Tomkins, S. (2011) ZX81: Small black box of computing desire, *BBC News Magazine*, March 11, 2011, available at: http://www.bbc.co.uk/news/magazine-12703674.
10. Polsson, K. (1981) Chronology of personal computers, 1981, available at: http://www.islandnet.com/~kpolsson/comphist/comp1981.htm.
11. Feder, B.J. (1984) British race is on in microcomputers, *New York Times*, February 27, 1984, available at: http://www.nytimes.com/1984/02/27/business/british-race-is-on-in-microcomputers.html?pagewanted=2.
12. ZX Spectrum, available at: http://en.wikipedia.org/wiki/ZX_Spectrum.
13. Williams, G. (1983) Microcomputing, British Style. *BYTE*, 8:1.
14. Birth of the IBM PC, available at: http://www-03.ibm.com/ibm/history/exhibits/pc25/pc25_birth.html.
15. Hayes, T.C. (1982) Winners of gamble on small computer. *New York Times*, December 4, 1982.
16. Commodore 64–1982, available at: http://oldcomputers.net/c64.html; Shaw, D. Commodore 64 turns 30: What do today's kids make of it? *BBC News Technology*, August 1, 2012, available at: http://www.bbc.co.uk/news/technology-19055707.
17. Reimer, J. Total share: Personal computer market share 1975–2010, available at: http://jeremyreimer.com/m-item.lsp?i=137.
18. The Commodore 64 – The best selling computer in history, available at: http://www.commodore.ca/commodore-products/commodore-64-the-best-selling-computer-in-history/.
19. This figure comes from adding up the data in Reimer above, but there is considerable debate as to the true figure with reports all the way from 12.5 million to as high as 30 million. See Pagetable.com, How many Commodore 64 computers were really sold?, available at: http://www.pagetable.com/?p=547.
20. Computer history 1980–1990, available at: http://www.computerhope.com/history/198090.htm.
21. Owen, K. (1980) A computer for fireside chats. *The Times*, April 11, 1980.
22. The data are based on surveys and should not be taken too literally. Sources are: UK; Schmitt, J. and Wadsworth, J. Give PC's a chance: Personal computer ownership and the digital divide

in the United States and Great Britain, LSE, available at: http://eprints.lse.ac.uk/20086/; Earliest dates from start of personal computer availability and sales figures. US, US Census Bureau, Computer and Internet use, Appendix table Oct 2009, available at: http://www.census.gov/hhes/computer/, and early figures from *The New York Times*, June 17, 1982 and May 11, 1983.

23. Schmitt, J. and Wadsworth, J. Give PC's a chance: Personal computer ownership and the digital divide in the United States and Great Britain. London: London School of Economics.
24. Schmitt, J. and Wadsworth, J. (2004) Is there an impact of household computer ownership on children's educational attainment in Britain? CEP Discussion Paper No 625, March 2004, London School of Economics.

18

Computers Take Over the Workplace

One of the most feared expressions in modern times is: 'The computer is down.'

Norman Ralph Augustine

In the spring of 1978, Dan Bricklin was a student at Harvard Business School studying for his MBA. As part of the course they had to do calculations on case studies that could be time consuming and tedious. He daydreamed of a calculator that had a big head-up display like a fighter plane and a mouse to move the calculator around on the display so that he could punch in a few numbers or do some calculations. By the summer, while riding his bike, he decided he wanted to pursue this idea and create a real product to sell after he graduated.[1]

Of course, the ambition had to be curtailed when he came to actually try to implement it but he started to make something that would test the concept. It was based on a long-known idea of the 'spreadsheet' which was used for calculations, but automating it in a manner that was easy to use by people who were not accountants.

He recruited his friend Bob Frankston to turn his ideas into more usable code and then teamed up with Dan Fylstra, who ran the Personal Software company, to sell the result. Fylstra named it Visicalc to emphasize that it was a simple way to make calculations visible as well as automating them. As the code had been written for the 6502 microprocessor it could run on the Apple II computer that was now available.

By the autumn of 1979, they had the product completed and started shipping copies. Here they were lucky in their timing in that the Apple II now was available with floppy disk drives so that the product could be supplied on a floppy disk rather than the unsatisfactory tapes that had been used previously.

As Steve Jobs of Apple was to say later, it was this spreadsheet program that really drove the Apple II to the success it achieved.[2] It changed the computer market. As Chuck Peddle, who created the PET computer, put it: 'From 1976 to 1978 we were solving the problem of selling computers to people who *wanted* computers; we were satisfying the home market. In 1979, with the introduction of Visicalc, we began to sell computers to people who *needed* computers; we saw the beginnings of the business market.'[3]

J.B. Williams, *The Electronics Revolution*, Springer Praxis Books,
DOI 10.1007/978-3-319-49088-5_18

For anyone in business who wanted to do some financial modeling Visicalc was a godsend. They could sit with the computer on their desk and try different scenarios simply by changing a few numbers. Gone were the days of endless tedious calculations or having to try to get the computer department to run a program again and again every time they wanted to see what happened with slightly different data.

As a result of its usefulness, around a million copies of Visicalc were sold in its short lifetime.[4] It was soon to be superseded by the more advanced Lotus 123, and later still by Microsoft's Excel. However, the stage had been set for mass programs that could be used on large numbers of small computers not even from the same manufacturer. This was a drastic change from the specialist programs that just ran on one type of large computer in a company.

Of course, computers can do more than manipulate numbers. By ascribing a number code to letters, they and thus text can be handled. As computer programing progressed from the raw codes to a 'language' to make the task easier there came a need to have a simple way of feeding this into the machine. As a result, text editors were developed to implement this. A huge variety were produced which reflected the many different types of incompatible machines that existed.

Microcomputers were based on a fairly small number of microprocessors, so the need for a variety of text editors was much reduced. This led to the desire to produce something better that could manipulate text in a more sophisticated way, suitable for writing letters and other office tasks. Some programs were produced that ran on large mainframe computers, but the costs involved ruled out almost all possible users.

The next stage was for the machine manufacturers to produce simpler computers dedicated to what was now called word processing functions. But again the costs were still too high for most users. It was inevitable that the burgeoning microcomputer business would try to tackle the same problem. Two early attempts at this were Electric Pencil and Easy Writer. The problems they faced were that the machines could only deal with lines of text and could not produce a complete image of what was required for printing. Also, there were a vast number of incompatible display and printing systems which meant that versions of the software had to be produced for all the possible combinations.

Gradually, as the operating system CP/M became some sort of partial standard, there was an opportunity to produce a product for a wider market. In 1979, a small start-up company, MicroPro, produced a word processing program called WordStar written by their brilliant programmer Rob Barnaby. It used key combinations to select its many functions but crucially it was a considerable step along the way to WYSIWYG (what you see is what you get) which is essential for a true word processor. With heavy marketing the program started to make a significant impact in the market and sales rose from half a million dollars in 1979 to 72 million in 1984.[5]

IBM began to see their peril as microcomputers running these spreadsheets and word processors could provide cheap alternatives for businesses and this had the potential to impact on their computer and office machine sales. They realized that they must get into that business, and rapidly. This led to the hurried development of the IBM PC and its appearance on the market in 1981 (see Chap. 17). Though it was not the most refined offering, it had the IBM logo and that made the difference. They started selling in large numbers and soon attracted 'clone' copies. The net impact was that these machines steadily took over virtually the whole market and established a de facto standard (Fig. 18.1).

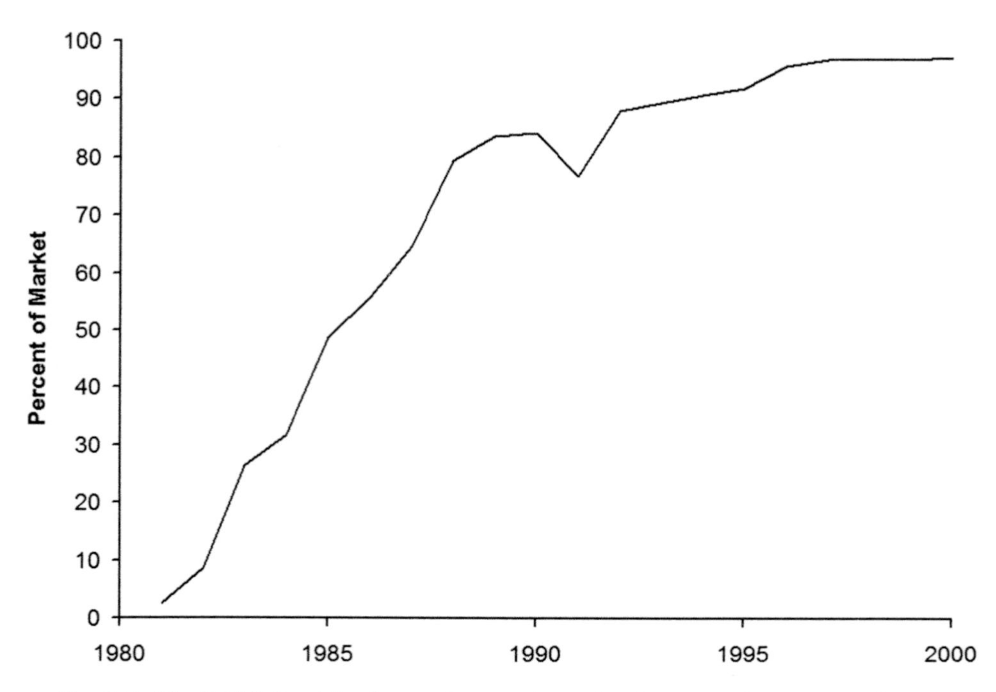

Fig. 18.1 The IBM PC and its clones take over the market for personal computers. *Source:* Author[6]

This changed the rules for the software businesses, who now had to have their offerings capable of running on IBM PCs. Although VisiCalc and WordStar converted their programs, it was the new entrants, Lotus 123 and WordPerfect, written for this new contender that rapidly took over, selling in far greater numbers, and the previous front runners faded.[7] These new programs had added features but it was the mistakes of the original companies in not keeping up with the required changes that ultimately was their downfall.

Now there was a synergy between the hardware and the software. Users bought the machines so that they could run the software to do useful work. Word processors started to impact on the traditional typewriters. It was easy to have a personal computer where it was needed and use that to generate documents rather than send away a dictation tape to a typing pool. It had the effect of returning the work to a local point rather than a central facility.

Despite the progress the computers still were somewhat unfriendly to use. Commands had to be typed into a line to action a piece of software, and when operating most of those just consisted of letters. There were 40 lines with 80 characters in each. It took considerable ingenuity on the part of the software programmers to make this look even vaguely attractive.

Meanwhile, the hardware of the computers themselves was progressing steadily. In a curious phenomenon the price changed little from year to year.[8] What you did get was more and more performance for the money. The microprocessors around which they were based went from the 8088, to 80286, to 80386, to 80486 increasing in power and speed along the way. The memories increased dramatically. The storage went from a single disk drive to two, and hard drives appeared and steadily increased in capacity.

Over in California the Xerox Corporation, which had made a lot of money from their copying machines (see Chap. 19), had set up the Palo Alto Research Center (PARC). To protect their office business they had been looking into personal computers but, inspired by ideas from Doug Engelbart at the nearby Stamford Research Institute, they were working on something completely different.

On the computer screen were 'icons' and they could be selected by moving a 'mouse', a small device that was manipulated over a flat surface and moved a pointer on the screen. Clicking a button on the mouse caused the icon to be selected and this produced a pop-up menu, allowing the user to move smoothly from one program to another by opening 'windows'. There was absolutely no typing in of commands.[9]

Steve Jobs, of the rising Apple Computer company, paid Xerox a visit. His cheeky proposal was that, in exchange for being able to invest in shares in the Apple Company, they should disclose what they were working on relevant to the computer business. Amazingly, they went for the offer and showed Jobs and his colleague their icon and mouse system.

PARC had their concept of an expensive sophisticated system for the high-end user. Jobs had a very different vision—for a mass personal computer. One of the first things he did was to commission the design of a low-cost mouse. He wanted to be able to sell it for $15 instead of the $300 of the PARC device. Apple also took the Graphical User Interface (GUI), introducing the menu bar, pull down menu and the 'trash can'. They also added manipulation of the 'window' where a corner could be pulled to make it bigger or it could be dragged across the screen to move it. It was a giant step forward.

All this was gathered together in Apple's offering of the Macintosh computer which appeared in early 1984.[10] While it might have been a bit short of usable memory and only had a single floppy disk drive, the shock it produced with its GUI was considerable. As *BYTE* magazine said at the time: 'Whatever its problems and limitations the Mac represents a breakthrough in adapting computers to work with people instead of vice versa.' Those who saw it were immediately convinced that this was the way forward, but the limitations meant that, though successful, it never achieved a dominant market position. Its biggest drawback was that it wasn't an IBM and by using a different microprocessor it wasn't compatible.

The man who could see the opportunities was Bill Gates at Microsoft. As the company supplying the operating system for the IBM computer and clones they were in the ideal position to exploit the possibilities. He announced software that gave a similar sort of GUI as a spoiler even before the Macintosh was launched, but it took until 1985 for their Windows 1.0 to appear.[11] The graphics weren't particularly good but it had the enormous advantage that it ran on the IBM and clones which were now half the market for personal computers and rapidly swallowing the rest.

As the hardware improved new versions could be produced. A sort of arms race between the hardware makers and the software producers developed. As each generation of machines came out with greater speed and capabilities these were rapidly absorbed by the bigger software packages. While these combinations produced improved performance it meant that the users were being constantly tempted to upgrade their systems.

In 1987, Windows was replaced by version 2.0, and in 1990 by version 3.0 where the quality of the graphics was much improved. After that came Windows 95, 98 and 2000 where the version number was replaced by the year when it appeared which only accelerated the trend towards rapid obsolescence. As this sequence progressed the machines became more and more capable of real work.

This new platform gave fresh opportunities for software and the company positioned to exploit this was again Microsoft. Their 'Word' word processor and 'Excel' spreadsheet were designed to run under the Windows environment and steadily took over from WordPerfect and Lotus 123. In 1990, Microsoft added a presentation program called PowerPoint, originally developed for the Macintosh, which they had obtained by buying the company that made it. They now had a considerable grip of the software market. The Wintel combination of Windows software running on machines using Intel processors (the IBM clones) was now virtually in complete control.

The Apple Computers, much loved by some of their supporters, hung on but largely in niche areas such as desktop publishing where the particular attributes of the machines or the available software gave an advantage.[12] Elsewhere the various 'IBM' clones had established a more-or-less standardized environment. As each generation appeared it had to be compatible so that older software was able to run on the hardware. Too great a variation from the standards of existing software could annoy the users greatly.

The increase in the numbers of computers in use was staggering. Once things got going in the 1980s there were soon many millions. Sales increased from 800,000 in 1981 to 12 million in 1992. It became the norm for the average office desk to be occupied by a machine, and it was nearly always some sort of IBM clone. In America the use in business always outstripped that in the home, except perhaps in the very early stages (Fig. 18.2).

In the UK the situation was rather different (Fig. 18.3). The home computer business had taken off more rapidly thanks to the efforts of Clive Sinclair and Acorn Computers as described in Chap. 17. With the IBM computer arriving two years later than in the US, and a greater caution for new things, the take-up in business was much slower. The result was that there were more computers in homes than in businesses until well into the 1990s when the GUIs made computers much easier to use. It was noticeable that newer firms were quicker to computerize than older ones.[13]

As the machines were relatively cheap it meant that virtually all businesses, rather than just the large ones, could afford to computerize. This became important as, for example, the Pay As You Earn (PAYE) tax system became so complicated that what used to be calculable by the use of the published tables now required a bookkeeping software package. As the word processing packages became common, the much better presentation of the documents they produced soon made it essential to use them instead of a typewriter that gave a less professional looking result. This was as well as the fact that it was much simpler to create and modify documents.

The appearance of the computer on the average office desk meant that computing was no longer the domain of the central computer and its priesthood. It was something for everyone. Also the word processor deskilled typing. The ease of correcting mistakes meant that anyone could quickly learn to type with two fingers. It was often quicker and easier for the person wanting to generate a document to type as they thought and so they didn't need a typist any more. Typing as a job virtually disappeared.

With the arrival of electronic mail (email) it became normal for even the managers to send their own messages, which again removed the need for typists and secretaries. Email

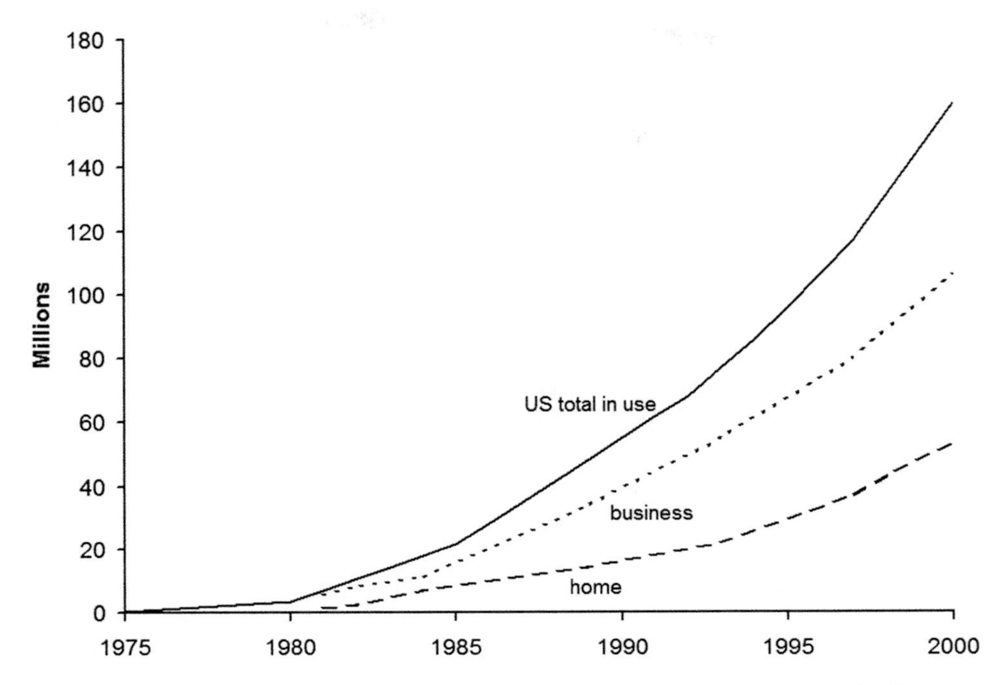

Fig. 18.2 The rise in the use of computers in the US, 1975–2000. *Source:* Author[14]

was often much more time efficient than telephoning—the recipient could deal with it when they came to their desk. Unfortunately, the advantages were soon negated by the ease with which multiple copies could be sent and the practice grew of sending copies to all vaguely relevant people. Thus, checking emails became a major task rather more onerous that opening conventional mail, but as a way of keeping a group of people informed it was very useful.

Email use started mostly in larger companies and slowly permeated down to the smaller ones.[15] Its adoption was subject to the same restraints as the telephone and required that both parties had the equipment. You cannot send to someone who doesn't use the system. However, from the mid 1990s firms increasingly found the need for it.[16] Within these companies only a minority of employees used it, who tended to be in the higher echelons of the organization.

Personal computers had not displaced the big central computer completely. In larger organizations where huge amounts of data were stored, whether for stock control, accounting, or the details of many small customers, the central computer still had a role. It was essential that there was a single database and any changes were made only to that one source. Originally many 'terminals' would have been used but these were steadily replaced by personal computers which could access the stored data, but also do other work such as word processing locally.

In a smaller organization which still had the need for a central data store, another personal computer, or something closely akin to it, could be used as a 'server'. Cabling

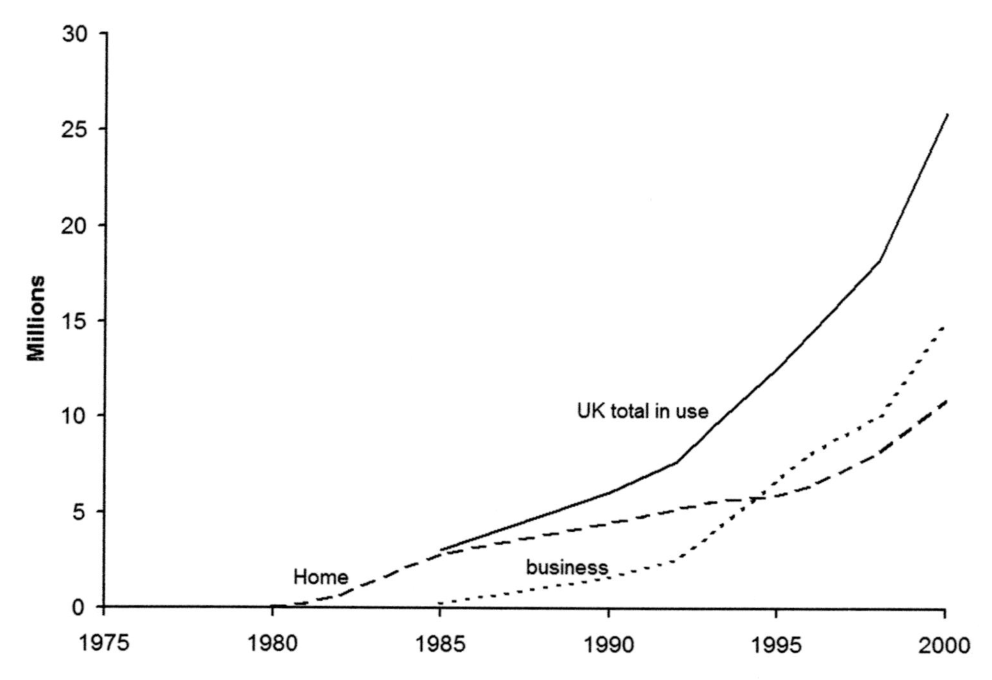

Fig. 18.3 The slower growth of business computing in the UK, 1975–2000. *Source:* Author (see note 14)

between the remote machines and this server meant that it would appear as though it was another disk drive. It still had the advantage of the single database but with the flexibility that work could be carried out on the local computer without needing the resources of the central machine and so running much faster.

Such systems were of great value in organizations such as insurance and other service companies where the person answering a telephone call needed to rapidly access all the information about that particular customer and answer the query.

It was only one step from there to connect the telephone system to the computer. Modern telephone systems can make the caller's number available and this is sent as a code even before the phone is picked up. This is used to find the customer's information so that it is ready for the adviser when they pick up the phone to answer the call. Up to this point, the technology was a boon in saving time and frustration and increasing the efficiency of the operation. Hence grew 'call centrers', where the computer could route the call to the next available adviser.

However, once connected to the telephone system the computer could also dial numbers. This is where these systems turned from being an advantage to the consumer to being a nuisance. The computer could be given a list of telephone numbers and, as each operator became available, the system would automatically dial the next number in the list. Unfortunately, at this extreme what had started out as an advantage had degenerated into a menace.

Thus, computers had taken over the office. Quite apart from the call centers which employed vast numbers of people, nearly all accounting functions had been computerized.

Any routine operation could be simplified and calculations carried out automatically. Those trading in the written word from solicitors to managers producing reports found that word processing greatly increased their output, though in some cases a little too much.

It wasn't just offices that were affected. Computing power didn't always appear in the form of a desktop computer. It could be hidden away inside some other machine. In this form it pervaded the factory. A classic example was machine tools. Where traditionally an operator had wound controls to progress cutters to shape a workpiece, now a computer system controlled the machine tool using a program containing a prerecorded sequence of coordinates. The system was known as Computerized Numerical Control (CNC).

CNC had the effect of moving the skill from the operation of the machine to the programming of it which, of course, was done on yet another computer. Once the program had been perfected and transferred to the CNC machine, large numbers of items could now be made with minimal operator intervention. All that was required was that someone ensured the machine tool was supplied with suitable material, and checked from time to time that all was well. One operator could now easily tend several machines, hence giving a great increase in productivity.

Another place where computing power crept in was with robots. The popular image is of 'mechanical men' but in reality they look nothing like that. Often they are just a large flexible arm with some sort of gripper on the end. Their function is to move items from one place to another or perform some sort of operation such as welding. A common use is on car production lines where many of the more boring or dangerous tasks have been taken over by the machines.

Once again a program contains the instructions for what movements are required and a computer hidden in the machine commands its operation to execute the desired sequence of moves. Of course, they will go on performing the sequence without getting tired, or bored, leaving human workers to do more complex tasks. Ideally, the machines should take over all the routine work, leaving people to supervise and ensure that supplies are there in good time and any difficulties are dealt with.

Even robots can take many forms, many of which probably wouldn't even be recognized as such. For example, the electronics industry uses large numbers of 'pick and place' machines whose task is to take components from the reels on which they are supplied and position them correctly on a printed circuit board. They can work at many times the speed of a human assembler and minimize the number of errors that occur in the process. The low cost of modern electronic devices would not be possible without them.

Most companies are now utterly dependent on their computer systems in one form or another. As a result, most employed people have had to come to terms with them. In the 1980s and particularly the 1990s it required a huge effort to train large numbers of people in the ways of these machines. As time has gone on, the machines have been adapted to the ways of people rather than the other way round.

Computing power has thus utterly changed the workplace for the majority of people. In response to any query the answer is nearly always; 'I'll look on the computer.' By the end of the century they had become a fact of life and were rapidly taking over. They had become so embedded that it was difficult to remember what life had been like before them. While it was easy to blame mistakes on the computer, and many cursed them from time to time, there was no possible way that society could function without them.

NOTES

1. Bricklin, D. The idea, available at: http://www.bricklin.com/history/saiidea.htm.
2. Bricklin, D. Steve Jobs talking about VisiCalc in 1996, available at: http://www.bricklin.com/jobs96.htm.
3. Peddle, C. (1984) Counterculture to Madison Avenue. *Creative Computing Magazine*, 10: 11.
4. Power. D.J., A brief history of spreadsheets, DSSResources, available at: http://dssresources.com/history/sshistory.html.
5. Bergin, T.J. (2006) The origins of word processing software for personal computers: 1976–1985. *IEEE Annals of the History of Computing*, 28@4, 32–47.
6. Data from Reimer, J. Personal computers from 1975–2004: Market growth and market share, available at: http://jeremyreimer.com/m-item.lsp?i=137.
7. Bergin T. J, (2006) The proliferation and consolidation of word processing software: 1985–1995. *IEEE Annals of the History of Computing*, 28:4, 48–63.
8. Bayus, B.L. and Putsis, W.P.,Jr. (1999) Product proliferation: An empirical analysis of product line determinants and market outcomes. *Marketing Science*, 18: 2, Figure 2.
9. Gladwell, M. (2011) Creation myth: Xerox PARC, Apple, and the truth about innovation. *The New Yorker*, May 16, 2011.
10. Knight, D. 25 years of Macintosh: Macintosh History, 1984 The first Macs, available at: http://lowendmac.com/history/1984dk.shtml.
11. A history of Windows, available at: http://windows.microsoft.com/en-gb/windows/history#T1=era1.
12. Bresnahan, T. and Malerba, F. (1997) Industrial dynamics and the evolution of firms' and nations' competitive capabilities in the world computer industry. Department of Economics, Stanford University 1997, available at: www.siepr.stanford.edu/workp/swp97030.pdf.
13. Chen, J.-C. and Williams, B.C. (1993) The impact of microcomputer systems on small businesses: England, 10 years later. *Journal of Small Business Management* 31:3.
14. Total use data from eTForcasts available at: http://www.etforecasts.com/pr/index.htm; and Computer Industry Almanac available at: http://www.c-i-a.com/pr1286.htm; home percentages from chapter 38 multiplied by number of households; business by subtraction. Some interpolation used.
15. Sillince, J.A.A., Macdonald, S., Lefang, B. and Frost, B. (1998) Email adoption, diffusion, use and impact within small firms: A Ssurvey of UK companies. *International Journal of Information Management*, 18: 4, 231–242.
16. Harindranath, G., Dyerson, R. and Barnes, D. (2008) ICT adoption and use in UK SMEs: A failure of initiatives? *The Electronic Journal Information Systems Evaluation*, 11: 2, 91–96, available at www.ejise.com.

19

From Clerks to Xerography: Copiers

Copying is the engine of civilisation: culture is behavior duplicated

David Owen

In 1777, Erasmus Darwin, grandfather of Charles Darwin, started investigating a way of copying letters or, more specifically, a method of making two copies at once.[1] Though he probably didn't know it, using a mechanical linkage between two pens, one of which was manipulated by the writer while the second produced the copy, wasn't a new idea. However, he pursued his thoughts to the point of making a working device in the following year.

He showed it to his colleagues in the Lunar Society, the club of Midland businessmen and inventors who met on the night of the full moon so that they would have light to ride home afterwards. Among those who tried his device was James Watt, of steam engine fame, who had a considerable problems producing multiple copies of instructions on how to assemble his machines in Cornwall.

Always one to go his own way, he thought he could do better and by the middle of 1779 he was writing to Darwin claiming that he had 'fallen on a way of writing chemically which beats your bigrapher hollow'. Clearly, an element of competition was driving him on, but to overcome the need for clerks to sit there making error prone copies was also a considerable incentive.

What Watt came up with was completely different. He wrote the original letter or document with a special ink that he developed with James Keir, another member of the Society. A sheet of dampened paper, which had not been treated with size, was then pressed on the original either with a roller or a press. This produced an image on the back of the sheet, but it was sufficiently transparent that it could be read from the front. As long as the copy was taken within around 24 h of the original being written, it worked reasonable well.

The copies were surprisingly good and in 1780 he was persuaded to patent the process.[2] He and Keir set up a company, James Watt & Co, using Matthew Boulton's money. By the end of the year they had sold 150 presses and the method became a standard way of copying documents for many years to come. A common arrangement was to have a 'letter copy book' where the letters would be placed between pages with oiled papers to protect the next page and the whole book put into the press. Once copied the originals were removed,

J.B. Williams, *The Electronics Revolution*, Springer Praxis Books,
DOI 10.1007/978-3-319-49088-5_19

blotted to dry them and sent to the addressee. The book then contained copies of all the correspondence tidily held together.

Ralph Wedgwood was not actually a member of the Lunar Society but, as Josiah Wedgwood's cousin and business partner, he was closely associated with it and a budding inventor himself. Trying to find a way of helping blind people to communicate he came up with a 'carbonated paper' which was placed between two sheets of ordinary paper. The user then wrote with a stylus guided by wires to keep to the lines. The written document then appeared on the lower sheet of paper. In 1806, he patented the method.[3]

The problem with this for ordinary copying was that the normal ink pens, whether quill or steel nibbed, didn't provide enough pressure for the method to work. However, a variant employed double-sided carbon paper which was used the same way, but now the top sheet was semi transparent and a stylus was still used.[4] Two copies were produced; the bottom one was treated as the 'original', and sent to the addressee if a letter, while the top sheet was used as a copy and could be read through the paper in the same way as the copies produced by James Watt's copy press.

Though the carbon paper was improved along the way it only became really useful with the coming of typewriters in the 1870s. The typewriter keys hit the paper with sufficient force to use a single-sided carbon paper, placed with its working side down between the two sheets of paper, and the copy was produced on the bottom sheet. The advantage was that the top copy was the original as produced by the typewriter and was not dampened and potentially degraded as in Watt's process.

Additionally, the copy was the right way around and so could use ordinary paper. It was possible to use additional sheets of carbon and paper to produce more copies at the same time though the results were fainter. Steadily it replaced the copy book process and remained the standard way of making copies of letters for the best part of a 100 years until plain paper copiers made it redundant. However, an echo of it still exists in the cc of email copies; this stands for 'carbon copy' and implies one that is less important than the 'top copy'.

Though printing could be used where large numbers of copies were required, there was a need for a simpler and cheaper method to produce perhaps 50 or 100 copies of a newsletter or circular. In about 1876, the Hectograph process was developed to answer this need.[5] A master was written or typed using one of the new aniline dye inks. This master was then placed face down on a tray of gelatine and most of the ink was transferred to the gelatine. Plain papers were then pressed on to the gelatine with a roller transferring the image to them. The copies gradually became lighter but 10s of reasonable copies could be obtained.

In 1923 a derivative of this, the spirit copier, appeared. Here a special paper was used to transfer ink on to the rear of the master as it was written or typed. This was then placed rear face up on a drum in the machine which was wiped with spirit to dissolve a small amount of ink which was then transferred to a fresh sheet of paper. The whole process was accomplished by the machine and the user only had to turn a handle to turn out multiple copies until the ink was used up. The most well known version of this was made by the American company Ditto. Later versions even had electric motors to power the process.

In the late 1870s and 1880s, a series of machines appeared which used a different principle. The idea was to make a stencil from which copies could be printed. Numerous inventors, including Thomas Edison, came up with methods of cutting the stencils but the principle was the same in that the writing needed to remove an impermeable layer.

When this was inked and a sheet of paper placed on the other side, the ink was squeezed through either with a press or roller.

The stencil system was good for perhaps as many as 1000 copies until the stencil started to disintegrate. At first, the stencils could only be handwritten, but by the 1890s it was possible to produce them on a typewriter which greatly improved the quality of the result. In the twentieth century, rotary versions of the machines appeared with electric motors that could turn out copies at speed. Famous names amongst these were Roneo and Gestetner.

In 1837, Sir John Herschel, son of William Herschel and himself an astronomer and scientist, started to investigate photosensive chemicals as part of his research into the new science of photography. He was really interested in the possibility of reproducing color as opposed to the monochrome of the existing methods. His intensive researches over the next 5 years led him to investigate vegetable dyes, but found them too unstable. This drove him back to inorganic chemicals, and in particular some of the compounds of iron.[6]

He found that if a paper was treated with potassium ferrocyanide and iron peroxide and allowed to dry, it was sensitive to light. If this paper was placed under a translucent master document and exposed to light the uncovered areas turned dark blue while the areas shaded by lines or writing remained white. Washing the paper with water removed any unused chemicals. He called this the cyanotype process.[7]

Because the chemical system was much too slow for ordinary photography the prepared paper could be handled for short periods in normal light. It required some minutes in bright light to get the exposure. This meant that it was convenient for the copying of documents or particularly drawings, but perhaps surprisingly it took many years to really catch on.

After the Marion company started producing preprepared paper in 1881 the system became popular for copying architectural plans and engineering drawings. Initially, to get the exposure in a reasonable time, the prints were placed outside in the sun, but later machines using arc lamps became the norm. These copies took on the name 'blueprints' after the colour of the background. The word has gone into general use but strictly it only applies to this early process which produced white lines on a blue background—a type of negative.

The drawback was, of course, that it was a negative and the white lines on a blue background are not comfortable to look at for any length of time. Despite this, it wasn't until 1923 that a better process was invented in Germany by a monk called Vogel, who licenced it to Kaller AG. By the next year his Ozalid process had reached Britain. The name came from the diazonium salts used and it was also known as the diazo method. In a similar fashion to the blueprint process, the paper is exposed to ultra violet light through the translucent master. But this time it is the exposed areas that react to light and go white while the unexposed parts under the lines or writing are unaffected. These turn dark blue when exposed to a developer, usually ammonia.

The great advantage is that the result is a positive with dark blue lines on a white background. The paper has to be stored in the dark, but can be handled briefly in normal light before passing though a machine with striplights or UV tubes. It comes out a minute or so later ready to be used. The only problem is that the machines, and the prints for some time afterwards, smell of ammonia or whichever developer is used. This produces a 'Dyeline' print and became the standard method for producing drawing copies during the middle part of the twentieth century.

In 1955, the Japanese Riken Optical Company Ltd., now known as Ricoh, developed the diazo process further and produced their model 101 which was a desktop copier intended for ordinary office documents rather than plans. They largely overcame the smell problems and the machines produced an almost dry copy. It was quite successful, with this and the succeeding models clocking up sales of a million units.[8]

It seemed an obvious idea to use photography to create copies, and before the First World War machines started to appear which were basically cameras with a developing system attached. Here again, a negative was produced which was not very satisfactory, but this could be passed through the system again and a positive produced. In the 1930s, the introduction of silver halide sensitive paper made 1:1 copies a practicable proposition. Despite this, the machines were large and cumbersome, and the paper with its silver content relatively expensive. As a result, though the word 'photocopy' entered the language the process was not much used.

In the 1950s, 3M in America came up with another copying process that used thermally sensitive paper. The advantage was that it was a totally dry process and a small and relatively cheap machine could be produced. The master to be copied was placed over the sensitive paper and infrared energy supplied to heat the lines on the master which would cause the same areas to darken on the copy. The downside was that any heat subsequently applied to the copy could cause the whole thing to darken and become unreadable. Also the copies had a tendancy to curl up. For short-term reference copies it was adequate, but not for permanent records.

The American Chester Carlson had a very difficult upbringing with a father wracked by arthritis and tuberculosis, and a mother who died when he was still in school. Despite that, he managed to get into the California Institute of Technology (Caltech) because he wanted to be an inventor. After graduating in 1930 he went to work for Bell Labs as a research engineer and a year later transferred to their patent department.[9]

In 1936, he decided to study law in evening classes, but as he couldn't afford expensive law books he would laboriously copy out large sections from library books by hand. It was then that he realized that all the copying processes like carbon paper didn't allow an existing document or book to be copied. It set his inventor's mind working.

His first thought was a photographic process, but if that was going to work one of the large photographic companies would already have done it. He therefore started to look in a different direction and searched in the library for ways in which light could affect matter. He soon alighted on photo-electricity, and particularly photoconductivity where conductivity of a material increases when light shines on it.

His idea was that if he could find a material that acted as a conductor when it was illuminated and as an insulator when it wasn't then maybe he had a starting point. If this material was applied as a thin layer to an earthed plate which was kept dark he could then electrostatically charge it. If it was then exposed to an image of what he wanted to copy the charge would leak away where it was exposed in the white areas and not for the black lines and writing. He would then dust the whole surface with a toner powder charged with the opposite polarity and it would stick to the charged image and could then be transferred to a clean sheet of paper, producing a copy.

That was the theory, but he was sufficiently convinced of its success that he made a preliminary application for a patent in 1937 and a full application in 1938 for the process which he called Electron Photography at the time but later was named Xerography.[10]

Fig. 19.1 Chester Carlson and a prototype of his copying system. *Source:* https://multimedia-man.wordpress.com/tag/xerox-model-a-copier/

However, his experiments at home didn't go well and he had to employ the young Austrian, Otto Kornei, and get a small laboratory before any progress was made.

The material they chose for the photoconductivity was a layer of sulfur and this was coated on to a zinc plate. On October 22, 1938 Kornei wrote the date and the place where they were working, Astoria, on a microscope slide. He turned out the lights, rubbed the coated plate with his handkerchief to give it a static charge, placed the slide on top and turned on a lamp for a few seconds. After he turned off the lamp, he removed the slide and dusted the plate with powder. There was a clear image and when Carlson pressed a piece of wax paper against the image it largely transferred to the paper. He was able to admire his first Xerographic copy (Fig. 19.1).

Though Kornei was taken to lunch to celebrate, he was unimpressed and soon left. Carlson soldiered on alone and spent the next 6 years visiting more than 20 companies such as IBM, General Electric, Eastman Kodak and RCA, trying to interest them in the idea. In 1944, things started to move when one man, Russell Dayton, saw the possibilities of the method, pointing out that the crude image Carlson had was the first copy he had seen which didn't need chemicals. This encouraged his employer, the research and development company Battelle Memorial Institute, to invest in the process.

In 1946, Joseph C. Wilson took over from his father as president of the Haloid Company. They made photographic supplies and Wilson was keen to diversify the business into something that wasn't dominated by powerful companies like Kodak, which was on the other side of the town. He came across information about Carlson's patents and by the next year had signed an agreement with Battelle to help Haloid manufacture office copiers based on Carlson's system.

The Model A, produced in 1949, was very unsatisfactory as it required numerous manual operations to make a copy and even an experienced operator would take 3 min or so. It might have sunk Haloid, and Xerography, if it hadn't found a niche market for producing paper masters for offset lithographic duplicators where the complexity and lack of speed didn't matter.

To make the process practical for Carlson's original concept of an office machine it needed to be automatic, and that required a system to time and control the operation. Electronics was required, and during the 1950s with the coming of transistors this became possible. There were further problems, particularly in producing a satisfactory toner, and it took Carlson's determination to keep Haloid from giving up. Finally, in 1959 they produced the Model 914, which was ready for shipping in the following year.

The machine was quite a brute and weighed more than a quarter of a ton, but it only needed to be filled with a pile of plain paper and then all the user had to do was press a button—it was completely automatic. It required no chemicals or special papers and the results didn't smell bad, curl up or turn brown. Where the developers had thought a heavy user might need 10,000 copies a month the first companies that bought them were producing 2000–3000 copies a day. Haloid had finally got it right, Carlson had been correct—it was a great success.

They had invented a new industry. A stream of new and improved models, such as ones small enough to fit on a desktop, followed over the next decade or so. The success could be measured by the fact the company name was changed to Xerox in 1961. Sales grew from $176 million in 1963 to $4 billion in 1975.[11] The annual number of copies produced worldwide rose from around 20 million in 1955, before Xerography, to some 2500 billion by the end of the century, virtually all produced by that process.

In Britain, the Xerox company had a joint venture with the Rank organization and set up Rank-Xerox, which was very successful in the 1960s and 1970s. The machines were often not sold but leased or 'placed' on some sort of complicated financial arrangement, sometimes just paying per copy which was an attempt to make the expensive machines affordable for smaller users. The machines were large and expensive and little effort was made to address the bottom end of the market (Fig. 19.2).

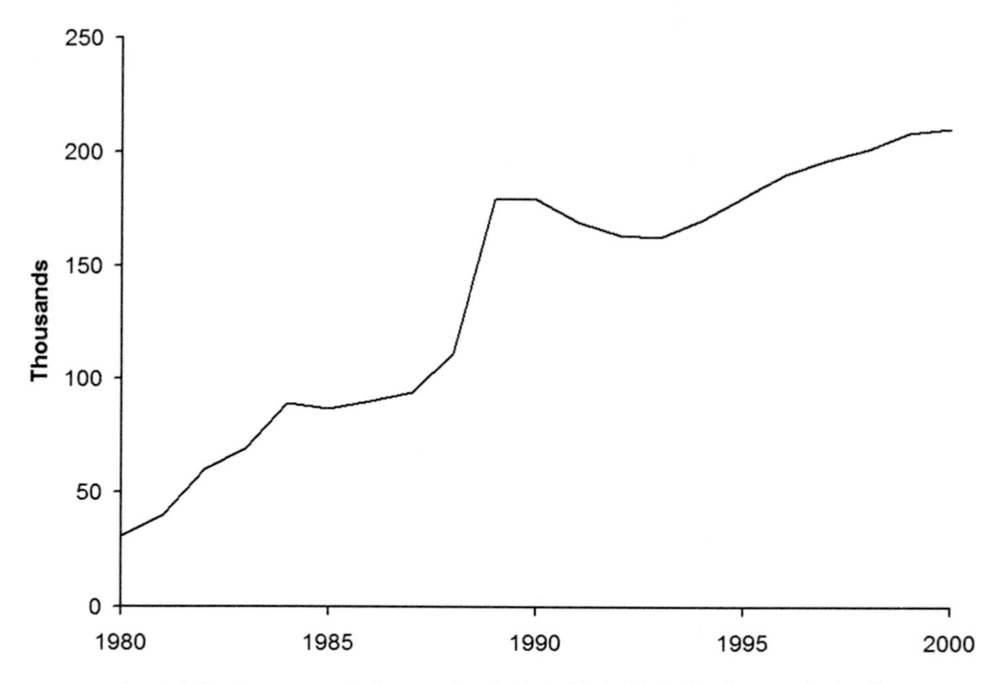

Fig. 19.2 Placements of photocopiers in the UK, 1980–2000. *Source:* Author[12]

This was a mistake as it left open an opportunity for Japanese firms to get into the market with smaller machines. They were always more consumer-oriented than the large American companies. In the early years, Xerox had had a licence to print money, as it had the market to itself and was making such huge amounts that it failed to see the competition coming. In the mid-1970s, a number of Japanese manufacturers, like Canon, Minolta, Sharp, Mita and Ricoh, instead of selling directly to customers created dealer networks for selling sophisticated inexpensive copiers to low-volume users.[13]

These companies steadily extended their ranges and made copiers affordable to a much larger potential market. The numbers being sold or 'placed' increased rapidly with only occasional hiccups due to depressed economic conditions. Meanwhile, Xerox began to suffer—their market share plummeted. Despite being the pioneers of the technology, they had failed to keep pace and had to go through a drastic reframing of their business, particularly to reduce their production costs.

By the mid 1980s, they had stopped the slide with the introduction of more competitive machines. Undoubtedly, the severe measures implemented saved the company, but the glory days were over. It was a lesson that no advantageous business position lasts for very long. It is bound to attract competition. Xerox got the message just in time and started to diversify, but despite considerable investment in research and development in their Palo Alto center, as described in Chap. 18, they were not good at exploiting its opportunities.

An example of this was the laser printer. Gary Starkweather was a young researcher working at Xerox's Webster research center, and in 1967 he asked himself why, instead of just copying an existing original, they couldn't produce one directly from the computer.[14] Other approaches to the problem could only type out the characters in a document using a 'daisy wheel' or other arrangements derived from typewriters; alternatively they used pins like the dot matrix which could print a picture but only crudely as the pins could never be close enough together for a high resolution image. Gary Starkweather wanted something that could do both and at high resolution.

The management didn't think it was a good idea, but Gary went on with it quietly. By 1969, he had a prototype which was made by modifying one of the company's copiers, but still he couldn't get any support. Fortunately, in 1970 the Palo Alto research center was set up and he was able to transfer there where he found a very different attitude. By 1971, he had built a laser printer which could take output from a computer and print it xerographically.

Instead of having the imaging system of the copier that took the image of the master and displayed it on to the photoconductive drum, he used a laser to write the image. The scan of this was achieved much like an early television scan by sweeping the laser beam side to side using a mirror drum, and then the main photoconductive drum rotating to achieve the scan in the other direction. The rest of the machine was basically the same as the copier.

The scanning system meant that the printer could put a spot on any point on the page, but in many cases the printer is simply required to print characters in various fonts. It required a considerable amount of software to convert the character code into the required pattern on the page. It took the coming of microprocessors during the 1970s before this was really practicable. Thus it was only in 1977 that their first product, the Xerox 9700, was finally produced. They had taken so long that they were beaten to the market by IBM, not known as the fastest mover.

The 9700 was a very large, fast machine and that was the sort of product that Xerox liked. Starkweather, though, could see the possibilities of smaller personal laser printers. Once again Xerox could not, and in the end he left the company. However, the baton was taken up by the Japanese camera and optics company Canon. They developed a low-cost desktop 'print engine' which they marketed for a while before developing the process further.

As they lacked experience in selling to computer users Canon made arrangements with two American companies, Apple and Hewlett Packard, for them to sell the system under their own names. Thus, in 1984 Hewlett Packard produced the LaserJet and the following year Apple came out with their Apple LaserWriter.[15] It took until 1990 before the prices dropped below $1000, but the laser printer became well established as a quality means of printing from a computer.

In parallel, another technology that could produce a complete image was developed. That was inkjet technology where fine droplets of ink were fired at the paper, and by moving the print head from side to side and the paper up then the whole image could he written. The ink droplets were either produced with heat causing a small bubble which would expel the drop, or with a tiny piezo-electric element.[16]

By using multiple heads containing different colored inks, and ever more complex software, a color image could be produced. This was developed further, sometimes using more than the normal four colors (black, magenta, cyan, and yellow). These systems were capable of producing images of photograph quality, which meant that digital images stored in the computer, like the user's photos, could be printed.

This laid down the challenge to the laser printer makers to produce a color version. At the cost of even more complexity they succeeded, but the machines were expensive and difficult to maintain, and though they appeared before the end of the century, and had a niche, they were not a mass market item, unlike the inkjet printers.

In a sense, things had come full circle. Now documents were usually generated on the computer and then, if required, were printed out. It meant that, like the original copying methods, copies were produced at the same time as the original. The difference, of course, was that the digital image was still there in the computer's memory and another copy could be produced at any time. However, the recipient of a paper version would still need a Xerox-type copier if they need a further copy.

With the price of inkjet printers dropping it was a fairly simple matter to add a scanning system and produce a complete copier. These were handy to have on a desktop where they could be used as the computer's printer or scanner as well as for the copying function. This further reduced the domination of xerography.

Though the completely paperless office still remained a dream, the coming of electronic mail (see Chap. 22), and hence the ability to send documents electronically, reduced still further the requirement to make copies of paper originals. Nevertheless, the numbers of copies made still continued to rise. No one could imagine a world in which you couldn't simply put a document in a machine and press a button to get a copy.

NOTES

1. Uglow, J. (2002) *The Lunar Men*. London: Faber & Faber, p. 306.
2. Rhodes, B. and Streeter, W.W. *Before photocopying: The art and history of mechanical copying 1780–1938*. Delaware: Oak Knoll Press, p.25; Proudfoot, W.B. (1972) *The Origin of Stencil Duplicating*, London: Hutchinson, p. 23.
3. Barnes, A. Ralph Wedgwood: Pioneer of Office Copying, available at: http://www.search.revolutionaryplayers.org.uk/engine/resource/default.asp?resource=2627.
4. Lawrence, K. The exciting history of carbon paper!, available at: http://kevinlaurence.net/essays/cc.php.
5. Antique copying machines, available at: http://www.officemuseum.com/copy_machines.htm.
6. Herschel, J. F.W. (1842) On the action of the rays of the solar spectrum on vegetable colours, and on some new photographic processes. *Philosophical Transactions of the Royal Society of London*, 181–214.
7. Andrew, J.H.(1981) The copying of engineering drawings and documents. *Transactions of the Newcomen Society*, 53:1, 1–15.
8. Ricoh. Ricopy 101: Forerunner of office automation, available at: http://www.ricoh.com/about/company/history/2000/ricopy101.html.
9. Owen, D. Making copies. *Smithsonian Magazine*, August 2004.
10. US patent 2221776.
11. Boulton, W.R. Xerox Corporation: Surviving the competitive crisis, Case studies. Auburn University, Montgomery, Alabama, available at: http://www.auburn.edu/~boultwr/.
12. Data from: The UK Market for Photocopiers, 1984, Gowling Marketing Services; Facsimile Machines and Office copiers, Special Market Report, 1990, Economic Intelligence Unit; Photocopiers and Fax machines, 1994, Keynote Report; Photocopiers and Fax machines, 2000 Market report, Keynote.
13. Boulton, W.R. The plain paper copier industry, Case studies. Auburn University, Montgomery, Alabama, available at: http://www.auburn.edu/~boultwr/.
14. Dalakov, G. Laser printer of Gary Starkweather, available at: http://history-computer.com/ModernComputer/Basis/laser_printer.html.
15. Yarin, P. History of the laser printer. MIT Media Laboratory, available at: http://alumni.media.mit.edu/~yarin/laser/laser_printing.html.
16. Le, H.P. (1998) Inkjet history: Progress and trends in ink-jet printing technology. *Journal of Imaging Science and Technology*, 42: 1, 49–62, also available at: http://www.printhead911.com/inkjet-history/.

20

Shrinking the World: Communication Satellites

A hundred years ago, the electric telegraph made possible—indeed, inevitable—the United States of America. The communications satellite will make equally inevitable a United Nations of Earth; let us hope that the transition period will not be equally bloody.

Arthur C. Clarke

In 1865, the French science fiction writer Jules Verne produced a new publication in his 'voyages extraordinaires' series of adventure tales.[1] This one was called *De la terre à la lune* (*From the Earth to the Moon*) and featured a gun 900 ft long that shot a capsule containing three intrepid travellers on their journey.[2] Five years later, he produced a sequel *Autour de la lune* (*Trip Around the Moon*) giving their adventures once they had reached their target. Needless to say, despite the aura of scientific calculation, the gun required would have needed to be impossibly large. It wasn't practicable.

Jules Verne was, however, very popular; his books were translated into more than 140 languages and had an enormous effect on many who looked at the problems of travel beyond the Earth in a more sound, scientific way. One of these was a Russian, Konstantin Eduardovich Tsiolkovsky, who though he worked as a schoolteacher, used his spare time to scientifically investigate the possibilities of flight both in air and in space (Fig. 20.1).[3]

After years of careful calculations, in 1898 he wrote an article *Investigating Space with Rocket Devices* which laid out many of the principles of using rockets for launching space ships into orbit. He described the basic rocket equation and showed why it was better to use multistage rockets to lift craft out of the Earth's atmosphere.[4] He submitted it to *Nauchnoye Obozreniye* (*Science Review*) who accepted it but were unable to publish it because they were shut down. It eventually appeared in 1903, but though influential, didn't have a wide circulation abroad.

One of his proposals was a 'space elevator'.[5] The idea was to have a 'celestial castle' in space with a cable connecting it to the Earth. He realized that if this castle was at approximately 35,800 km above the surface of the Earth it would rotate at the same speed and would appear to hover over a fixed point.[6] He understood that the speed at which a satellite

J.B. Williams, *The Electronics Revolution*, Springer Praxis Books,
DOI 10.1007/978-3-319-49088-5_20

Fig. 20.1 Konstantin Tsiolkovsky, Hermann Oberth, and Robert H Goddard. *Source:* http://en.wikipedia.org/wiki/Konstantin_Tsiolkovsky#/media/File:Tsiolkovsky.jpg; http://en.wikipedia.org/wiki/Hermann_Oberth#/media/File:Photo_of_Hermann_Oberth_-GPN-2003-00099.jpg; http://en.wikipedia.org/wiki/Robert_H._Goddard#/media/File:Dr._Robert_H._Goddard_-GPN-2002-000131.jpg

orbits depends on the distance from the surface and at this magic height this would be geosynchronous.

The idea is hopelessly impractical but, had it been possible, it would have been useful to have the elevator go up the cable and hence lift items into space. Tsiolkovsky, like so many of those investigating the possibilities of space travel, also wrote science fiction. For people like him there was a very thin line between serious proposals, dreams, and science fiction.

In Germany there appeared another such dreamer. His name was Hermann Oberth and he was born, appropriately, in Hermannstadt, a Transylvanian town in the Austro-Hungarian Empire which is now Sibiu in Romania. Though trained as a doctor, in which role he served in German army during the First World War, his real passion was for physics, and particularly space travel.[7]

In 1922, he presented his doctoral thesis on rocketry, but it was rejected by his university in Munich. However, by the next year he had published his first version of *The Rocket into Planetary Space*, and continued to refine this up to 1929. He patented his rocket design and launched his first rocket in 1931. As an academic he was influential for an upcoming generation and in particular to a young student, Werner von Braun, who was to make his mark on German rocketry.

In America, also unaware of the work of the others, was Robert Hutchings Goddard. He too had read science fiction as a boy and become fascinated with space, and even wrote some himself later.[8] He studied physics, and by the time of the First World War he was studying rockets and conducting tests on small solid fuel prototypes. After the war, he wrote his paper ‘A Method of Reaching Extreme Altitudes’, published in 1919. In this he defined 11.2 km/s as the ‘escape velocity’ necessary to get free from the Earth.

He received a great deal of ridicule for his ideas, so he continued his work on perfecting rocket components in secret, supported by a few philanthropists who believed in what he was doing. He filed 214 patents for rocket components and managed to launch a number of rockets that were at least partially successful. During the Second World War, he was involved in booster rockets to aid the takeoff of heavy aircraft. Sadly, when the country finally became interested in rockets after the war, he wasn't there to help as he died in 1945.

In 1928, another book appeared in German, *Das Problem der Befahrung des Weltraums* (*The Problem of Space Travel*) written by Hermann Noordung whose real name was Hermann Potocnik.[9] He was from another piece of the multiethnic Austro-Hungarian Empire, this time Pula in today's Croatia. The book described the detail of a space station which was intended to be at a fixed point in the sky in geosynchronous orbit. He died of tuberculosis the following year, so his work was not widely translated though its existence and some information about it did reach a wider readership.

All this fascinated a young Englishman, Arthur C. Clarke, who was a member of the British Interplanetary Society between the wars whilst also writing science fiction stories on the subject. During the Second World War, he became an RAF technical officer working on radar systems, and from these interests he obtained a sound understanding of the background which led to his paper, Extra-terrestrial Relays, which was published in the *Wireless World* magazine in 1945.[10]

In this paper, he described the use of a geostationary satellite as a means of relaying radio signals, particularly over the oceans. While relatively long wavelength radio signals can bounce off the ionic layers of the atmosphere and hence travel great distances, the higher frequencies needed for television transmission, or to carry a large number of telephone conversations, will not do this, and are limited to line-of-sight transmission. Thus, where it was impractical to use repeater stations, such as over the sea, there was no way of transmitting television pictures. Clarke's solution was to put a relay station in space which was in line-of-sight to ground stations on each side of the ocean.

His paper showed that, by having three stations at suitable positions around the world, over the Equator, a more-or-less complete coverage of the Earth could be achieved. He evisaged these as manned space stations, which would have been necessary with the vacuum tube technology of the time, as the tubes would have needed to he replaced regularly. He could see that the wartime V2 rockets could be developed further and might one day be capable of lifting the required parts into space. He was aware of Noordung's work as he quotes it in his references.

It is often claimed that Arthur C. Clarke invented the geostationary satellite concept. As can be seen, this isn't so; the idea dated much further back, but what he did work out explicitly was the possibility of the communications satellite, though he didn't claim originality even for that. He laid out the basic science in a clear and unambiguous way even though, as he well knew, this was not yet practical. Because of his wartime radar work, Clarke was aware of the progress in the high frequency electronics that would be necessary in the satellites to amplify the weak signals arriving up from the Earth and retransmit them back down again. The key device was called a traveling wave tube, which was another in the family of vacuum tube devices in which streams of electrons were manipulated in magnetic fields in a similar manner to the cavity magnetron.

Rudolf Kömpfner was born in Vienna before the First World War, and only survived the difficult times after it by being evacuated to Sweden.[11] Though keen on physics he trained as an architect, qualifying in 1933. These were difficult times for Jews in Austria so he managed to come to London where he worked his way up in the building firm belonging to his cousin's husband until he became managing director.

However, he was still interested in physics and spent a good deal of his spare time in the Patent Office Library and reading books and journals. He invented and patented a device which he called the Relayoscope in which, in a way related to a television camera, a light image falling on a nonconducting photoelectric grid controlled the flow of an electron beam. This work was to be important later.

When the war intervened, Kömpfner was taken away and interned on the Isle of Man as an enemy alien. It took some 6 months to convince the authorities that he was a stateless person and to be released. His work on advanced electron devices had been noticed and he was sent to the Physics Department of Birmingham University. In one bound he went from prisoner to the heart of the wartime research effort on radar, as this was the place where Randal and Boot invented the Cavity Magnetron (see Chap. 5).

The faith put in him was not misplaced as he went on to invent the travelling-wave tube while trying to make a better klystron amplifier for radar receivers. His fundamental idea was the continuous interaction of an electron beam with an electromagnetic wave of the same velocity travelling along a helix. It was ingenious and, even more importantly—it worked. From this device a useful low-noise microwave amplifier was developed.

In America at Bell Labs, John R. Pierce was aware of Kömpfner's work as he was interested in the same subject. He undertook theoretical studies, but had not done the practical work that had led Kömpfner to the device. He recommended that Kömpfner was offered a job, but it took until 1951 for the latter to obtain his visa and join the famous Bell Labs. So capable did he prove that he became director of electronics research in 1955, director of electronics and radio research in 1957, and associate executive director, research, Communication Sciences Division in 1962.

Unaware of Arthur C. Clarke's paper, Pierce became interested in the possibilities of unmanned communication satellites, and in 1955 he published a paper 'Orbital Radio Relays' which was published in *Jet Propulsion*, the journal of the American Rocket Society.[12] He examined a number of possibilities, not just synchronous satellites, pointing out that it could well be far cheaper to obtain a large number of telephone channels this way than with transatlantic cables. While interesting, no higher power was prepared to invest in this.

All this changed in 1957 with the launch of the Sputnik satellite by the USSR. With the satellite bleeping over their heads, what frightened the Americans was that this might mean that the Russians had the capability of launching an intercontinental missile which could threaten their security. They didn't know that the USSR was not capable of doing that yet, but only had the technology to place a small satellite into orbit.

The launch of Sputnik and its successors galvanized various American organizations into action, and led to the setting up of the National Aeronautics and Space Administration (NASA).[13] In 1958, the US launched its own first satellite, Explorer 1, and began a serious space program. The commercial companies started jockeying for position, but most interested in communications was AT&T, the main US telephone organization and the parent company of Bell Labs.

In 1959, another paper appeared, this time under both Pierce's and Kömpfner's names. It was called 'Transoceanic Communication by Means of Satellites'.[14] It discussed the various possibilities but saw as a first step a 'passive' satellite which would be a large, balloon-type device which could be used to bounce radio signals back to the Earth the other side of an ocean. The next stage could then be an orbiting satellite which actively retransmitted the radio signal and then possibly to a synchronous satellite.

What was different about these proposals compared to all the previous ones was that the two authors were both directors of divisions in Bell Labs and had available all the electronics technology that was necessary to make this a reality. What they needed was a rocket to launch it. In the meantime, they began serious design work on the electronic systems, both on the ground but also what would be required of an 'active' satellite.

In 1960, after an initial failure, NASA had a successful launch of a 30 m (100 ft) metalized balloon that inflated itself in orbit, named ECHO 1.[15] Bell Labs worked on their 'earth station' at their laboratories at Holmdel in New Jersey and were able to achieve good voice communication between there and a NASA facility in Goldstone, California. While it could be made to work the signals were very weak, and it showed that an active satellite would be a better option.

They started work on an active satellite, to be put into a low Earth orbit and called Telstar. This was to be a trial with the view of launching a whole constellation of 50 or so satellites in random orbits at around 9300 km (7000 miles) height, which could allow coverage of virtually the whole Earth. AT&T were keen on this idea, despite the cost, as it could be used to bolster their monopoly on international telephone communications to and from the US.

In April 1961, the USSR launched Yuri Gargarin into orbit, claiming the moral prize as the first country to achieve a man in space. This motivated the new President Kennedy to make a commitment to put a man on to the moon within a decade.[16] What is less well known is that he also committed the United States to spending 50 million dollars to accelerate the use of space satellites for worldwide communications. What he didn't say publicly was that the government was determined that AT&T should not achieve a monopoly of these worldwide communications.[17]

As a result, NASA contracted with one of their rivals, RCA, to build an experimental communications satellite for low Earth orbit to be called Relay. However, they did agree to launch Bell Lab's Telstar as long as the company paid the costs. AT&T built a ground station featuring a large horizontal horn aerial in Andover in Maine, while at Pleumeur-Bodou in France the French Post Office built a near copy. In Britain, the Post Office built their own design of station at Goonhilly Downs on the Lizard peninsula in Cornwall, with a simpler dish aerial which was to become the pattern for future installations (Fig. 20.2).

On July10, 1962 the first Telstar was successfully launched on top of a Thor-Delta rocket. The satellite was roughly a sphere 88 cm (34.5 in.) in diameter weighing 77 kg (170 lb); figures which were dictated by what the rocket could carry.[18] It could support 60 two-way telephone channels or a TV channel that could only operate in one direction at a time. That channel was soon in service, with the French station picking up American transmissions. The British station was a little behind due to a misunderstanding about the polarity of a signal which was quickly rectified.

There then followed exchanges of short television programs from each side of the Atlantic which generated huge interest, particularly in Europe. Never before had viewers been able to see live pictures from the far side of the ocean. Of course, they were unspectacular and the UK's newspaper,'The Times', sniffily reported that 'the Americans were delighted because it showed them what they expected to see in Europe: it showed the

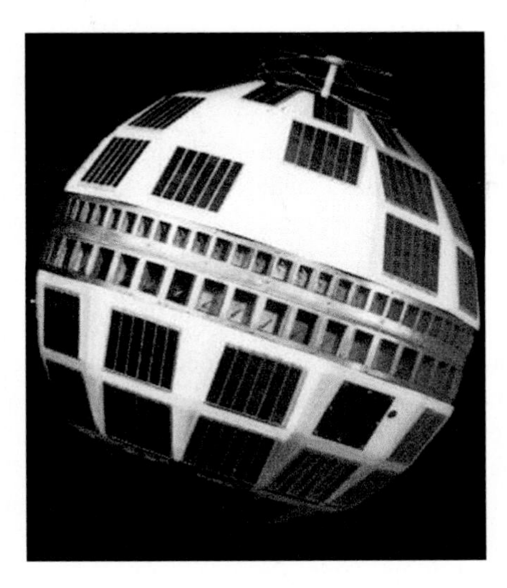

Fig. 20.2 Telstar communications satellite. The solar panels for power are arranged around the sphere. *Source:* http://en.wikipedia.org/wiki/Telstar#/media/File:Telstar.jpg

Germans being industrious, the Italians being artistic, and the British being ceremonious.'[19] They felt that the new marvel had reduced communication to an exchange of clichés.

The available time for transmission was only about 19 min on each two-and-a-half hour elliptical orbit as the satellite passed over and was in view from both ground stations. It was thus inevitable that any television program would be strictly curtailed which isn't the conditions for quality. The satellite had been intended as a trial and in that it was successful though it did start to misbehave after a few months. A great deal was learned about operating in space, and it proved without any doubt that satellites could be launched to handle transocean communications.[20]

NASA decided on their own experimental program and carefully chose someone other than AT&T to build their satellite. On December 13, 1962 they launched the RCA-built Relay satellite which was bigger and heavier than Telstar. It was also launched into a somewhat higher orbit. Despite malfunctions it was able to transmit television programs across the Pacific as well as the Atlantic using some of the same ground stations. It was thus able to transmit some pictures of the 1964 Tokyo Olympics to America and then again to Europe.

These satellites had been in relatively low Earth orbits of 1000 km up to 7000 km which was much easier for the launch rockets to reach than the synchronous orbit at 35,784 km. In addition, to use the synchronous orbit would require the satellite to be maneuvred into position over the Equator, and even when in the correct position it would need adjustment from time to time to keep it correctly on station. It was thus a far more difficult proposition and most thought that it was beyond the existing technology.

To compound matters, the greater distance from the Earth meant that the signals were weaker. There were thus greater demands on both the receiving and transmitting parts of the device. The advantages, on the other hand, were great. It would provide continuous coverage of some 40% of the Earth from a single satellite and there would be no interruptions as it orbited. The ground antennae would not need to track the satellite and there would be no interruptions while the signal was switched to a different orbiting station as the first one went out of range.

At the Hughes Aircraft Corporation, Harold A. Rosen was convinced it could be done. In 1959, together with his colleagues Donald Williams and Thomas Hudspeth, he started the design of just such a satellite.[21] Not only did it need the superior electronics for the receiver and transmitter, it needed to carry fuel for small control jets to make the position adjustments it would need. All this was required to be in a smaller and lighter package than that used for the low orbit satellites so that the rocket could lift it to the required height.

Rosen and his colleagues managed to convince NASA that they had something and received a contract to make three Syncom satellites compared with the two each of the Telstar and Relay devices. It was realized that this was riskier and that at least one might be a failure. The first Syncom was launched on February 14, 1963 and during the maneuvre to lift it from a temporary to the synchronous orbit communications were lost and never re-established.[22]

A few months later, on July 26, 1963 they tried again with a slightly modified design. This time they were successful, and by August 16 they had inched Satcom 2 into a position over the Equator at 55 degrees west in the middle of the Atlantic.[23] The orbit wasn't perfectly aligned with the Equator and, as a result, the satellite moved north and south in a figure-of-eight orbit, though staying more-or-less in the same longitudinal position. This was quite adequate for communication with the Earth stations, particularly as they had stearable antennae.

Communications were successfully made across the Atlantic, but the bandwidth on the satellite was not sufficient for a full television channel. However, it proved that a synchronous satellite was practicable. By the time Syncom 3 was launched on August 19, 1964, the rockets had advanced and this meant that it could be placed into a true geostationary orbit at 180 ° on the Equator over the Pacific.[24]

After satisfactory trials it was used to relay some of the television pictures from the 1964 Olympic Games from Tokyo to the US, thereby demonstrating that Harold Rosen's faith in the geostationary satellite concept. Early the next year it was handed over to the American Department of Defence. The war they were waging in Vietnam meant there was a large requirement for communications with south Asia which could conveniently be provided by Syncom 3.

Meanwhile, the American government and Congress passed the Communications Satellite Act in August 1962. This set up a semi-commercial company called Comsat. Its function was to provide international telephone and television channels via satellite. The organization now had a problem as to whether to choose a group of satellites for a medium orbit or a geosynchronous one.

Comsat put out a request for proposals, and in early 1964 it received four. Three of them were for medium orbits while Hughes, with the success of Syncom 2 and with Syncom 3 well advanced, proposed a geostationary one. Without committing themselves to a final decision, Comsat decided to accept the Hughes proposal as an 'Early Bird', which was to have 240 telephone circuits and a television channel.

It was rapidly realized that Comsat needed to recruit international partners, particularly from Europe, and the organization metamorphosed into the International Satellite Corporation

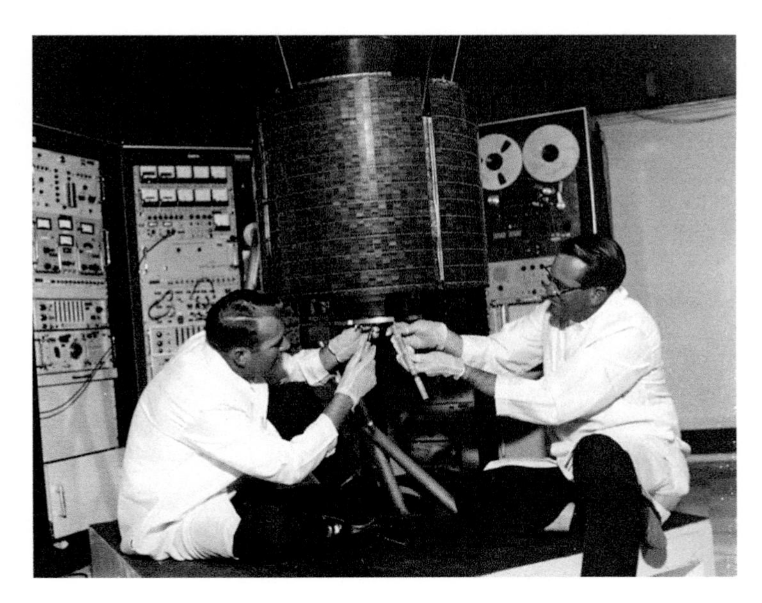

Fig. 20.3 Engineers working on the Early Bird (Intelsat 1) satellite. *Source:* http://en.wikipedia.org/wiki/Intelsat_I#/media/File:INTELSAT_I_(Early_Bird).jpg

(Intelsat). This was well in place by the time Early Bird was launched on April 6, 1965 and led to its alternative name of Intelsat 1 (Fig. 20.3). It was very similar to Syncom 3 but used commercial frequencies for the communications rather than military ones.

By the beginning of June, Intelsat 1 was in position at 28 ° west over the Atlantic and had passed its initial tests ready to be put into commercial operation. Overnight, it almost doubled the number of telephone channels and provided a permanent television channel that could be used at any time and not just when the satellite was passing overhead. It had cost some $7 million to put up, which was a fraction of the cost of a transatlantic telephone cable.

The primary objective of those promoting its use was for telephone circuits. For them, the one television channel was merely an extra. However, by the beginning of May the broadcasters were starting to make use of it to show live programs straight from America. With only one channel this had to be shared between the various European countries and also with the American television networks, so the amount of time that any one TV company could use it was strictly limited.

President Johnson took the opportunity of the new satellite channel to broadcast a message to Europe on the 20th anniversary of VE day on May 8, 1965.[25] He was concerned to counter the programs that the Russians were able to transmit using the Eurovision network celebrating the same event. It did produce somewhat of a storm as by custom he should have obtained permission of the various governments to speak directly to their peoples.

Before long, the broadcasters began to find how to use this new resource. It was quite expensive, so using it just for the sake of it soon dropped out of fashion.[26] Its real use was for transmitting up-to-date information, particularly news items, and live TV came into its own. The idea that live news could come from other parts of the world, and not just from your own country or nearby, began to take shape.

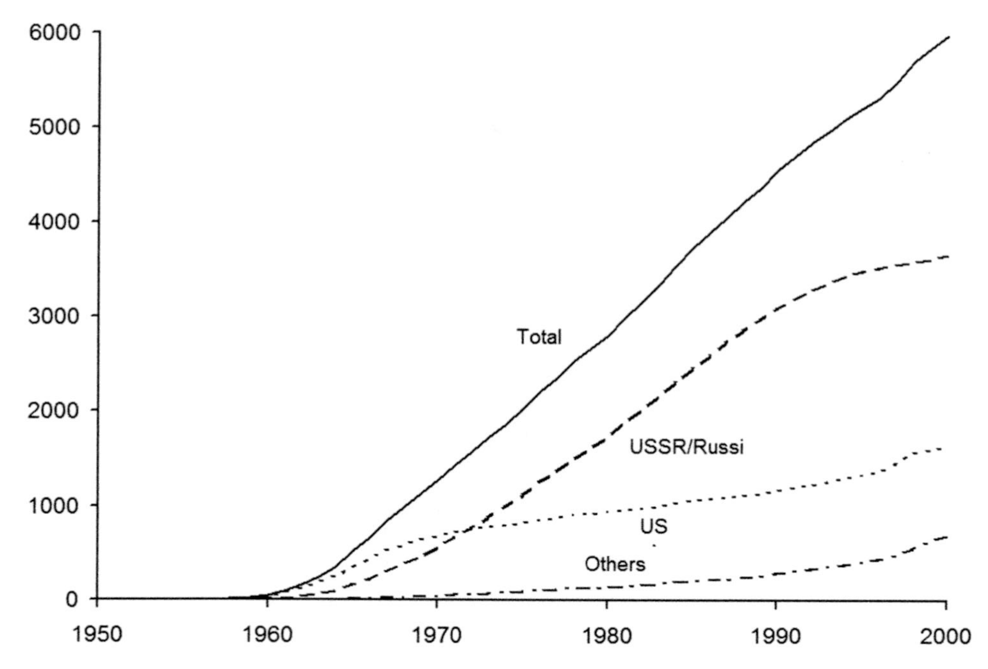

Fig. 20.4 Cumulative satellite launches, 1950–2000. Many are no longer operational or have broken up. *Source:* Author[27]

Intelsat were just getting started and by early 1967 had a satellite over the Pacific providing telephone circuits and a television channel to Japan.[28] Next was a replacement for Early Bird over the Atlantic. One of the drawbacks of the synchronous satellite was that it needed fuel to power the small thrusters to keep it on station. The amount of fuel that it could carry determined the lifetime of the satellite, because when it was exhausted the satellite tended to drift away from its desired position.

In 1969, another Intelsat was placed over the Indian Ocean, thus giving coverage to almost all the populated parts of Earth. The only areas difficult to cover were the far North and South. This turned out to be just in time, as less than 3 weeks later pictures of the Moon landing were beamed to the whole world. Some 500 million viewers from all parts of the globe were able to watch it live. Worldwide television had come into its own.

The launching of satellites of all types came thick and fast (Fig. 20.4). The military, of course, were in the van using them for communications as well as surveillance in various forms. It was much safer to fly a camera over an area of interest from space than to use a plane which might get shot down and cause a diplomatic incident. The commercial world used them largely for communications but also for looking at weather patterns, providing an enormous amount of data simply and hence improving forecasting.

An obvious need, which could be met only with satellite communication, was to keep in touch with ships at sea. In 1979, the International Marine Organization set up the International Marine Satellite Organization (Inmarsat) to provide two-way communication and, importantly, a distress system to call for help when in difficulties far out to sea.[29]

In 1982, the first satellite, Inmarsat A, was deployed by the Royal Navy over the South Atlantic. It was the beginning of a worldwide system that meant ships were always in touch and had the capability to call for help when in trouble.

On a ship it is easy to have a reasonably sized dish aerial though there is a complication of keeping it pointing at the satellite, but the problems were soon solved. Once the organization had their satellites and communications system in place it could be used for other purposes. All that was required was a dish aerial and the electronic systems. While these were not very small there were many situations where this was not a problem and communications were vital.

Once more communications satellites were available, the broadcasters started using what they called Satellite News Gathering (SNG). This consisted of a truck with a satellite dish on the roof and the necessary transmitters and receivers inside. Cameras operated a short distance away, connected their signals to the truck and then via the satellite link to their studios which could be many thousands of miles away. This system proved its worth in the British war with Argentina over the Falkland Islands in 1982, and then for succeeding conflicts where reporters could make dispatches from out-of-the-way places direct to their studios.[30]

Some broadcasters had even greater ambitions for satellites, and that was to use them to broadcast directly. There were two areas of interest: in developed countries where the airwaves were already congested, particularly in Europe; and the other extreme in thinly populated areas where there were considerable difficulties supplying an adequate service to the more outlying districts. This last was of concern in Canada and Australia, and for Indonesia with its multiplicity of islands and often difficult terrain.

The receiving stations for the first satellites used large dish aerials to extract the weak signals, which was quite impossible for a domestic situation. The design of the satellite attacked this in a number of ways. First, larger arrays of solar cells produced more power which meant that larger transmitters could be used. Better aerials could focus the signal to a smaller area.[31] Instead of sending the signal in all directions, it could be sent only towards the Earth, producing something like a ten times improvement in the signal at the Earth's surface. However, if the transmission aerial focused the beam on to a single country a further order of magnitude increase could be achieved.

These improvements started to make it look feasible to direct broadcast, but there was still a problem at the receiving end. The transmission from the satellite was at such a high frequency that it required specialized components to receive it. Certainly the signal wouldn't go through the sort of coaxial cable used for normal TV reception. The solution was to use a small parabolic dish and then mount a device called a Low Noise Block (LNB) at its focus. Often the dish was offset from a true parabola so that the LNB was not in its line of sight and the dish was mounted in a more vertical position so didn't fill with rainwater.

The LNB consisted of a small waveguide to channel the signal into the electronics where it used the heterodyne technique of beating the signal with a local oscillator to convert it to a lower frequency. With some amplification also provided the signal could now be fed along a coaxial cable to the receiver set top box. This could then pick out the required channel and send it to the television set. Surprisingly, with a good LNB design, there was just enough performance for the system to work in practice.

The Japanese, with considerable problems of distribution of TV signals in their mountainous country and scattered islands, were the first to look at the possibilities.[32] In 1978, they arranged for NASA to launch their experimental satellite and it began direct broadcast trials. Six years later, in 1984, they launched the first direct broadcast satellite to provide a service to consumers. They had solved all the problems and it allowed them to easily provide a service to the outer islands as well as the main part of Japan.

The next step was made by British Satellite Broadcasting (BSB) in 1987 who contracted Hughes to build them a satellite to be named Marco Polo 1.[33] The plan was to have sufficient transmission power so that the receiving aerial was only a 35 cm dish. The first satellite was launched on August 27, 1989, but the service didn't start until April the next year.[34] It was beset by difficulties and was losing money heavily.

What BSB hadn't bargained for was that at the end of 1988 a Luxembourg-based company, Société Europienne des Satellites (SES), also launched their Astra satellite for direct broadcast to Europe. Newspaper tycoon Rupert Murdoch saw his opportunity and leased channels so that he could set up his Sky television company. He used a simpler system with a larger dish aerial but, importantly, it only required a 'set top' box rather than the new television which was needed for the BSB system. Despite this, a year down the line he was losing money as well.

His next step was to seek a merger between the two companies, and the government let him do it. The result was B Sky B which rapidly became Sky as the Marco Polo satellites were sold off and the more complex BSB system dropped. All that remained of their offering were one or two channels but only available via the Sky system. Murdoch had won and after a while began making handsome profits.

In a few years, many other countries had their own systems, direct broadcast became common and the number of channels multiplied. The main effect was to produce channels aimed at particular interests, whether sport or home improvements. It became common to have hundreds of channels at the flick of the remote control rather than the few that had been available on the terrestrial systems. A side effect was that it was often possible to obtain channels from other countries and in other languages. It was useful for students, but even more so for homesick expatriates!

One of the dreams of the satellite pioneers was to be able to communicate directly with telephones. This was largely overtaken by the cellular phone business (see Chap. 21) but there are places on the planet where it is either not practical or impossible to build the necessary network of base stations that they need. That was sufficient for some entrepreneurs to attempt to build a satellite phone service.

The problem is that the requirements are even more severe than for direct broadcast as the aerial needs to be even smaller and cannot be a dish aimed at the satellite. It seemed that the only way to make this work was with low Earth orbit satellites. Of course, at that level they would be orbiting rapidly and it would require a whole constellation of satellites, with the ability to switch from one to the next as they came into view, to provide a service.

Two companies, Globalstar with their 48 satellite constellation and Iridium with 66, attempted to build systems and started launching before the end of the century.[35] Both ran into financial problems as it proved difficult to sign up customers fast enough to fund the enormous cost of putting up all those satellites. After financial restructuring, both survived, but these systems were never the great success that had been planned for them.

Other terrestrial systems had taken most of the market, and there wasn't a lot left that needed their expensive offering.

Yet another area opened up with navigation. What started as a way for the military to aim their missiles accurately metamorphosed into a universal system. A whole fleet of satellites allowed several to be in sight at the same time and from their signals positions could be calculated. This is explored in detail in Chap. 28.

Space exploration has thus always been intricately involved with electronics. It is needed for communications but also computing power is required to calculate orbits and flight paths for the space vehicles. These needed to advance hand-in-hand with the rocketry. On the other side, communications have been greatly enhanced by the ability to send high bandwidth signals across oceans. Hence, live television pictures can be had easily from the other side of the world for news and sporting events.

While moonshots and men walking in space gained the publicity and the glamor, it was the more humble communication and weather satellites cramming that valuable synchronous band around the Equator that had the greatest impact on society (Fig. 20.5). Quietly, but completely, they changed the world. We now expect to know what is happening even in the far flung parts of the planet, and it has brought us all closer together. Sadly, that knowledge has not overcome all prejudices and, as Arthur C. Clarke suspected, many wars still take place.

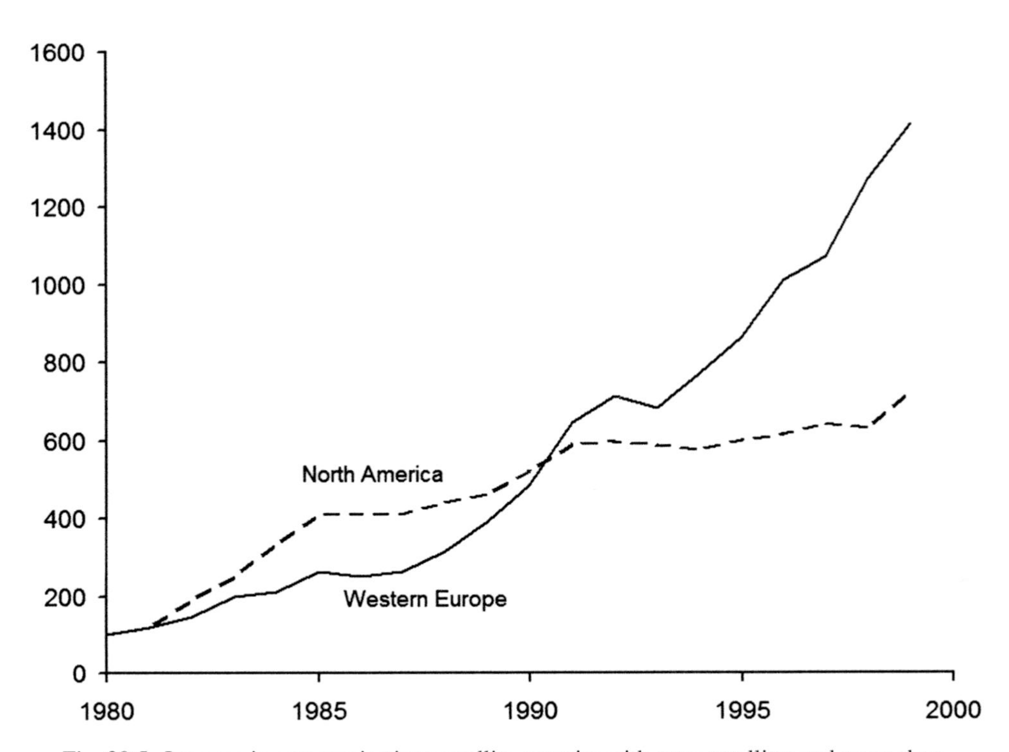

Fig. 20.5 Increase in communications satellite capacity with more satellites and more channels on each, 1980–2000. *Source:* Author[36]

NOTES

1. Evans, A.B. Jules Verne. *Encyclopaedia Britannica.*
2. Verne, J. From the Earth to the Moon, p. 18.
3. The exploration of space: Konstantin Eduardovich Tsiolkovsky, available at: http://www.century-of-flight.freeola.com/Aviation%20history/space/Konstantin%20E.%20Tsiolkovskiy.htm.
4. Darling, D. Biography Tsiolkovsky, Konstantin Eduardovich (1857–1935), available at: http://www.thelivingmoon.com/45jack_files/03files/Tsiolkovsky_001_Intro.html.
5. Bellis, M. Konstantin Tsiolkovsky, available at: http://inventors.about.com/library/inventors/blrocketTsiolkovsky.htm.
6. The exact figure for the height of the geosynchronous orbit varies a little between various sources. It is also roughly 22,300 miles.
7. NASA, Hermann Oberth, available at: http://www.nasa.gov/audience/foreducators/rocketry/home/hermann-oberth.html.
8. The exploration of space: Robert H Goddard, available at: http://www.century-of-flight.freeola.com/Aviation%20history/space/Robert%20H.%20Goddard.htm.
9. An English translation together with background information can be found at: http://istrianet.org/istria/illustri/potocnik/book/index.htm.
10. Clarke, A.C. Extra-terrestrial relays. *Wireless World*, October 1945, also available at: http://lakdiva.org/clarke/1945ww/1945ww_oct_305-308.html.
11. Pierce, J.R. Rudolf Kompfner: 1909–1977, A biographical memoir, available at: http://www.nasonline.org/publications/biographical-memoirs/memoir-pdfs/kompfner-rudolph.pdf.
12. Pierce, J.R. (1955) Orbital radio relays. *Jet Propulsion*, 25:4, 153–157.
13. Whalen, D.J. Billion dollar technology: A short historical overview of the origins of communications satellite technology, 1945–1965, NASA, available at: http://history.nasa.gov/SP-4217/ch9.htm.
14. Pierce, J.R. and Kompfner, R. (1959) Transoceanic communication by means of satellites. *Proceedings of the IRE*, 47:3, 372–380.
15. Cavagnaro, M. The early satellites, NASA, available at: http://www.nasa.gov/missions/science/f-satellites.html.
16. Kennedy, J.F. Speech to congress, May 25, 1961, text available at: http://www.jfklibrary.org/Asset-Viewer/Archives/JFKPOF-034-030.aspx.
17. Glover, D.R. NASA experimental communications satellites, 1958–1995, NASA, available at: http://history.nasa.gov/SP-4217/ch6.htm.
18. Martin, D., Anderson, P., and Bartamian, L. (2008) Satellite history: Telstar. *SatMagazine*, available at: http://www.satmagazine.com/story.php?number=511938650.
19. *The Times*, August 11, 1962, p. 9.
20. Telstar 1, NASA SP-32 Volume 1, June 1963, available at: http://articles.adsabs.harvard.edu/cgi-bin/nph-iarticle_query?bibcode=1963NASSP..32..739D&db_key=AST&page_ind=6&plate_select=NO&data_type=GIF&type=SCREEN_GIF&classic=YES.
21. Gregersen, E. Harold Rosen. *Encyclopaedia Britannica.*
22. Syncom 1, available at: http://nssdc.gsfc.nasa.gov/nmc/spacecraftDisplay.do?id=1963-004A.
23. Syncom 2, available at: http://nssdc.gsfc.nasa.gov/nmc/spacecraftDisplay.do?id=1963-031A.
24. Syncom 3, available at: http://nssdc.gsfc.nasa.gov/nmc/spacecraftDisplay.do?id=1964-047A.
25. The room-size world. *Time*, May 14, 1965, Vol. 85, Issue 20.
26. *The Guardian* (1959–2003); May 29, 1965.
27. Data from: http://satellitedebris.net/Database/LaunchHistoryView.php.
28. Our history, available at: http://www.intelsat.com/about-us/our-history/1960-2/.

29. We ensure you are never beyond reach, Inmarsat, available at: http://www.inmarsat.com/about-us/.
30. Rouse, M. Satellite News Gathering (SNG), available at: http://searchnetworking.techtarget.com/definition/satellite-news-gathering.
31. Pritchard, W.L. and Ogata, M. (1990) Satellite direct broadcast. *Proceedings of the IEEE*, 78:7, 1116–1140.
32. Milestones: First direct broadcast satellite service, 1984, Engineering and Technology Wiki, available at: http://ethw.org/Milestones:First_Direct_Broadcast_Satellite_Service,_1984.
33. Marco Polo satellites, available at: http://www.astra2sat.com/bsb-british-satellite-broadcasting/marco-polo-satellites/.
34. British Satellite Broadcasting: The full responsibility, available at: http://www.terramedia.co.uk/reference/documents/BSB.htm.
35. History of the handheld satellite phone, Globalcom, available at: http://www.globalcomsatphone.com/phone-articles/history-of-the-handheld.
36. Data from: Mattock, M.G. The dynamics of growth in worldwide satellite communications Capacity, RAND, available at: http://www.rand.org/content/dam/rand/pubs/monograph_reports/2005/MR1613.pdf; W Europe Figure A4, and US Figure A1.

21

Personal Communicators: Mobile Phones

Three objects were considered essential across all participants, cultures and genders: keys, money and mobile phone.

Jan Chipchase, Nokia.

Almost from the moment that Marconi began a wireless service, artists and writers began to conceive of personal communicators. The idea was always a small unit that the user could carry around to receive messages and, better still, send them. A cartoon by Lewis Baumer from Punch in 1906 foresees personal boxes that could receive telegraphy messages by wireless, with the users sporting aerials on their hats. The cartoonist hasn't had the imagination to realize that the transmission could be of voice as this hadn't been invented at the time (Fig. 21.1).

By 1926, Karl Arnold in Germany was drawing mobile telephones being used in the street.[1] Quite soon the idea was appearing in the children's books of Erich Kästner. After the Second World War they were in novels by Robert Heinlein, and in an essay in 1958 Arthur C. Clarke imagined 'a personal transceiver, so small and compact that every man carries one', even though that wouldn't appear for some 20 years.[2] Perhaps the most famous example from film is 'Star Trek', where the voyagers can make the call on their handheld devices to 'beam me up, Scotty'. The reality was a long way behind.

In Germany, experiments began with wireless telegraphy even before the First World War on the military railway between Berlin and Zossen.[3] After that war, with improvements in electronics, further trials began on the relatively short Berlin-Hamburg line, this time with telephone connections. This resulted in a service in 1926. It used the long wave and so the antennae on the train were 192 m long, being carried along the tops of the carriages. This coupled to a wire run along the telegraph poles alongside the railway. Inside, two toilet compartments were converted into the telephone booth and a space for the operator.

Naturally, this service to make and receive calls from a moving train was only available to first-class passengers.[4] The bulk of the calls (some 90%) were from the train to landlines rather than the other way around, which is hardly surprising as it was essentially a portable telephone box.[5] The size and weight of the equipment, and particularly the antenna, made

J.B. Williams, *The Electronics Revolution*, Springer Praxis Books,
DOI 10.1007/978-3-319-49088-5_21

Fig. 21.1 Early idea of personal communicators using wireless telegraphy. *Source:* Punch Dec 26 1906

the system practical only on a train. Though inefficient, it worked surprisingly well and remained in service until the Second World War.

In other countries things weren't as advanced. Mobile radios were used to call police cars in Detroit in 1921, and two years later New York police had two-way communications to theirs.[6] By early 1940, some 10,000 US police cars and fire trucks had such systems. However, these were merely mobile radios and were not connected to the public telephone system.

In Britain in 1920, Sir Samuel Instone of the Instone Air Line had a conversation by wireless radio from a house in London to an aeroplane in flight to Paris.[7] Marconi had largely built his business on mobile communication to ships, but again this was not telephony connected to the public telephone system. By 1934, a short-wave radiotelephone system was operating to coastal ships. Of course the military were using mobile radios, particularly in tanks.

As the likely combatants geared up for the inevitable clash that became the Second World War, they all saw the need for portable radio systems. There is considerable doubt as to who invented the 'walkie talkie', as it became known, due to the inevitable secrecy that many countries imposed on their rearmaments program. What is clear is that, by around 1940, a number of armies had portable radios which could be carried by a soldier in a backpack.[8] Later, Motorola produced a handheld device known as the 'handie talkie' for the American army.[9]

These wartime sets were not telephones. They were radios, with each group operating on a common frequency. Normally, all sets listened and could only transmit when a switch was pressed. In addition, there was a protocol that the users had to follow to avoid chaos by ensuring that only one tried to transmit at any one time. Other disadvantages were that they were very bulky and heavy, their range was only a few miles and, unless encrypted, the enemy could also listen in.

The war had advanced electronics technology for these portable sets and for radars, so it was natural that, once it was over, consideration would once again be given to the possibility of mobile telephones. Quickest off the mark were Bell Labs who tested a system in St Louis, Missouri, in 1946. It was a 'simplex' system in that it used a single radio channel and the user had to press a button to talk. This accessed an operator who could connect the call to the fixed telephone network. As there were only three available radio channels the system was very limited in its number of users.

The next year, a 'highway' system began operating along the route from Boston to New York. Slowly this spread to other cities, but the severe limitation of the number of channels held the system back.[10] There was a considerable unfulfilled demand with many potential customers on waiting lists. However, it wasn't until 1956 that the Federal Communications Commission (FCC), in charge of the radio spectrum, agreed to make more channels available.

In Germany, the Federal Post Office rounded up all the bits of existing systems and created the A network which was the first national mobile phone network and by the end of the decade was the largest in the world. By 1968, it covered about 80% of West Germany but, due to the high costs and limited frequencies, it only had some 10,000 users. Like other systems at the time, the receiver units were heavy and only suitable for use in cars.

Britain started a system in 1959 in South Lancashire, and this was soon followed by one in London. Again, this was a limited system with the capacity for 300 subscribers which was reached within 6 months of opening.[11] Again, this was manual and calls were connected by the operator.

In the 1960s, the US started upgrading their systems to higher frequencies and more channels and automatic operation were introduced. In Germany and Britain, the systems were upgraded in 1972 and automatic features added, but the number of users only grew slowly due to the channel limitations. These were systems only for the very few who could justify the high costs, such as diplomats and business leaders. It was something of a status symbol to have a radio telephone in one's car.

All these systems suffered from the same fundamental drawbacks. The radio spectrum was crowded with users, and the regulators were unwilling to licence large amounts of it for telephone purposes as one channel took up the same amount of frequency spectrum as a broadcast radio channel.

The coverage area was determined by the power of the transmitters. For the base station this was not a problem as it was simple to have a powerful device and a high antenna. However, the mobile units could not do this. The practical size of the antennae was very limited, as was the transmitter power before the vehicle's batteries were exhausted. Sometimes the networks had a single base transmitter but several receivers spread around its working area. This helped to some degree but was only a partial solution.

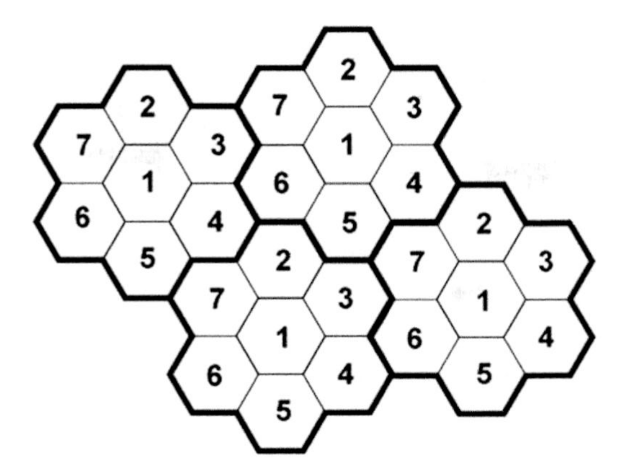

Fig. 21.2 Pattern of mobile phone cells. Those with the same number would use the same frequency. *Source:* Author

As early as 1947, Bell Labs had been examining the problem. In December of that year, D.H. Ring issued a memo outlining a new approach. Instead of trying to have as large a coverage area as possible, the transmitter powers should be relatively small and, particularly with the use of higher frequencies, the coverage area would not be great. If this area was then surrounded by six others operating on different frequencies, then they would not interfere with each other.

The clever part of the idea was that the sets of frequencies could be reused by other groups surrounding the first. There would always be two cells between ones on the same frequency. By designing the receivers to separate the higher power signal from the weaker, only the intended signal would be received. This was the basis of the 'cellular' system, though in practice the cells were not really hexagons as in Fig. 21.2. The advantage was obvious in that the capacity of the system was enormously increased by the reuse of the frequencies. Also, by keeping the cells small the transmitter powers in the mobile units could be reduced, hence lowering the battery drain.

The problems were all in the detail. Somehow the receiver needed to know which cell it was in and thus the frequency to use. If the user moved from one cell to the next his phone needed to know this and be able to switch instantly to the new frequency in that cell. The central system would need to know where the subscriber was at all times in case someone called his phone. The complexity was simply beyond the technology at the time, even for Bell Labs, though they didn't give up on the idea.

What was needed was the development of transistors and integrated circuits to get the size down, and of computing technology to be able to deal with the complexities of controlling the operation of the phone and of the network. It wasn't until the late 1960s that these began to come into place, though there was still some way to go. In 1969, a cellular payphone service was set up on the Metroliner trains between New York and Washington. The trains could easily carry the bulky equipment and provide the necessary power.

Now various groups were looking at the possibilities, and in particular Martin Cooper who worked for Motorola in the US. He had a very clear concept that people didn't want to talk to buildings or to cars, they wanted to communicate with other people. What this meant was that a person should be able to carry a mobile phone around, and not require a vehicle to cart it about.

Cooper's group worked on this concept, and by 1973 they had built a prototype unit. To demonstrate its capability they set up a base station on a building in New York and Cooper made calls from the street, including one to Joe Engel, his counterpart at Bell Labs.[12] While this was just a publicity stunt it moved the concept along. Whereas Bell Labs and AT&T had been fixated on car phones, Motorola was now concentrating on truly portable devices.

It took until 1977 for Bell Labs and AT&T to have a working prototype cellular network, but they were beaten to an operating system by Bahrain who introduced a two-cell system the following year.[13] The Americans soon followed with a cellular network trial in Chicago which had over 2000 customers, but because of squabbles between potential operators and the FCC the American systems didn't progress for some years.

By the end of 1979, the Japanese had stolen a march and had a 13-cell system working in Tokyo.[14] Soon systems were springing up in the Nordic countries, Australia, Saudi Arabia and Mexico. Finally, in 1983 the cellular system in the US began to expand to other cities and the number of users increased rapidly. Britain launch its system with two separate operators, Cellnet and Vodaphone, in 1985. Cellular systems quickly appeared in many other countries.

Many of the phones were still car based, but Motorola launched their DynaTAC 8000X handheld phone in 1983. Though smaller and lighter than the prototype that Martin Cooper had used a decade earlier, it was still bulky and gained the unflattering nickname of 'the brick'. These were expensive at just under $4000 in the US which limited the market, but despite this high price they became the thing to have for those who could afford it and wanted a status symbol.

The advancement that had finally made these systems practical was the arrival of the microprocessor whose computing power could carry out the complicated procedures for 'handing off' when the phone passed from one cell to another and deal with the issues of keeping track of where it was. The arrival of integrated circuits, which meant that the radio could easily jump from one frequency to another, was also essential.

These systems all used analogue signals for the speech channels, but due to a lack of coordination they were not compatible. (The exception was the Scandinavian countries which collaborated with each other.) This was to prove a particular drawback as the manufacturers could not get real economies of scale because they had to make different devices for each country.

Despite the fact that these systems used two channels, one for the outgoing speech and another for the reception, there were more channels available in the new high-frequency bands that had been opened up. This, coupled with the efficient reuse of frequencies by the cellular system, meant that a vastly greater number of users could be accommodated. Whereas the final precellular system in Britain eventually reached 14,000 subscribers, the numbers on the new system reached 100,000 by the autumn of 1986, and 200,000 less than a year later.[15] This was despite the high costs and that most of the phones were still used in cars.

As the decade progressed the price and size of phones gradually fell, increasing the number of subscribers further. There was no doubt now that there was a very large demand for mobile phones if they could be made sufficiently small and cheap.

There was one worry: that the conversation was a simple analogue signal which could, in principle, be picked up by a radio receiver tuned to that frequency. In fact, there were two infamous examples involving members of the British Royal family colloquially known as 'squidgygate' and 'Camillagate'. However, as the two sides of the conversations would have been on different frequencies, and the tapes of these had both halves of the conversation, they could not have been what they were claimed, i.e., off-air reception and it is possible there was something more complex going on.

Despite their slightly earlier start in the US, the take-up of mobile phones there and in the UK, in terms of percentage of the population, largely kept pace during the 1980s. There was not the usual delay between them. Those involved with the supply, such as regulators, makers and operators, were aware that even these improved networks were in danger of being saturated and started to look at what should come next.

In Europe, the error of having incompatible systems was recognized and the European Conference of Posts and Telecommunications (CEPT) established a technical committee known as the "Group Special Mobile". This was the origin of the initials GSM, though later the meaning was changed to Global System for Mobile telecommunications.

There was tremendous wrangling about frequencies and the precise standards, and a number of competing systems were trialled, but everyone understood the value of reaching a common standard. Finally, in 1987 consensus was reached and countries started to sign the memorandum of agreement. The system was so thoroughly worked out that the signatories rapidly spread from Europe to the Arab states, to Asia-Pacific regions and Africa. It had reached 120 signatories by 1995.

What they had agreed on was a digital system, and it was replacing a mixture of systems, there was no temptation to try to bolt it on to the existing analogue ones. They were able to start again and design a new, really efficient system with all the knowledge that had been gained in a series of countries.

Using digital techniques meant that the effective 'dead space' in much of speech could be utilized and a number of users could use the same radio channel by sending their information in bursts in different time slots. By this and other techniques they were able to get something like 5–10 times as many effective simultaneous users with the same number of frequencies as the analogue systems. In fact, many countries also opened up a second band of frequencies around 1800 MHz, roughly double the main 900 MHz band that most used.

As the handsets needed to be able to work on various frequencies it was a simple matter to hop from one to another during the conversation. This made it considerably more difficult to monitor a call, giving a level of security much greater than the analogue systems. This idea had been invented much earlier in the century but a curious person was a joint holder of a patent for frequency hopping as it was called. This was the film star, Hedy Lamarr. In 1942, she and composer George Antheil filed a claim which was intended to give secure communications to radio-controlled torpedoes.[16] Though her day job was as a film actress, her hobby was inventing.

Security was not the main reason for using frequency hopping. A particular frequency in a locality could be subject to poor radio propagation. By hopping around, with the transmitter and receiver keeping in synchronism, the good and bad frequencies could be shared out, giving everyone a more consistent service. Though the frequency hopping gave better security, it was further enhanced by an encryption system making it almost impossible to eavesdrop.

Another feature of GSM was the ability to send a short text message known as SMS. These could be sent very rapidly on the system and, as a result, the operators made them quite cheap. They became very popular, particularly amongst the younger generation.

The first GSM systems started to appear in a number of European countries in 1991. After that they spread rapidly, despite the operators having to effectively build new networks consisting of large numbers of base stations. In populated areas these needed to be quite close together to get enough capacity, though they could be further apart in places where usage was lower. The phones were designed to alter the power of their transmitters depending on the circumstances so that it could be kept to a minimum, hence saving battery power.

There had initially been considerable doubts as to whether the available technology was capable of making handsets at acceptable size and cost. However, integrated circuits, and particularly microprocessors, were advancing at an enormous rate. Of particular interest were the ARM cores described in Chap. 15. These could be put on to a chip with the other circuitry to convert to and from analogue signals. What was particularly valuable was the small amount of power they consumed while still undertaking the complex operations needed.

With the GSM system in use in a number of countries the number of handsets needed was high enough for the set makers to put great efforts into reducing the cost of their products. Of course, as they did so the market size increased, which led to further cost reductions and so into a virtuous circle. A particular beneficiary was an obscure Finish company called Nokia, which soon began to dominate the business.

However, in America things were rather different. At the time when decisions needed to be made about the second generation there was a Republican president and the attitude was that the market should decide. As a result, no guidance on the system to be used was forthcoming, unlike the first generation when the government had insisted on everyone using a compatible system. The result for the second generation was that there were two incompatible competing systems. In addition, GSM was used in some areas effectively making a third incompatible system. Users could not move from one to another without a completely new phone.

While Europe was rapidly converting to a digital system, America was largely hanging on to analogue. One of the US systems was a digital version of the existing analogue one and could be interworked with it using the same frequencies. Hence, there was not the pressure to convert the users so that the analogue service could be shut down and the frequencies reused. The other US system was quite different, and though more efficient in its use of the frequency spectrum, needed a completely new network.

It was noticeable that, despite the early domination of much of the technology by the Americans, it was the GSM system which spread to most countries other than the US and Japan. In 1999, 89% of the mobile phones in Europe were GSM while in the US 60% were still analogue with the rest split between the three digital systems.[17] As can be seen from Fig. 21.3, the take-up of mobile phones in the latter half of the 1990s was much faster in Britain than in America. This was also true for most European countries.

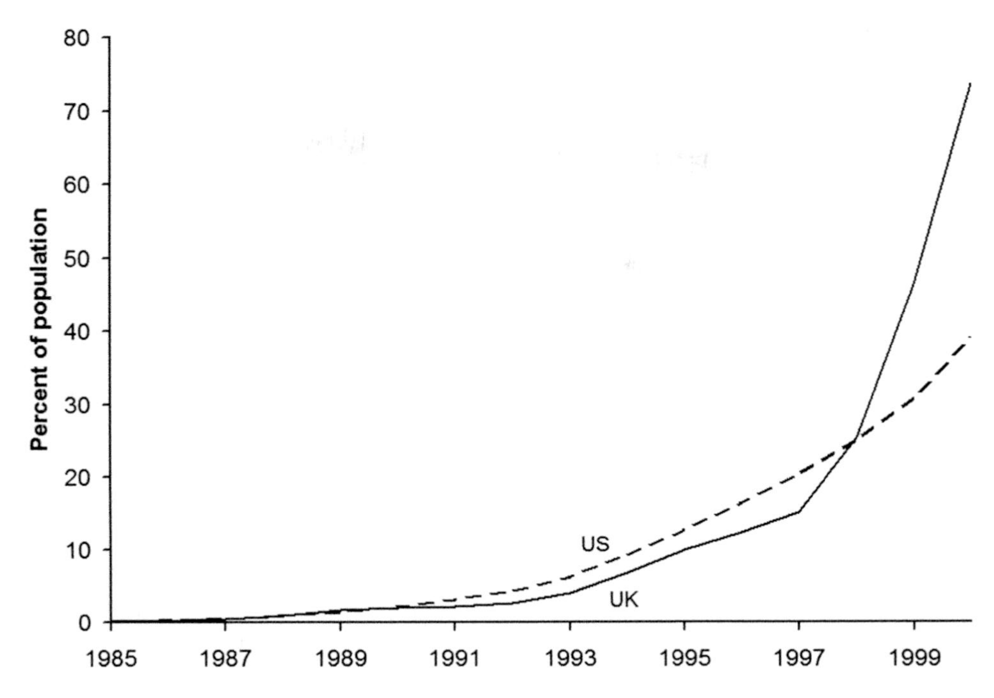

Fig. 21.3 The rise of mobile phone subscriptions in the US and UK, 1985–1999. *Source:* Author[18]

Where most innovations normally had a faster take-up in the US than elsewhere, it was different for mobile phones. It stemmed from a number of differences between the systems. The ability to roam in other countries was undoubtedly a benefit of GSM. However, one early decision in the US had a drastic effect, though it seemed quite sensible at the time, and that was to intermix the telephone numbers of fixed and mobile phones. There was thus no way of knowing what sort of phone you were ringing.[19] The problem was that it was much more expensive to call a mobile phone. The telephone companies' solution to this was to charge the receiving phone for the call, rather than the caller paying. This put people off buying mobile phones as they could run up bills just by receiving calls. In Europe, mobile phones have a distinctive code so the users know immediately whether it is a mobile or not, and the charging is, as normal, to the caller.

One of the biggest factors for the difference in take-up was that in around 1996 or 1997, European operators introduced the prepaid phone. With this, the user didn't run up a bill to be paid later; they loaded credit on to the phone, and when this was used up they paid for some more. What it meant was that youngsters could have phones and it was impossible for them to run up large bills which the parents would have to pay—they could only use the credit on the phone. Ownership amongst the younger generation rocketed, particularly when the supermarkets started a price war for these prepaid phones.[20]

What the kids particularly liked was to send text messages to each other which they could do cheaply, so not using up very much of the credit. The American systems didn't have either prepayment (difficult with the charging for receiving calls) or text messages. The result was that growth was much slower.

Around the turn of the century, mobile phone subscriptions in Britain passed the 25 million mark and were increasing at the rate of a million a month.[21] The phones were now small, easy to use and light to hold in the hand. With the network operators often recovering some of the cost of the phones in their charges for calls, the phones themselves could be obtained very cheaply.

This time Europe had got it right. The advantages of a standardized system were so great that GSM was now the dominant system in the world, with about 70% of the over 600 million mobile phones on the planet using it. By 2001, there were 340 networks in 137 countries.[22] It showed the advantages of cooperation in that all aspects of the system were subject to considerable scrutiny during the development process, and that pressuring people to work together to define a system and then compete afterwards was much better than leaving it to the market.

The computing power both in the system and the phone meant that it was fairly easy to add further features. 'Voice mail', which is basically the same as an answering machine for taking a message, is available on most systems. On the phone, a list of regularly-used numbers can be compiled with the name of the person attached. This is further enhanced when a call is received. The phone will show the number of the caller, but if the number is in the internal list then the name will be displayed instead.

Often there are further improvements, such as the ability to change the ring tone that announces when the phone is being called. With many sets new tunes could be downloaded—for a fee—by calling certain numbers. Clocks and other gadgets were also often included; these could be added cheaply because they were mostly a matter of adding software to the microprocessors which, once programmed, cost nothing in the production of the phone.

Of course, the industry was still looking ahead. Now that mobile phones had become a mass device the regulators were less reluctant to allocate more radio spectrum for their use. Many governments found a good source of income by selling chunks of spectrum at high prices to the intending operators. These were wanted in order to provide a next generation, known as 3G. The displays on the phones were made larger and these new networks had a larger bandwidth so that digital data could be downloaded faster which made connection to the internet practical.

This was all in the future as these networks only started to come on stream as the new millennium began. With large displays and abundant onboard computing power the phones began to be computers in themselves. They also started to accrete cameras, motion sensors and all sorts of extras, turning them into the universal handheld device and a 'must have' for almost every person.

It is difficult for those who have grown up in a world with mobile phones to understand what it was like without them. If a person wasn't at home or in their place of work there was no way of contacting them, or in most cases of being able to find them. If an adult let their child out of their sight, they had to trust them to reappear. There was no way of knowing where they were, or the child to contact them if in trouble.

Although the mobile phone began as a business tool its use rapidly changed as more and more people obtained them. It allowed families and other social groups to remain in touch with each other and, as a result, it began to change society. The constant contact between people strengthened bonds which previously could only be achieved by proximity, which in a modern world was not always possible.

Perhaps the impact was even greater in the developing countries where the costs of installing a comprehensive fixed telephone system were just too great. Here a mobile system could be installed far more cheaply, and for many that new ability to communicate was a liberation. It boosted business as farmers or fishermen, for example, could find out prices to get the best for their produce. In fact, for all forms of trading it became essential, hence boosting the economies.

Altogether, this fusion of telephone, battery, radio and computing technologies has gone far beyond the expectations of those who dreamed it up. It has found its way into every corner of society and become ubiquitous. It has gone beyond the simple communicating device beloved of science fiction to enable the users to obtain and exchange information and not just talk to each other.

NOTES

1. German satirical magazine *Simplicismus*, reproduced at: http://en.wikipedia.org/wiki/History_of_mobile_phones.
2. Ramirez, V. Pocket phone, available at: http://www.technovelgy.com/ct/content.asp?Bnum=595; Lee, A. Can you guess what the world's first text message said?, *Daily Express*, December 30, 2014, also available at: http://www.express.co.uk/life-style/science-technology/549261/Vodaphone-s-first-mobile-phone-call-New-Years-Day.
3. Öffentlicher beweglicher Landfunk (Public countrywide mobile radio) available in German at: http://www.oebl.de/A-Netz/Rest/Zugfunk/Zug1926.html.
4. Informationszentrum Mobilfunk, The development of digital mobile communications in Germany, English version available at: http://www.izmf.de/en/content/development-digital-mobile-communications-germany.
5. Deutsches Telephon Museum (German Telephone Museum), von 1900 bis 1999 (from 1900 to 1999), available in German at: http://www.deutsches-telefon-museum.eu/1900.htm.
6. Wheen, A. (2000) *Dot_dash to Dot.com*. New York: Springer-Praxis, p.164; Bellaver, R.F. Wireless: From Marconi to McCaw, University as a Bridge from Technology to Society. IEEE International Symposium on Technology and Society, 2000.
7. Freshwater, R. UK telephone history, available at: http://www.britishtelephones.com/histuk.htm.
8. Anderson, L.H. The first walkie-talkie radio, available at: http://www.repeater-builder.com/motorola/pdfs/scr300.pdf.
9. University of Salford. Walkie talkies, car telephones and pagers, available at: http://www.cntr.salford.ac.uk/comms/walkie_talkies_radio_telephones_pagers.php.
10. Young, W.R. (1979) Advanced mobile phone service: Introduction, background, and objectives. *The Bell System Technical Journal*, 58:1, 1–14.
11. Harrison, F.G., and Bishop, P. (1995) Cellular radio – Just ten short years. 100 Years of Radio Conference Publication 411, 5–7 September 1995, Institution of Electrical Engineers.
12. Shiels, M. A chat with the man behind mobiles. BBC News, available at: http://news.bbc.co.uk/1/hi/uk/2963619.stm.
13. Wheen, A. (2000) *Dot_dash to Dot.com*. New York: Springer-Praxis, p. 168.

14. Oetting, J. (1983) Cellular mobile radio- An emerging technology. *IEEE Communications Magazine*, 21:8.
15. Retrowow. 80s mobile phones, available at: http://www.retrowow.co.uk/retro_collectibles/80s/mobile_phone.html.
16. Wheen, A. *Dot_dash to Dot.com*, p. 171.
17. Selian, A. 3G Mobile licensing policy: From GSM to IMT-2000 – Comparative analysis. International Telecommunications Union series of Telecommunication Case Studies, available at: http://www.doc88.com/p-5177148469990.html.
18. Data from the World Bank, available at: http://data.worldbank.org/indicator/IT.CEL.SETS.P2/countries/1W?page=2&display=default.
19. Tanenbaum, A.S. (2003) *Computer Networks, 4th Edition*. Upper Saddle River, NJ: Pearson Education, p.153, Chapter 2.6 The mobile telephone system, available at: http://authors.phptr.com/tanenbaumcn4/samples/section02_06.pdf.
20. Braggs, S. Mobile phones – timeline, available at: http://www.mobilephonehistory.co.uk/history/time_line.php.
21. Mobile phone subscribers break 25 mil barrier, available at: http://www.telecompaper.com/news/mobile-phone-subscribers-break-25-mil-barrier--210348.
22. Lacohée, H., Wakeford, N. and Pearson, I. (2003) A social history of the mobile telephone with a view of its future. *BT Technology Journal*, 21:3, also available at: http://dm.ncl.ac.uk/courseblog/files/2010/03/a-social-history-of-the-mobile-telephone-with-a-view-of-its-future.pdf.

22

Going Online: The Internet

The Internet is based on a layered, end-to-end model that allows people at each level of the network to innovate free of any central control. By placing intelligence at the edges rather than control in the middle of the network, the Internet has created a platform for innovation.

Vinton Cerf.

Pavel Baranov was born in 1926 in Grodno, Poland (now in Belarus), but 2 years later his parents moved to the US where he took the name Paul Baran.[1] Though his father owned a small grocery store Paul's interest was in electrical engineering and in 1949 he received his degree. He worked in the infant computer industry for some years before obtaining his masters degree in 1959, after which he went to work for the RAND organization which carried out research for the US military.

With the cold war in full swing there were considerable worries about the vulnerability of the communication systems to nuclear attack. Baran was set the task of finding a method of producing a survivable and reliable network. He realized that a normal telephone network, with its exchanges where everything came together, was very vulnerable. In a series of papers written from 1961 to 1964 he described a mesh network with many nodes and no central control. Through this small messages would be sent by any available route. Unfortunately, at the time little notice was taken of these ideas.

Donald Watts Davies was born in Treorchy, South Wales in Britain, in 1924, but his father died shortly afterwards so the family moved to Portsmouth and he was educated there.[2] He then gained both a physics and mathematics degrees at Imperial College in London. This gave him an entry into the early work on computers with Alan Turing at the National Physical Laboratory (NPL) as described in Chap. 10. He was part of the team that completed the ACE computer after Turing left.

By 1963 he had risen to be technical manager of the advanced computer techniques project and was particularly interested in communication between the machines. In 1965, he had an inspiration about the best way to do this. He realized that the messages were what he called 'bursty'—they consisted of short bursts of information followed by relatively long periods when nothing happened while either the computer or the user digested it. This was very inefficient use of a dedicated connection.

J.B. Williams, *The Electronics Revolution*, Springer Praxis Books,
DOI 10.1007/978-3-319-49088-5_22

His crucial counterintuitive concept was not to send the complete message, which could be of any size, but to break it into short packets of fixed length which contained the address and information on their position in the complete message. These would then be passed by any route from node to node through the network. At the receiving end the packets would be reassembled into the message. He documented his thoughts, but it wasn't until March 1964 that he gave a presentation of his ideas to interested parties. There an attendee from the Ministry of Defence drew his attention to Paul Baran's work.[3] Though they had started from very different points, and were trying to solve very different problems, the solutions they had come up with independently were almost the same.

This was completely unlike the telephone system, where a connection is made between sender and receiver, and more like a postal system where something is broken down into pieces and each piece is sent addressed to the recipient and then reassembled there. What it meant was that the network was highly resilient and it could carry many times the traffic that a telephone connection could achieve.

At each node the packets had to be temporally stored before being sent on. The small size of the packets meant that the amount of storage needed was small and known. Also the delays to the messages through the system were minimized. Another advantage was that there was no central control and hence any method could be used to send on the packets. The NPL set to work to design such a system to connect all their internal equipment but, having to start from scratch, it took several years to build.

The launch of the Russian Sputnik satellite in 1957 had galvanized the Americans and one of the results was the setting up of the Advanced Research Projects Agency (ARPA) to try to ensure that they were not caught on the hop again. Obviously interested in communications, in 1962 they employed J.C.R Licklider, unsurprisingly known as 'Lick', to further explore his ideas on human–machine interaction. He propounded the idea of an intergalactic network which would connect people to the available computing power and programs.[4] This was very influential at ARPA even after he moved on in 1964.

By 1966, ARPA were under pressure from their contractors for more computing resources and decided that they should build on Lick's ideas and construct a network to connect all the computers at ARPA-funded research centers.[5] A young man, Larry Roberts, with a background in computer communications, was recruited to manage this project. During that year, Roberts met all the scientists working on ARPA-funded projects. They were very resistant to his ideas as it would consume a considerable amount of the available computing power just to run the network but, by the following year he was ready to share some of his ideas.

At an ACM conference at the beginning of October 1967, Roberts presented a paper on 'Multiple computer networks and intercomputer communication' which was an outline plan for what he had in mind. At the same meeting, Roger Scantlebury, who was running the NPL plan to implement Donald Davies' packet switching network, presented a paper of their detailed work. It was a revelation to Roberts, particularly as Scantlebury also referred to Baran's work.

After the meeting, Wesley Clark came up to him and said what he needed was to use small computers alongside the main machines to handle all the communications. Roberts adopted this idea and called them Interface Message Processors (IMPs).

A meeting was held between the three groups involved in this packet switching network idea, ARPA, RAND and the NPL, to discuss the detail.[6] Now all the basic building blocks were in place for them to start to build what would become known as the ARPANET. On Roberts desk was a well-thumbed copy of the detailed NPL report.[7] By 1968, he issued a specification to contractors to quote to build the IMP units. To ensure a reasonable response it was sent to 140 companies, but many didn't take it seriously and only 12 responded.[8]

The eventual winner was a small company called Bolt, Beranek and Newman (BBN). At first sight this seemed a strange decision, but the company was well known to ARPA and was full of people from the Massachusetts Institute of Technology (MIT). The trust wasn't misplaced as the first IMP was ready by August 1969 and installed at the University of California at Los Angeles (UCLA). In September, UCLA connected their host computer to the IMP and the first node of the ARPANET was ready.

This was an appropriate place to start as the person in charge there was Leonard Kleinrock, who had contributed much of the early theoretical work around computer networks. He had analyzed sending messages through them, but had not taken the final step to breaking up the messages into packets.[9] His contribution to the beginnings of the internet is thus slightly contentious.

Soon the next IMP arrived at Stanford Research Institute (SRI) and further ones at the University of California at Santa Barbara and the University of Utah. They were rapidly connected to their host computers and to leased telephone lines. The ARPANET was born. After a few teething troubles these organizations were able to access the computing power of all the others.

As the IMPs were based on commercial minicomputers, most of the work was in software and in particular protocols. In any sort of communication medium there has to be some rules to prevent chaos. When someone is speaking, generally others wait until they have finished and then have their turn. On a single radio channel, most people are familiar with the use of 'over' when they finish speaking, though in practice it is more complicated with the need to identify who is speaking.

For the network arrangements such as the ARPANET the whole thing completely depended on the protocols. Unlike a telephone system there were no exchanges or central controllers and thus each IMP only knew what to do with a packet of data when it read the instructions. These had to conform to a strict protocol or it would become confused.

In October 1972, APRA decided that the whole concept of their net needed some publicity.[10] It was still suffering from sniping by the telephone lobby which thought all this packet switch business was nonsense. They set up a demonstration at the International Conference on Computer Communications (ICCC) in a Washington hotel and invited every relevant person that they could think of.

In the center of the room, they had one minicomputer, which was a modified form of their IMP and designed to connect computer terminals with 40 terminals connected to it. Visitors were able to sit at one of the terminals and run demonstration programs on computers connected to the network on the other side of the country. It was such an enormous success that the participants had to be ejected after midnight each day or they would have stayed the whole week. It gave the ARPANET a boost and by the following year 23 computers were connected.

For a long time, users of one computer had been able to send messages to each other. Ray Tomlinson at BBN had the idea that it would be useful to be able to send similar messages over the ARPANET. He cobbled together the program that sent messages on his computer with a program for sending a file over the net and produced a crude email program. Larry Roberts took this and turned it into a usable tool; the idea of electronic mail was born.

By 1970, the NPL had their network running, but it wasn't until the following year that it was fully operational with 60 connections across the extensive NPL site. It seemed sensible to Larry Roberts that they should link the ARPANET and NPL networks together, but Donald Davies was completely unable to persuade the government to fund this.[11] The timing was bad as Britain had just applied to join the European Economic Community, and things European were in and things American were out.

However, Peter Kirstein at University College London (UCL) agreed to have a go, thinking that funding would be easier for a university to get than for the NPL, which was basically a government department. It turned out to be almost as difficult as neither the Science Research Council (SRC) nor the Department of Industry (DOI) would support the project. However, he managed to find a way. ARPA had a satellite connection to a seismic array in Norway, and for historical reasons there was a cable link from this to the UK.

Kirstein was able to get the Post Office to support the cost of the connection to Norway, Donald Davies managed a to get small amount of funding, and ARPA agreed to loan the interface processor which was one of the terminal variety known as a TIP. On this shaky foundation they agreed to go ahead. Kirstein had one advantage which made the project interesting to ARPA, which was that he already had a connection to the largest computer in the UK at the Rutherford High Energy Laboratory.

They had to construct interface computers as they couldn't put any software on the main machine, but in July 1973 they had the link running. By 1975, some 40 British academic research groups were using the link and Kirstein had stable funding. The DOI was even boasting about the connection of its Computer Aided Design center (CADC) to the ARPANET. In February 1976, the Queen formally opened the link between RSRE (Malvern) and the US, which of course went via UCL. Once something is successful everyone jumps on the bandwagon.

As time went on, the shortcomings in the existing protocols were becoming apparent. In 1972, Larry Roberts enticed Robert Kahn from BBN to come and work at ARPA to work on network communications.[12] Kahn soon became convinced that the architecture of the network needed to be more open so that any network could communicate to any other without the need for changes and only needing a 'gateway'. It was important that no changes should be needed to a network to connect it to others.

The following year, Vinton Cerf joined Kahn and they set to work to define a refined protocol (Fig. 22.1). The first version of what became known as Transmission Control Protocol (TCP) was written that year and fully documented by the end of 1974. ARPA then contracted BBN and UCL to implement versions to run on different host computers and it was demonstrated operating not only over ARPANET but a satellite and a packet radio network in 1977.[13]

They aimed for as much flexibility as possible while maintaining the openness to other networks and in 1978 it was split into two protocols: a slimmed down TCP and an Internet Protocol (IP) to deal with the different layers of requirements. The whole thing

Fig. 22.1 Vinton Cerf and Robert Kahn in later life. *Source:* https://en.wikipedia.org/wiki/Bob_Kahn#/media/File:CerfKahnMedalOfFreedom.jpg

was migrating to a multilayered protocol model and eventually reached seven layers. These were handled at different levels in the system. The lowest ones dealt with the physical moving of the packets while at the top were the actual applications that the user saw.

TCP/IP was gradually accepted as a standard, and that the ARPANET should be switched over to it. This was far from simple as everyone needed to change at the same time. Finally, after extensive planning, the system was switched over on January 1, 1983. It then became possible to separate the military network from the more civilian side of ARPANET.

One of the problems was that the hosts on the system were all known by long numbers of the form 123.45.6.78 which was fine for the computers but not easy for humans. In 1983, Paul Mockapetris at the University of California, Irvine, wrote the first implementation of what became known as the Domain Name System (DNS). This had the disadvantage that it need some sort of central look-up to convert from the human-friendly names to the numbers, hence cutting across one of the fundamentals of the net. However, this was largely overcome by having domains for each country and separate ones for commerce and academia. These are the familiar .com, .co.uk and .ac.uk or .edu endings.

The idea of networks connecting universities and research facilities was now well established. A great deal of the traffic was as email as this was a very convenient way of communicating with colleagues at other institutions. It had the advantage of the letter in that it could be dealt with when the person was available, and the immediacy of the telephone. The problem was that networks were growing up for specialized groups.

Britain set up the JANET network to connect universities and the American National Science Foundation (NSF) followed in 1985 with the NSFNET linking five supercomputer sites. Both countries insisted that the sites accept further connections so that the networks grew. This was so effective that by 1990 NSFNET had replaced ARPANET as the core of the Internet. It was much faster and ARPANET was now nearly 20 years old and showing its age.

The only problem was that, under the funding arrangements, the networks could only be used by academic institutions. It was clear that there was a demand outside this area, particularly for email, and a commercial email system was allowed to connect to it. In 1991, the definition of who could use the NSFNET was widened and finally, in 1993, it was allowed to be opened to full commercial use.

Then the first commercial Internet Service Providers (ISPs) began, who could connect users to the net for a suitable fee. It only took 2 years for this system to pay its way and the NSF could withdraw its financial support. At this point the infant Internet was on its own. It had no owner and still had the freewheeling attitudes stemming from its academic background. It grew exponentially as the number of hosts on the NSF system rose from 2000 in 1985 to more than two million in 1993.[14]

Emails, however, had a drawback. They could only send a text message which meant that drawings or other non-text information couldn't be included. Nathaniel Borenstein and Ned Freed teamed up to provide a method that could go beyond the simple ASCII text and cope with languages in other scripts, and also be able to attach documents or pictures.[15] Borenstein said that people kept asking him, 'Why do you care so much about putting media into email?' His reply was, 'I always said because someday I'm going to have grandchildren and I want to get pictures of them by email.' While people laughed at him at the time, it was very farsighted.

What Borenstein and Freed came up with was Multipurpose Internet Mail Extensions (MIME) which allowed other scripts but, crucially, the ability to attach documents or pictures to an email. This then needed to be approved by the Internet Engineering Task Force (IETF), which took some convincing but eventually incorporated MIME into the standards. While the email system worked reasonably well, trying to access information on a computer was a nightmare of incompatible programs. One place where this was becoming a serious problem was at the European Organization for Nuclear Research, more commonly known as CERN, the acronym of the French version of its name. Its research, both at their site in Geneva and at the various universities involved, generated vast amounts of data, but it was all on incompatible systems and there was no simple way of sharing and disseminating it.

In 1989, one man decided to tackle this. He was Tim Berners-Lee, an Oxford-educated Englishman who was both a physicist and software engineer.[16] His basic concept was to take the idea of hypertext and apply it to the Internet. Hypertext wasn't new, as an index can be considered a form of it, and it had been used to provide cross-references in electronic documents.[17] Berners-Lee's contribution was to apply the idea to linking documents on the Internet.

He wanted a system where information was all in a public space where anyone could access it without having to go through some centralized system. He also wanted to link text in one document with relevant information in another regardless of where it was stored in the network. To do this he realized that the documents needed to be in a standardized form which could incorporate the hypertext links.

To do this, each document needed a file name so it could be located, and this he called the Uniform Resource Locator (URL). The pages would be written in Hypertext Markup Language (HTML) and so could always be displayed with a suitable piece of reading software. The URLs took the form http://www.google.com, where the http specifies that this was the Hypertext Transfer Protocol and the :// separates the rest of the web address. By convention, this always accesses the 'index' page.

To enable the site to have more pages their addresses used the same entry address and then used /nextpage in the same way as computer file directories. Thus there was a very simple system which could produce easily readable pages anywhere on the Internet and they could be read from anywhere else. The existing ability of the Internet to move packets of data, without knowing what they were, meant that this system could easily be spliced onto it.

Berners-Lee's first proposal to the CERN management, put forward in 1989, received no reply. They probably didn't know what to make of it. A year later he resubmitted it, but this time he decided to go ahead without formal approval. By later in 1990, he had produced a crude browser/editor which he called WorldWideWeb that could decode and display the contents of an HTML file. He had also created the software for the world's first web server.

The prototype demonstrated the principles but was too complex for most CERN workers, so Berners-Lee produced a simpler browser. During 1991, other servers started to appear and the CERN one was getting 100 request for pages (hits) a day. A year later this was 1000 hits a day and by 1993 requests had reached 10,000 per day. Realizing it was a success, CERN released the software free as a service to the world. It was a decision that was to have very far-reaching consequences.

The next issue was how to view these new web pages. The National Center for Supercomputing Applications in Chicago released the first version of their Mosaic browser which, crucially, ran on all those personal computers which were now starting to appear. It could be downloaded over the Internet and within a few months a million users had done so. Eventually, those behind the development set up Netscape and in 1995 launched the commercial Netscape Navigator. Microsoft at last saw the importance of the net and replied with their Internet Explorer. Now web browsers became regarded as a normal piece of software to have on a personal computer and so the web really began to take off.

The problem with the increasing amount of information appearing on the web was how the user could find out what was there. Initially, a list of sites or a directory was sufficient but quite soon this became inadequate. How could users know what was there? One approach was to get site owners to register, but that depended on them knowing the registration site. Various attempts were made to overcome these difficulties but something more sophisticated was needed.

The idea was to use a 'web crawler' or 'spider' which rambled around the web following links and indexing pages that it found.[18] That solved the problem of what was there, but raised the further question of what to do with it. It is all very well to find all the pages with a particular word or group of words, but that isn't what the user wants. He needs the most relevant page, and that is where the real skill comes in.

The search algorithm is the key to the whole system and the one that really got this right was Google which was launched in 1998. Though the details are carefully guarded, some of the principles are known. A number of factors are used to produce a ranking and some of these are the number of links to a site, and the ranking of those links.[19] If multiple words are used then how close are these words to each other, and so on.

The result is an attempt to give the user the answer to what he really was looking for. Misspelt words are corrected, and synonyms used so that there doesn't need to be an exact match in order for the answer to still arrive. Though there are considerable numbers of search engines which work on similar principles, Google still rules the roost.

As was seen in Chap. 18, the arrival of the IBM personal computer and its clones produced a de facto standard for small computers, and they were being used in very large numbers, initially in offices but later in homes. This meant that it was relatively simple to produce programs to monitor emails or surf the web. The only other device needed was a modem to connect the computer to a telephone line. The large demand soon drove down the price and it became a mass market item.

As clever ways were found to squeeze faster signals down the ordinary telephone wires, the speed of modems steadily increased which made the web in particular more useful. To begin with, only very simple sites were acceptable as otherwise they just took too long to download over the slow communication channel. At the same time, the backbone of the Internet was constantly being upgraded in both speed and extent.

The result, of course, was that more and more people became connected. Like telephones there is a clear 'network effect'—as more and more users have email addresses and larger numbers of web sites appear then the potential user has an increased number of reasons to join in. As usual, things took off more rapidly in America, but in Britain some quarter of the population was online by the end of the century, with the numbers rising fast (Fig. 22.2). Considering the equipment that was needed and also access to a telephone line, it was extraordinary how rapidly it grew.

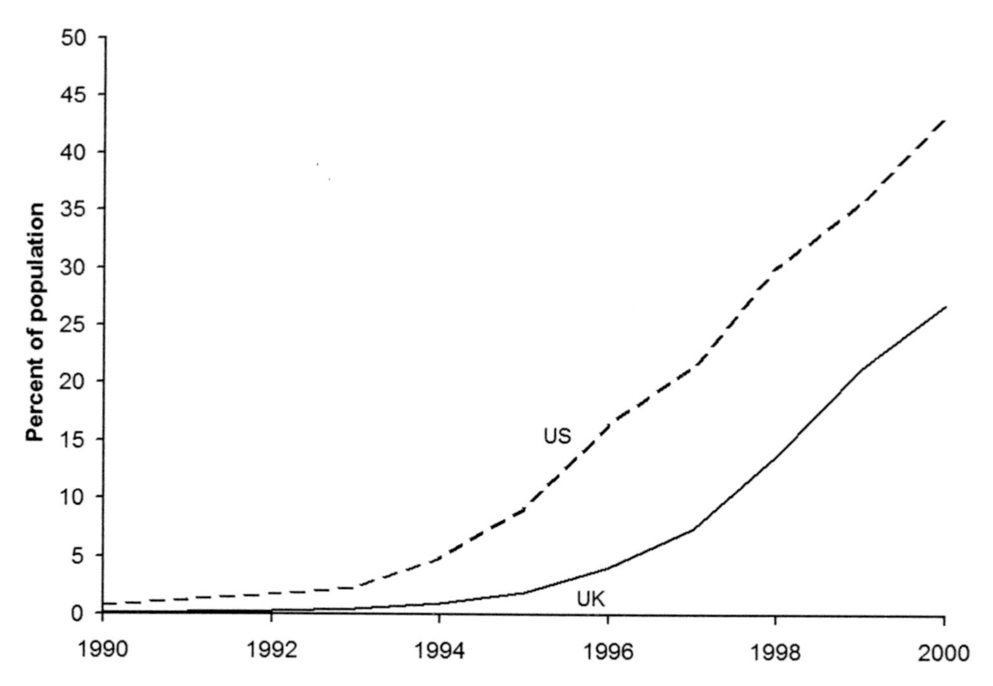

Fig. 22.2 Percent of the population in the US and UK with access to the internet, 1990–2000. *Source:* Author[20]

The real need was for much faster connections. The problem lay in that telephone systems were optimized for speech signals which only require a frequency range up to around 3.5 kHz. However, research had shown that the basic wiring to individual premises was capable of considerably more than this. A clever system was devised where a normal telephone conversation took up the low frequency end and then wide band channels were fitted in above this.[21]

The system was called Asymmetric Digital Subscriber Line or ADSL. The asymmetric part referred to the arrangement where data could be downloaded at faster speed than it could be uploaded. This matched the typical Internet use reasonably well. Surprisingly, this could all be shoved down an ordinary telephone pair of wires. Some small boxes were required at the customer's premises to separate the signals plus a special modem, and naturally more equipment at the exchange. The system worked as long as the user wasn't too far from the exchange.

By the late 1990s, the telecommunications bodies were ready to offer what was now called 'broadband' to the public. It appeared in 1998 in America, and the following year it was introduced by British Telecom or BT as it was now known. At first, the speeds were in the order of 512 kb download which was some ten times faster than what was being achieved with dial-up modems working over ordinary telephone lines. As time went on, these speeds steadily increased for those users reasonably near an exchange.

It wasn't only the increase in speed that changed, it was the charging system. The dial-up modems used the normal telephone line and were charged as such. Though this was normally a local telephone call and relatively cheap, all the time the user was online he was clicking up cost. With the broadband there was a fixed charge per month and then the system could be used for as long as was required without incurring extra costs. That boosted use further.

The convergence of computing and telecommunications technologies had produced something greater than either of them. There were two arms to this: a quick and simple means of passing messages which was email; and then a universal reference source which was well on the way to containing far more information than the greatest of libraries—the World Wide Web.

Though these were only just beginning to get into their stride at the end of the twentieth century, the potential was already visible. The web was exploding at a tremendous rate and email opened up ways of keeping in touch which paid no attention to distance. It was just as easy to communicate with the other side of the world as with someone down the road. It complemented television in bringing the world closer together.

NOTES

1. Dalacov, G. Paul Baran, available at: http://history-computer.com/Internet/Birth/Baran.html.
2. Dalacov, G. Donald Davies, available at: http://history-computer.com/Internet/Birth/Davis.html.
3. Campbell-Kelly, M. (1988) Data communications at the National Physical Laboratory (1965–1 975). *Annals of the History of Computing*, 9:3/4 .

4. Leiner, B.M., Cerf, V.G., Clark, D.D., Kahn, R.E., Kleinrock, L., Lynch, D.C., Postel, J., Roberts, L.G., and Wolff, S.S. (1997) The past and future history of the internet. *Communications of the ACM*, 40;:2, 102–108.
5. Dalacov, G. Larry Roberts, available at: http://history-computer.com/Internet/Birth/Roberts.html
6. Kristula, D. The history of the Internet 1957–1976, available at: http://www.thocp.net/reference/internet/internet1.htm.
7. Pelkey, J. An interview with Donald Davies, available at: http://www.historyofcomputercommunications.info/Individuals/DaviesDonald/InterviewDonaldDavis.html.
8. Pelkey, J. A history of computer communications 1968–1988, Chapter 2, Networking: Vision and packet switching 1959–1968, Intergalactic vision to Arpanet, available at: http://www.historyofcomputercommunications.info/Book/2/2.8-The%20RFQ%20and%20Bidding68.html.
9. Davies, D.W. (2001) An historical study of the beginnings of packet switching. The Computer Journal, 44:3, 152–162.
10. Pelkey, J. A history of computer communications, 1968–1988, Chap. 4, Arpanet: 1969–1972, The beginnings of computer networks, available at: http://www.historyofcomputercommunications.info/Book/4/4.12-ICCC%20Demonstration71-72.html.
11. Kirstein, P.T. (1998) Early experiences with the ARPANET and INTERNET in the UK, History of the UK ARPANET/Internet links, 28 July, 1998, available at: http://nrg.cs.ucl.ac.uk/mjh/kirstein-arpanet.pdf.
12. Dalacov, G. TCP/IP, available at: http://history-computer.com/Internet/Maturing/TCPIP.html.
13. Wheen, A. (2000) *Dot-dash to Dot.com*, p.135.
14. Zimmermann, K.A. (2012) Internet history timeline: ARPANET to the World Wide Web. *Livescience*, June 4, 2012, available at: http://www.livescience.com/20727-internet-history.html .
15. Brodkin, J. (2011) The MIME guys: How two Internet gurus changed e-mail forever. *Network World*, February 1, 2011, available at: http://www.networkworld.com/article/2199390/uc-voip/the-mime-guys--how-two-internet-gurus-changed-e-mail-forever.html?page=3 .
16. NNDB, Tim Berners-Lee, available at: http://www.nndb.com/people/573/000023504/.
17. Wheen, p.141.
18. Wall, A. History of search engines: From 1945 to Google today. *Search Engine History*, available at: http://www.searchenginehistory.com/.
19. Wheen, p.148.
20. Data from World Bank, Data, Internet users (per 100 people), available at: http://data.worldbank.org/indicator/IT.NET.USER.P2?page=5.
21. Knagge, G. Digital Subscriber Loop - DSL and ADSL, available at: http://www.geoffknagge.com/uni/elec351/351assign.shtml.

23

Glass to the Rescue: Fiber Optics

It is my firm belief that an almost ideal telecommunication network . . . going a long way toward eliminating the effects of mere distance altogether, can be made—but only, at least as now foreseeable, by optical means. I believe too that within and between the highly developed regions of the world such a technological revolution . . . will be demanded by the public by the start of the next century. It will be a real adventure, and lead too, to truly stimulating new challenges in human behaviour.

Alec Reeves, John Logie Baird Memorial Lecture, May 1969

Ever since the rise of the telegraph in the nineteenth century, the demand for communication over distances has become greater and greater. Telegraph signals can be as slow as they like, but the user's impatience caused them to be speeded up. Telephones needed fast signals—more bandwidth—and so on. Television required more than radio and so the need for ever more sophisticated means of communication became apparent as the twentieth century progressed. Developments in electronics kept pace until after the Second World War.

It was soon found that there tended to be a trade-off. A signal could be sent with a wider bandwidth but usually it needed to be amplified or 'repeated' every so often to restore it to a reasonable size. Thus the 'goodness' of a transmission system was often expressed as the product of the maximum signal frequency and the distance between repeaters—the bandwidth times the distance. This had risen from around 10 bits/sec.km in the early days of the telegraph to something like 1000 for telephones in the twentieth century.[1]

Whereas telephone signals had been sent down a pair of wires, this became increasingly difficult as the traffic increased. It meant that massive cables containing thousands of pairs of wires needed to be laid between towns. The next step was to have a coaxial cable (similar to a television aerial cable) and send a large number of telephone signals modulated onto different carrier frequencies along it. Despite the need for frequent repeaters it still had a better goodness factor, which had increased by about two orders of magnitude.

The next innovation was to use microwave links. Radar work during the war had pushed electronics technology so that frequencies in the GHz region were possible. The only problem was that at these short centimeter wavelengths the signal would only go in a

J.B. Williams, *The Electronics Revolution*, Springer Praxis Books,
DOI 10.1007/978-3-319-49088-5_23

straight line. This was overcome by erecting tall towers and aiming the antenna from one to the next. As long as these were line-of-sight then the arrangement would work.

Again, multiple channels could be fitted into the signal communication medium. By now, that could mean television channels as well as telephones, and the beginnings of a demand for data transmission. The microwave links produced another increase in the goodness factor to around a million, and though this was developed further to another couple of orders of magnitude, gradually the limits began to be reached. The question was: where should the industry look for further improvements?

The natural line of development would be to go to ever-higher carrier frequencies on the microwave links. However, as the frequency was increased it began to be affected by clouds, rain and fog. The lower frequencies, at longer wavelengths, are unaffected but attempts to go higher than 10 GHz ran into this problem. The solution, the telephone companies thought, was to put their signals inside a special sort of pipe known as a waveguide.[2]

Bell Labs started working on this in America, while in Britain the Post Office took up the idea. A commercial company, Standard Telephones and Cables, was also interested at their research laboratories Standard Telecommunications Laboratories (STL). Despite being part of the International Telegraph and Telephones Corporation (ITT), STL was based in Harlow in England, where research was controlled by the shrewd Alex Reeves.

Though he investigated the waveguide technology, Reeves soon felt it wasn't the right way to go. Though it could work if the guide was absolutely straight or went around very gentle curves, in practice tiny movements in the ground could cause small kinks that produced large losses of the signal. Besides, though it might be possible to lay such nearly straight waveguides in the US, it was totally impractical in crowded Britain.

Like many others, Reeves began to consider the possibilities of light. Here the wavelength is around a micrometer (μm)—actually visible light lies between 0.76 and 0.38 μm—which is 100,000 times smaller than the 10 cm of the microwave band they had been using. Hence the bandwidth, and also the possible number of channels, increased by that factor which meant that it was very tempting. The trouble was that they only had to look out of the window to realize how affected it could be by the smog and drizzle that were common in Britain. The question then was: could it be made to go through some sort of 'waveguide'?

In 1841, Daniel Colladon, a professor of physics at the University of Geneva, was trying to demonstrate fluid flows and the behavior of water jets. Finding this difficult to see, he arranged sunlight to pass through his tank and into the stream of water. Surprisingly, the light stayed within the water stream, following its curve downwards. The light was bouncing off the sides of the stream and staying within the jet by total internal reflection (Fig. 23.1).

Also in the 1840s in France, Jacques Babinet had demonstrated the same phenomenon. Though rather less interested in it than Colladon, he remarked that it also worked in glass rods, even following a curve in them. The glass at the time was not very transparent so this was of little interest. However, by the 1880s Collodon's basic principle was being used for spectacular displays of fountains where the colored light was fed into the jet of water and illuminated it as it sprayed upwards, producing a magical effect.

It took until the 1930s before it was realized that if the glass rods were made very thin—in other words, a fiber—then this would still work and transmit the light from a small point. Glass is quite easy to pull out into thin strands when heated to the correct

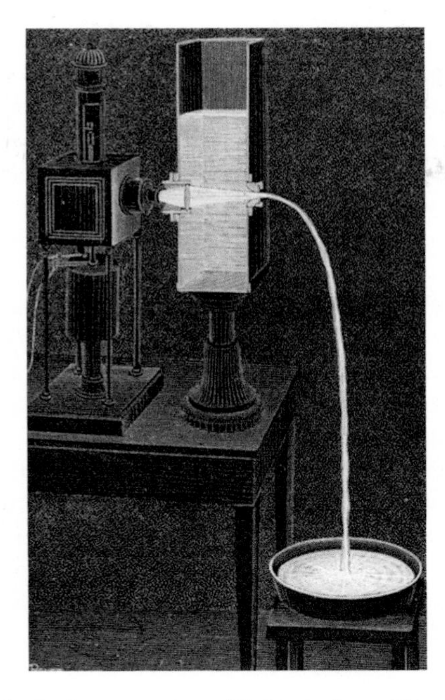

Fig. 23.1 Daniel Colladon's demonstration of light staying within a water jet, using an arc lamp in this version. *Source:* Hecht J, City of Light, p14

temperature. The idea was that if a bundle of fibers was used then each would carry the light from one spot and the bundle would transmit the whole of an image bit by bit. The trouble was that very little light appeared at the other end and everyone lost interest.

It took until after the Second World War for the mistake to be realized. For total internal reflection, essential for transmission of light along the fiber, it must be surrounded by a material with a lower refractive index. If the fiber was in air then there was no problem but, in a bundle, the fibers would be touching and this condition wouldn't be fulfilled. In addition, movement of the fibers over each other would cause scratching, further degrading the internal reflections. It was no wonder that little light came out of the far end.

In an issue of the respected journal *Nature* in early 1954, two letters appeared which moved fiber optics along again. The first was from professor van Heel at the physics laboratory of the University of Delft in Holland. He had experimented with bundles of fibers and had found the importance of surrounding each fiber with a material of lower refractive index.[3] However, this idea may have originally come from a suggestion by the American Brian O'Brien.[4]

The other letter, printed immediately below, was from Harold Hopkins and Narinder Kapany at Imperial College in London.[5] They had experimented with a much larger bundle of fibers with the idea of using it as an endoscope for peering inside the human body. In reality, they had only made a bundle 4 in. long and had thus avoided many of the real problems. They didn't seem to understand the importance of cladding the fibers which van Heel had pointed out.

Van Heel and Hopkins soon became involved in other fields and Kapany went to America where he made academic contributions to the subject of fiber optics. One person who had been inspired by these letters was the South African Basil Hirschowitz who trained as a gastroenterologist in London. He too went to America, to the University of Michigan. He saw the possibility of using the technique to make a gastroscope that was really flexible and could be passed easily down the throat into the stomach.[6] Perhaps they could even make an endoscope flexible enough so that it could be used to examine the intestines and other parts of the body.

At the university, Hirschowitz was allowed to pursue the idea and he recruited the help of optics professor Wilbur Peters and a young student Lawrence Curtiss to do the practical work. Although they had only a small budget they were soon drawing fibers, but then they had to relearn the importance of cladding. Hirschowitz and Peters thought that a lacquer coating should be sufficient but it didn't work well.

Curtiss was convinced that a layer of glass of a lower refractive index would be better, and while his collealgues were at a conference he shrank a glass tube onto the glass for the fiber and then pulled the composite material out into a thin fiber. It worked far better then any previously-made fibers. Only later was it discovered exactly how lucky he had been to get the combination precisely right at this first attempt. The way was now clear to make the bundles that they sought.

Hirschowitz immediately applied for a patent covering a gastroscope while Curtiss applied for one for glass-clad fibers.[7] The first gastroscope was tested in early 1957. A second, only 0.8 cm thick and a meter long, was demonstrated by Hirschowitz at a meeting of the American Gastroscopic Society in Colorado Springs in May. He then went looking for a manufacturer to make commercial versions.

American Cystoscope Makers Inc. agreed to manufacture them on the condition that they had assistance from the team, which in practice meant Curtiss. It still took time to source supplies of glass and production machinery much more sophisticated than the crude homemade equipment that had been used in the laboratory. Eventually, in 1960 Hirschowitz was able to test the first commercial device.

The new flexible fiber gastroscope was so superior to the semirigid lensed instruments the profession had been using, that other doctors soon adopted them. By the late 1960s they had almost completely replaced the original devices and were becoming important tools in investigations of many internal conditions.

Elsewhere short lengths of fiber bundles were used for many optical purposes. One application was to connect between two image intensifiers so soldiers could see in the dark. One tube had a curved output and the other a flat input. A suitably-shaped 'plate' of fibers could do the job that conventional optics could not. Fused bundles could also invert an image or split it into two. IBM used them to illuminate the holes in punched cards as part of their readers.[8] What linked all these uses was that the lengths of the fibers were very short.

This was the position when Alex Reeves at STL put together a team to examine the possibility of using light as a communications medium, spurred on by the huge potential bandwidth available. Other research groups had looked at this issue but soon abandoned it because glass just absorbs too much light. Though a sheet of window glass looks clear, if viewed end on it looks green and it is impossible to see through it. Any significant thickness absorbs far too much of the light.

In the early 1960s, STL decided to abandon the millimeter waveguide. This meant that Toni Karbowiak, well versed in the mathematics of waveguides, needed to look at other areas. Karbowiak had come from Poland to fight with British forces in the war, had stayed and earned his doctorate in the country.[9] He set his small group to examine whether it was possible to build an optical waveguide.

Unfortunately, Karbowiak's concept for the thin film optical waveguide turned out to suffer from the same problems as the millimeter ones: the signal didn't like going around bends, and leaked out.[10] Karboniak was lured away to an academic post in Australia, and a young man of Chinese extract, Charles Kao, was promoted to head the group.

One proposal doing the rounds was to have a succession of lenses in a 'light pipe' which could keep an optical signal going in the correct direction. Bell Labs investigated this, but STL rapidly rejected it as impractical. Once again it required great accuracy and the signal didn't like bends. It was not going to be any use in Britain.

After considering the other options Kao, and his colleague George Hockham, started to focus on the waveguide properties of optical mediums, and particularly glass. Everyone knew that glass absorbs most of the light in a length of more than a few meters, and Kao thought that if he could get 1% of the light out after a kilometer then that would be usable in Britain. The trouble was that the glasses he could obtain were so bad that he was down to the 1% after only 20 m.

Other research groups had given up, but Charles Kao did not. He asked the simple question: what was the theoretical loss figure for glass? Finding that no one seemed to know the answer, he set about investigating it himself. He and Hockham examined all the possible loss mechanisms in glass and came to the startling conclusion that there was no fundamental barrier to achieving the sort of target they had set themselves. In theory, the losses could be much lower than this and it would then be possible to see through hundreds of meters of glass.

The problem was, as many had suspected, that there were too many contaminants in the glass. For most purposes these don't matter; they were often added deliberately to lower the melting point of the glass and make it easier to handle. What Kao and Hockham realized was that a very pure glass was needed, but that was totally outside the experience of the glassmakers. However, ultra pure silicon had been outside everyone's experience, but the semiconductor industry had mastered this. In 1965, they wrote up their results, which were published as a paper in June the following year.[11] Not only did they claim that a fiber with sufficiently low loss was theoretically possible, but they showed that it needed to work in a single mode. This means that the fiber must be thin enough so that only a single mode of reflections occurs it otherwise the signals take multiple paths with different lengths and so broaden any pulse sent along it, cutting the possible bandwidth.

The difficulty with single mode is that the fiber diameter needs to be of the same order as the wavelength, which would be around 1 μm. This is really too fine to be practical, but Kao and Hockham came up with an elegant solution. The size of the fiber for a single mode depends on the difference in refractive index between the fiber and whatever is around it. By having a cladding with a refractive index just 1% lower than the core fiber, the size could be considerably bigger and of a practical dimension.

In one bound they had laid out the possibility of making light pass along thin fibers of glass and being able to realize some of that vast increase in bandwidth which was theoretically available. Also, they had shown that the fiber should be able to go around reasonable bends. There was only one problem: could anyone make glass that was pure enough to realize this dream?

Charles Kao was quite convinced and traveled the world trying to 'sell' the idea of thin glass fibers for communications. What he realized was that, unless some of the big players such as Bell Labs became involved, it was never going to take off. Despite his efforts there was not a lot of interest except in the UK where the Post Office Research Department,and Southampton University working for the Ministry of Defence, started to investigate it seriously.

Kao and his colleagues set about the difficult task of actually measuring the losses in various samples of glass. This loss is normally expressed in decibels (db) which is ten times the logarithm of the ratio. Thus the reduction to 1/100th that they were seeking was a loss of 20 db. A thousandth would be 30 db and so on. They carried on measuring samples of glass until they found one that only had a loss of 5 db at a useful wavelength around 0.85 μm which is just outside the visible range.

When these results became known the world started to take notice. What had seemed an impossible target suddenly became within reach. At this point, Bell Labs put more effort into the subject. One of the few glassmakers who started to take this seriously was Corning. They appointed Robert Maurer to research pure glasses.

Maurer had a simple philosophy which was to try something different from everyone else. Corning were a relatively small company, and just following the herd was a certain way to get crushed, but if they went their own way they might succeed. He asked himself a simple question: what was the purest glass? The answer, he knew, was fused silica—pure silicon dioxide. Everyone else turned away from this because of the difficulties of working with its very high melting point (1600 °C). Maurer decided to look at it, particularly as the company had some experience of working with it, and a suitable furnace.

He assembled a team and they worked steadily, gradually inching forward. Their system was to use a pure silica cladding and a core doped with some titanium dioxide. This wasn't ideal as the core should be of the purest glass but they needed the titanium dioxide to raise the refractive index that bit higher than the cladding.

After many trials, in 1970 Corning finally produced a quantity of good fiber. The loss was mesured at 16 db/km, comfortably below Charles Kao's target of 20 db/km. However, the material was brittle and not very practical, which gave Corning two dilemmas: whether to publicize the result and how to stay ahead of their competitors, who they assumed were not far behind. They didn't realize that they were a long way ahead.

Very coyly, Maurer announced the results at a conference on 'Trunk Telecommunications by Guided Waves' in London run by the Institution of Electrical Engineers (IEE).[12] The conference was dominated by the exponents of millimeter waveguides and there were few contributors on fiber optics, so Maurer didn't create a splash and only a very few realized the significance of his guarded remarks.

Of those who did, one was Stew Miller from Bell Labs and, importantly, others were the British Post Office and, of course, STL. These last two both invited Maurer to bring samples to be tested in their labs. Maurer took up the invitation of the Post Office as they were a potential customer, and returned with a small drum of fiber. Very cautiously, he let them measure the loss which confirmed his figure.

However, a piece of the brittle fiber broke off and became lost in the detritus on the floor. Though he searched for it, he couldn't find it. As soon as he was gone they swept the floor and soon discovered it and had it analyzed. At STL, they used a different technique of immersing the fiber in liquids of different refractive indexes to find one where the signal no longer went through the fiber. They couldn't find one and so they were sure the material must be pure silica.

The Post Office were even reasonably sure that there was also titanium dioxide present, and so despite Maurer's best efforts the secret was out. Of course they didn't know how it had been made, but it still pointed the way. At last it had been proved possible to fulfil Charles Kao's dream, though there was still work to be done on Corning's brittle fiber.

The appearance of low-loss fiber galvanized the Post Office, STL, Bell Labs and other companies around the world to redouble their efforts. Nevertheless, Corning were in the lead and within 2 years they had switched to germanium instead of titanium for the core doping and solved their brittleness problems. They had also reduced the losses to around 4 db/km and thought that 2 db/km was possible. This was much better than Charles Kao's 20 db/km target and proved that his theory on the ultimate limit was correct.

As the fiber work appeared to be coming along nicely, attention turned to the other parts needed to make an optical telecommunications system. There were problems of aligning light sources and detectors to the very fine fiber, and how to inject and collect the light. At around 0.85 μm, where there was a low loss point in the fibers, it was possible to use silicon diodes as detectors because they were sensitive in this region. The light source was more of a problem.

It was possible to use a Light Emitting Diode (LED) but it was not very bright and produced a range of wavelengths (or colors) which were not coherent. For a really effective system, like a radio signal it needed to be a single wavelength or frequency. Fortunately, in 1960 the laser had been invented. This is a light source that uses a cavity, usually between a pair of mirrors to pump up the light signal and produce a strong bean of tightly focused coherent light.

The early lasers only worked at very low temperatures and produced pulses of light, making them not very useful for fiber communications. A couple of years later, when practical LEDs appeared, the search was on to produce a laser from them. By 1964, a number of groups had succeeded, but the devices could only produce very short pulses or they would destroy themselves.

It was found that a more complex structure, with multiple junctions, would work. However, it took until 1970 for the semiconductor processing technology to catch up and this sort of device to be able to be constructed. In that year, two workers at Bell Labs, Mort Panish and Japanese physicist Izuo Hayashi, succeeded in making a device which would produce a continuous output at room temperature. American researchers were even more motivated when they discovered that a group in Russia under Zhores I. Alferov had also achieved this.

Other groups soon caught up, including STL, but all of their devices suffered from very poor lifetimes. They would work for a while and then gradually go dim, or would simply stop working after a very short time. Steadily over the next few years the various problems were solved and the lifetimes of the laser diodes increased to usable levels. At last, there was a practical source of light to use with the fibers. It was particularly useful in that the diodes produced a very narrow beam which was convenient for connecting to the thin fiber.

In the summer of 1975, lightning struck the antenna for the two-way radio communications system used by Dorset police in England, destroying the electronics equipment. While this was unlucky for the police, it was a stroke of luck for Standard Telephones and Cables (STC), STL's parent company. The chief constable wanted a system that wouldn't be affected in this way, and was recommended to look at fiber optics and talk to STC. The reasoning was that, unlike electrical connections, lightning doesn't pass along optical fibers so another strike would not destroy the replacement equipment.

STC had supplied some equipment for trials, but no commercial systems. The chief constable didn't care that it was new; he only wanted it to work and trusted STC to make it. Within a few weeks he had a system up and running and STC could boast of the first commercial operational fiber optic system in the world.

Meanwhile, Bell Labs were running a trial at Atlanta, In May 1976 they reported the success of this, which was more convincing to the sceptics than STC's police radio system. AT&T, Bell's parent, was finally convinced and in early 1977 started feeding the thin fiber optic cables into ducts in Chicago, taking a fraction of the space of the copper cables but supplying a greater capacity. On April 1, the first test signals were put through them and they were in full service 6 weeks later.

Bell Labs were horrified to find that General Telephone, the second largest US telephone company, had beaten them to it. They had put in a fiber optic connection between Long Beach and Artesia in California on April 22, though it was not as advanced as AT&T's. In Britain, the Post Office was also running trials near its new research center at Martlesham Heath, steadily increasing the speeds until they were much faster than the American systems. On June 16 they started using the system for real telephone traffic. It was clear that fiber optics had finally arrived.

All these systems worked at around 0.85 μm where the lasers and detectors were relatively easy to make, but the losses in the fiber were theoretically higher than slightly longer wavelengths, and practical fibers were getting close to those limits. What researchers had discovered was that at around 1.3 μm the losses in fibers were even lower, but this was beyond the capabilities of the lasers.

The Japanese now got in on the act. By 1978, they had a new laser that could work at 1.3 μm and fiber with losses below 0.5 db/km, greatly under Charles Kao's target, and capable of having repeaters at much greater distances. By August of that year, NTT had a system running which, in addition to the other advantages, was also running at even higher speeds and hence greater capacity. This encouraged others to switch to the longer wavelength and take advantage of the increased repeater spacing.

Soon organizations were moving to a third generation at an even longer wavelength at 1.55 μm where the loss was in the region of 0.2 db/km, which was at levels that could only have been dreamed of earlier.[13] This allowed the spacing between repeaters to be as much as 50 km. Not only that, but the dispersion at this wavelength was very low so even higher modulation frequencies could be used, giving greater throughput of information. The bandwidth multiplied by repeater distance 'goodness' factor was going up and up (Fig. 23.2).

All these schemes required electronic repeaters where the light signal was converted to an electronic one, amplified, cleaned up and then converted back again to light for the next length of fiber. There were those who thought that was inefficient. What was required was an optical amplifier. It happened that such a thing existed in the form of a laser without the

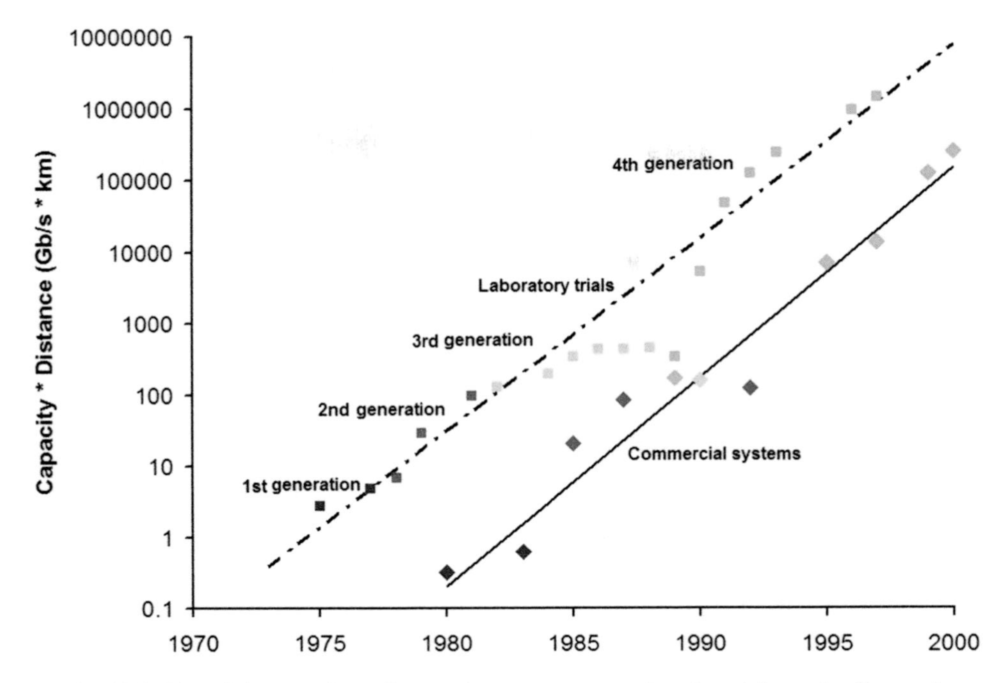

Fig. 23.2 Rise of the capacity × distance between repeaters 'goodness' factor for fiber optic cables. *Source:* Author[14]

mirrors. Suitable sections of fiber could be 'pumped' with a laser at a different wavelength and those within its working band would be amplified.[15]

This fourth generation had a further advantage in that it amplified signals over a range of wavelengths, meaning that a number of wavelengths or 'colors' could be sent along the same fiber, thus packing even more information through it. By using the same sort of techniques as had been used to get more information through the channels for mobile phones, as described in Chap. 21, and the highest frequencies the electronics could achieve, as well as the multiple wavelengths, the amount of data passing through the thin fiber could reach 1000s of Gb/s (Terabytes).[16] This was getting close to the 10,000 improvement from the microwave systems that had seemed theoretically possible.

There was a further advantage to the fibers. As long as a good fiber cable was laid it could be upgraded simply by changing the equipment at each end so that it used higher frequencies or more advanced arrangements to pack more on to it. All around the world, contractors were laying fiber cables and towing them through existing ducts, knowing that these could easily be upgraded without the enormous expense of continually digging up roads.

One of the areas that needed this greater bandwidth was between countries, such as across the Atlantic. The first conventional TAT1 cable had had 36 circuits, but by 1976 TAT6 added 4000 to the 1200 that already existed. Another 4000 was soon added in TAT7, but represented the limit of what copper cables could achieve. Satellites could do better but users didn't like the delay as the signal went out to the satellite and back. The first fiber cable TAT8, in operation at the end of 1988, added 40,000 circuits—a huge jump and a thousand times greater than the original TAT1.

During the 1990s, a whole network of high capacity links was built around the world. It culminated in 1997 in the Fiber optic Link Around the Globe (FLAG) system which linked the UK to Japan taking in 12 counties.[17] It joined the Atlantic cables to the Pacific ones around the world the other way, hence producing a complete network circling the globe. In practice, it wasn't a single cable but a complex arrangement with a good deal of redundancy to produce a secure network.

A measure of the phenomenal growth rate of fiber optical cables can be seen from the fact that in 1978 the total of all of it in the world amounted to 1000 km.[18] However, by 1984 there were 400,000 km in America alone. In 1988, the TAT1 cable was 6000 km, but the FLAG system amounted to 28,000 km. By the end of the 1990s there was a scramble to lay even more fiber optic cable, the total length reaching some 120 million km.[19] There was some danger that the over exuberance would greatly exceed the demand for telephone traffic which was rising but only steadily at some 10% per year, particularly after the 'Picturephone' championed by AT&T failed to interest the public.

However, there was another phenomenon which was threatening to take over—the Internet. The exponential rise of the Internet made huge demands on the communications network. For a brief period in 1995/1996, Internet traffic was doubling every 3 or 4 months. With the digital traffic rising faster than that from the telephone it was inevitable that by the late 1990s it should exceed it.[20] Fiber optic cables had arrived in the nick of time so that the growth of the Internet was unhindered by the bandwidth available in the communications network.

With the switch over to broadband from dial-up modems, this demand took a further jump. The available capacity and the Internet access provision got into a virtuous circle as broadband speed steadily rose. Fiber started to creep out from the communications backbone ever closer to the actual user.

Without the drastic increase in capacity brought by fiber optic cables, the Internet would have been strangled, and the instant communications that are so entangled with the modern 'one world' would have never developed in the way they have. Reliance on copper cables, and perhaps satellite communications, would have formed a brake on progress. Fortunately, glass came to the rescue, thanks largely to the determination and foresight of Charles Kao, who received a Nobel Prize in 2009 'for groundbreaking achievements concerning the transmission of light in fibers for optical communication'.[21]

NOTES

1. Agrawal, A.P. Fiber-optic communications systems, p.3, available at: http://www.qsl.net/kb7tbt/manuals/Ham%20Help%20Manuals/FIBER-OPTIC.pdf.
2. Hecht, J. City of light: The story of fiber optics, p.85, available at: http://image.sciencenet.cn/olddata/kexue.com.cn/upload/blog/file/2010/10/201010311703539253.pdf. This chapter relies heavily on this book.
3. Van Heel, A.C.S. (1954) A new method of transporting optical images without aberrations. *Nature*, January 2, 1954.

4. Hecht, p.48.
5. Hopkins, H.H. and Kapany, N.S. A flexible fibrescope, using static scanning. *Nature*, January 2, 1954.
6. Hecht, p.61.
7. Basil I. Hirschowitz, US Patent 3,010,357, Flexible light transmitting tube, filed December 28, 1956; issued Nov. 28, 1961; Lawrence E. Curtiss, US Patent 3,589,793, Glass fiber optical devices, filed May 6, 1957, and eventually issued June 29, 1971 after bitter opposition.
8. Hecht, p.74.
9. Hecht, p.88.
10. Kao, C.K. (2009) Sand from centuries past: Send future voices fast. Nobel Prize lecture, available at: http://www.nobelprize.org/nobel_prizes/physics/laureates/2009/kao_lecture.pdf.
11. Kao, K.C., and Hockham, G.A. (1966), Dielectric-fibre surface waveguides for optical frequencies. *Proceedings of the IEEE*, 113:7, 1151–1158.
12. Kapron, F.P., Keck, D.B., and Maurer, R.D. (1970) Radiation losses in glass optical waveguides. Conference on Trunk Telecommunications by Guided Waves, Institution of Electrical Engineers, London, 1970, pp.148–153.
13. Davis, C.C., and Murphy, T.E. (2011) Fiber-optic communications. *IEEE Signal Processing Magazine*, 28:4, 147–150.
14. Data from: Agrawal, p.6 and Kogelnik, H., (2000) High-capacity optical communications: Personal recollections, Table I and Fig 6. IEEE Journal on Selected Topics in Quantum Electronics, 6:6, 1279–1286.
15. ITU, Optical fibres, cables and systems, ITU-T manual 2009, available at: http://www.itu.int/dms_pub/itu-t/opb/hdb/T-HDB-OUT.10-2009-1-PDF-E.pdf.
16. Wheen, A. *Dot-dash to Dot.com*, p.100.
17. Welsh, T., Smith, R., Azami, H., and Chrisner, R. (1996) The FLAG cable system. *IEEE Communications Magazine*, 1996, 34:2, 30–35.
18. Ledley, K. Optical ribbon fibre in today's networks, ACFIPS, available at: http://www.acfips.org.au/Resources/PageContent/Files/optical-ribbon-fibre-resource-vi.pdf.
19. McWhan, D. (2012) *Sand and Silicon*. Oxford: Oxford University Press, p.127.
20. Beaufils, J-M. (2000) How do submarine networks web the world? *Optical Fiber Technology*, 6, 15–32.
21. Nobelprize.org, Charles Kao, available at: http://www.nobelprize.org/nobel_prizes/physics/laureates/2009/kao-facts.html.

24

Towards Virtual Money: Cards, ATMs and PoS

Money plays the largest part in determining the course of history.

Karl Marx, *Communist Manifesto*

Banking operations in themselves are reasonably simple. It is the sheer number of these transactions and the need to do them all within the same day that cause the difficulties. It was natural that the banks should look to mechanization as it became available. Despite being rather conservative by nature, these pressures meant that they were some of the earliest organizations to investigate computerization.

Traditionally, banks did all their accounting within the branch. When a transaction took place over the counter it then had to be added or subtracted from the totals for the day, but also be 'posted' to the individual's own account. A check (cheque) had to be 'cleared', ultimately sending it back to the bank branch on which it was drawn. By the 1950s, the increasing level of business was beginning to overwhelm the system.

Therefore, it was natural for the banks to start looking at the potential of computers as they became commercial, and the British ones did so in a herd. Though they had all spent a considerable time looking at the issues, quite suddenly in 1961 and 1962 a number of them opened their first computer centers. Barclays was first, soon followed by Martins Bank, with the other main players not far behind.[1]

All these systems merely automated the procedures that had been carried out in the branches, but now in a central computer. In practice, only a few branches were involved. As an example, Barclays' staff typed in the information on a machine in their branch, and this was converted to a punched hole paper tape. This was read by a tape reader which sent the data along a dedicated telephone line to the computer center. There it was converted back to a punched tape, checked and then fed to another tape reader which fed the information into the computer. Some other banks physically took the tapes to the computer center, avoiding the use of telephone lines.

All this creating of tapes was very inefficient, but at the time the computer storage was minimal and they couldn't store the data in any other way. Altogether, the system didn't make the savings that the banks hoped, but it did reduce the load in some branches where

J.B. Williams, *The Electronics Revolution*, Springer Praxis Books,
DOI 10.1007/978-3-319-49088-5_24

increasing business was causing difficulties as there was simply no more room to introduce the calculating machines.

In terms of the operation of the bank, it changed nothing. The information was fed to the computer center at the end of the working day. The machines worked in the evenings and as late as necessary to ensure that all the information was back in the branch by the opening time in the next morning. The early computers were ideal for this 'batch' operation.

One of the biggest problems that the banks had was the processing of checks. The difficulty is that, once received in a branch, the check must be processed, but that isn't the end of the story as it must be returned to the bank and branch where it was drawn. As the volume of checks increased this became a considerable burden. Also at the end of each day the banks had to settle between themselves the difference between all the payments and receipts.

A first step in dealing with this problem was made in 1955 by the Bank of America, which despite its name was a Californian bank. They worked with the Stanford Research Institute to introduce a check processing system called ERMA. The key feature of this was that the bank's sort code and the account number were preprinted on the check in a magnetic ink which allowed partial automatic processing. By 1960 the system was in full operation (Fig. 24.1).

It didn't take very long for British banks to follow suit, but first they had to standardize the size and type of paper used between themselves, and by 1962 they had reached agreement. Once they had done this they too began using the magnetic ink system and almost straight away the Westminster Bank opened its computer center for check clearing. Other main clearing banks soon followed.

For a long time, financial organizations had been looking for an alternative to cash and checks to make it easier for customers to pay in shops or restaurants. One of the first effective schemes was Diners Club, launched in 1950.[2] This could only be used in restaurants that supported the scheme but, despite this, the number of users grew from 200 to 20,000 in its first year. This was what was called a charge card scheme. After their meal, the user 'paid' by presenting the card and effectively the bill was sent to Diners Club. At the end of the month Diners Club recovered the total for the month from the user.

In 1958, American Express also launched a charge card, but Bank of America went one further by allowing the user longer to pay off the account but charging them interest for the privilege. This was a 'credit' card. The following year, American Express changed their cards from the usual cardboard to plastic and by 1963 this had reached Britain.

Bank of America's credit card was progressing and in 1965 they expanded the scheme by licencing their card to other banks. The following year, Barclays in Britain produced their Barclaycard which was based on the Bank of America system. At this stage, to make a payment the card was placed in a small machine which imprinted the card number on a set of three slips. One was retained by the retailer, one was handed to the customer, and one went to the bank for processing. Although the bank was still handling paper, these schemes were usually computerized from the start and handled by a single processing center.

1 2 3 4 5 6 7 8 9 0

Fig. 24.1 The 14 magnetic ink characters used on cheques so that they could be read by machine. *Source:* https://en.wikipedia.org/wiki/Magnetic_ink_character_recognition#/media/File:MICR_char.svg

Another area of interest to banks was the possibility of dispensing cash out of banking hours. In the UK, this was particularly driven by the bank union's desire for their members not to work on Saturday mornings. The banks resisted this as it would make it very difficult for many fulltime workers to ever get to a bank. An unsuccessful attempt was made in America in 1959 with a machine manufactured by Luther G. Simjian and installed in New York; it was hardly used and removed after 6 months.[3] The real development took place in Britain.

It was a race between the banks and their respective cash machine developers as to who could introduce the first machine. It was won by Barclays Bank at their Enfield branch in June 1967 with their machine made by De La Rue.[4] This used a voucher, which the user had to obtain in advance from the bank. It had a collection of holes in it which the machine read to check for a correct voucher. An electronic signature in the form of a 6-digit number validated the operation, allowing the user to get the cash. The voucher was then handled within the bank's system just like a check.

A month later, the Westminster Bank had their first machine in operation at their branch next to Victoria railway station in London. This was made by Chubb, the safes company, but the electronic system within it was the work of Smiths Industries. A design engineer called James Goodfellow had been given the task. The solution he came up with was to use a credit card-sized plastic card with punched holes in it and get the user to validate it by using a 4-digit code. The important point was that this was related to the number punched into the card by a complex process which could be checked in the machine. He obtained a patent for what became known as a PIN number.[5] The cards were retained by the machine and later returned to the customer for re-use.

Barclays had insisted on an exclusive arrangement with De La Rue and this meant that most other banks turned to the Chubb machine when they introduced their own cash dispensers. The exception was the Midland Bank which, insisting on exclusivity, turned to a small company called Speytec and installed their first machines the following year.

The basic difficulty with all these was that they had no knowledge of the state of the customer's bank account when the cash was dispensed, so the vouchers and cards could only be given to trusted users. Though the processing of the cash withdrawal was still a manual operation it did mean that customers could obtain cash out-of-hours, thus allowing the banks to begin closing their branches on Saturday mornings.

In the late 1960s, the British banks were concentrating on the fact that the UK currency was to be decimalized, changing from pounds, shillings and pence to pounds and new pence, due on February 15, 1971. The task of converting every one of the millions of accounts was a huge incentive to being fully computerized by that date. In the event, only one bank succeeded in having all their branches converted in time and that was Lloyds Bank.[6] All the others still had branches where the conversion had to be undertaken manually.

With the arrival of the third generation of computers, particularly the IBM System/360 series, the banks could see the possibility of reaching their ultimate goal of having 'online' access to customers' accounts which would be updated immediately a transaction took place. This required that, instead of the computer proceeding in a stately way through the batch task it had been set, it needed to respond to 'interrupts' from terminals from the branches, disturbing what it was already doing.

This required programmers who were used to computing power applied to machinery rather than the world of the large mainframes used by the banks. The result was that everyone had considerable problems with the transition, which was compounded by some of the manufacturers being unable to supply reliable machines in the timescales they had promised.

The first real moves to online real-time systems occurred not in the high street banks but in savings banks. In America these were the East Coast Savings and Loans institutions, but the impetus moved to Europe and in Britain to the Trustee Savings banks. One of the driving forces was to be able to offer ATMs in institutions which could not offer overdraft facilities.[7] It was thus imperative to be able to check the customer's account before dispensing cash. At the same time, counter services updating passbooks could modify the central accounts instantly, hence reducing a great deal of bookkeeping work.

One of the reasons that the Trustee Savings banks and building societies could more easily convert to online systems was that their operations were simpler than those of the high street banks. However, Lloyds Bank had developed their computer systems in advance of their competitors and in 1972 they introduced the first online Automated Teller Machines (ATM).[8] Unfortunately, because of problems with obtaining sufficient reliable telephone lines to connect the ATMs to the computers, the ATMs often did not check the account before dispensing the cash, and simply stored the transaction for later processing.

Slowly, as the decade progressed, true online machines that could look at the account before dispensing the cash started to appear. The card had a magnetic strip containing an encoded form of the account and branch sort code that was returned to the user once the transaction had taken place. With this real-time connection the ATM could extend its services to include such things as giving the balance of the user's account.

An interesting development was the subtle change in the users' attitude to obtaining cash. If they had to exchange something for it, such as a check over the bank counter or a card which was swallowed by the machine, there was still this sense of 'paying' for the cash. With the returnable card there was somehow a feeling that the cash was 'free', even though they knew perfectly well that the amount was being deducted from their bank account.

At this point, the ATMs were tied to the particular bank or institution that operated them, but when in 1978 credit cards were standardized with the magnetic strip they became usable in the ATMs and the way was open for a wider use of the machines. In 1985, the LINK network was set up where building societies and other smaller financial institutions grouped together to set up an ATM network. The advantage was that their customers had access to a greater number machines than each individual organization could afford.

The key point was that the LINK organization ran the computer that coupled to the ATMs, only going back to the individual organizations for information. Soon there was also the MATRIX group and the high street banks started doing deals whereby customers could use the machines of the other institutions. Steadily the systems became interlinked. This also allowed the placing of ATMs in places other than bank branches, such as in supermarkets.

In the UK, the networks steadily became connected until finally there was a single network for the whole country, so that a customer could go anywhere and find a machine which would accept their card and allow them to withdraw cash. Though there were attempts to charge customers of a bank for using other organizations' machines this was

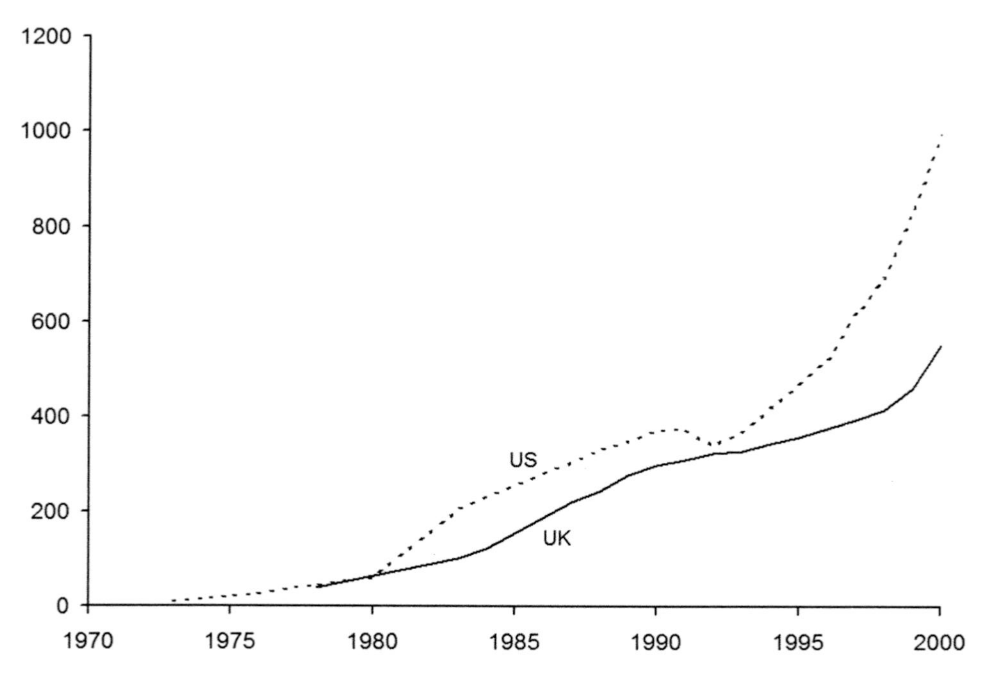

Fig. 24.2 The numbers of ATMs per million of population in the UK and US, 1970–2000. *Source:* Author[9]

resisted and soon dropped. However, in America the situation was very different, with a large number of competing networks which accounts for the larger numbers of machines per million of population (Fig. 24.2).[10]

With the on-line banking systems more-or-less under control by the 1980s the UK banks explored the possibility of customers making direct payments in retail outlets. At first a cooperative approach was followed, but competitive tendencies came to the fore. Really the banks would have liked to offer only their own systems, but the retailers obviously needed to be able to accept cards from any bank, so some degree of cooperation was necessary.[11]

Though they all had the goal of a cashless and checkless society, the banks and retailers continued to squabble until the mid 1980s. Then some 'trials' were held by individual organizations, but it wasn't until 1987 that they all started pointing in the same direction with the setting up of EFTPOS UK Ltd. Though the costs of implementing the system were great the potential savings were enormous.

One of the drivers was the deregulation of the building societies so that they could compete with the high street banks. The cosy cartel could no longer hold, and so it was essential to get the whole system operating. The details of debit cards were standardized but almost immediately rival systems of 'Connect' and 'Switch' cards broke out. The banks thus to some extent went their own way while at the same time publicly supported the EFTPOS UK Ltd. initiative.

Gradually this changed as the banks built up their customer bases. The only cooperation that remained was the need to 'clear' the payments between the organizations, thus losing the advantages of online systems. If it hadn't been for the Visa and Mastercard credit card networks the whole system would have degenerated into chaos. Nevertheless, slow progress was made, but a large percentage of transactions still involved cash.

By the early 1990s deals had been struck between the various schemes, and though there were different arrangements for Mastercard and Visa credit cards and Switch debit cards, they were all usable in the retailers' machines. The problems of the interworking were handled behind the scenes. In practice, many retailers were not actually 'on-line', with the transactions collected together by their own computers and sent to the banks either by tapes or electronically at the end of the working day. Only gradually were more 'real time' systems implemented (Fig. 24.3).

From the banks' point of view this was satisfactory; their costs dropped markedly as more transactions were handled by these electronic systems. They continued to try to phase out checks altogether, but some organizations were not big enough to use the expensive point-of-sale (POS) terminals, where checks continued to be essential. The potential loss was fiercely resisted and the banks backed away. They were reasonably happy with the situation as by 1998 debit card payment exceeded check payments.

By 2001, more than half retail spending was on plastic cards; customers had got used to them and rather liked them. It was simple to push a card into a slot and sign the slip. The pain of the money going was barely felt—it was somewhat unreal as though it wasn't

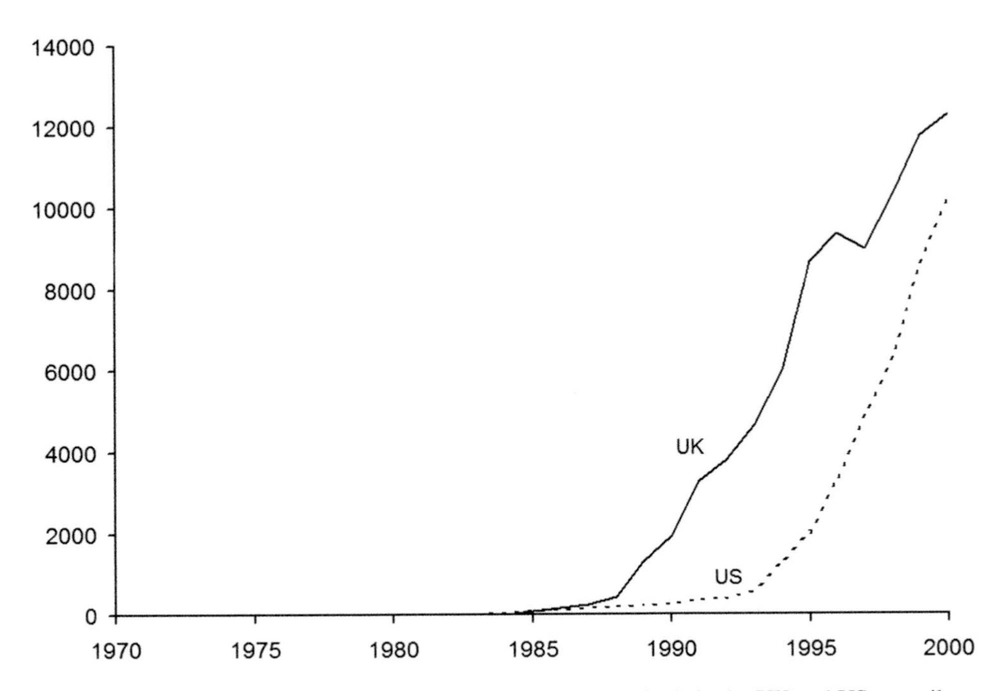

Fig. 24.3 The number of electronic point of sale (EPS) terminals in the UK and US per million population, 1970–2000. *Source:* Author (see note 9)

actual spending. Money was becoming virtual, and for some it was far too easy for things to get out of control and large debts pile up.

In the US, with its less integrated banking system, retail EPS terminals were much slower to be adopted. The American system lagged the UK by some 5 years. There the check was king and the take-up of electronic systems around the world was skewed by the slow take-up in the US.

The next step, chip and pin, only appeared in the UK at the beginning of the new millennium.[12] This was a more secure system in that, instead of the user having to sign a slip, they put in a 4-digit PIN number. This, of course, referred back to early ideas for security, and had only taken more than 30 years to come into full use.

The banks hate cash. It requires time-consuming counting and processing; it uses valuable counter time in the branches. Despite all their efforts to eliminate cash, people stubbornly clung to its use, and the amount used in day-to-day transactions refused to shrink at anything like the rate the banks would have liked. It was time for the banks to look for other ways to wean the customers off it, if they could.

In 1990, a scheme called 'cashback' was launched. Large retailers, such as supermarkets, offered the user the opportunity to have some cash when they used a debit card to pay for the goods. At first sight, it seems strange that the banks encouraged a greater use of cash, but they liked it. To them, it was 'free' as they were processing the transaction anyway, and it meant that the amount of cash passing across their counters was reduced. The retailers liked it because it reduced the amount of cash in their tills that would have to be processed. So something that was sold as a benefit to the customer was in the interests of the banks and retailers.

In 1995, there was another attempt to tackle the cash issue when the National Westminster Bank ran a trial for a smartcard called Mondex, which was a form of electronic cash.[13] Though the bank, and some retailers, liked it, the public was not enthusiastic and in the end the idea was quietly shelved. There was something about handling cash that people liked. You know where you are and how much you are spending and have got left. It was much easier to control your spending with real cash.

Another area that the banks were keen to automate was the paying of bills, such as to utilities. By the late 1990s, the World Wide Web was beginning to take off so the temptation was to use that. In 1999, Barclays were ready and launched their internet banking. This allowed the customer, should they so wish, to look at their account and make payments 24 h a day. It was sold as a great asset to the user but the advantage was mainly to the bank as the transaction was handled entirely by the computers.

Barclays were quickly followed by Lloyds Bank, but they had considerable problems getting it to work properly.[14] Once the dust had settled the other banks were forced to follow this lead. Some younger people took to this with enthusiasm but the growth of internet banking was slow. Like so many things to do with money, most people are quite conservative and continue doing the things they have always done.

Spending using cards and other automated methods grew steadily and by early this century these methods exceeded cash as the most popular method of payment.[15] The banks were still pushing these systems to try to reduce the amount of cash they handled across their counters, as this was a considerable cost. For the customers, there was a steady disconnection between the reality of cash and this virtual money which flowed about just as numbers in computers.

Thus the coming together of computers and communications has had a drastic effect on something as simple as paying for things. The ledgers that banks used to keep disappeared into their computers and transactions took place without anything tangible being seen. For some people, the downside was that it was easy to lose sight of how much they were spending, and as a result they piled up large debts. The banks, of course, were happy to lend to them. They had little understanding that this was dangerous, as time was to show.

NOTES

1. Martin, I. Centring the computer in the business of banking: Barclay's Bank and technological change 1954–1974. PhD thesis, University of Manchester; Lloyds TSB History available at: http://www.lloydstsb-ppi.co.uk/lloyds-tsb-history; Booth, A. (2004) Technical change in branch banking at the Midland Bank 1945–75. *Accounting, Business & Financial History*, 14:3, 277–300.
2. History of cards, available at: http://www.theukcardsassociation.org.uk/history_of_cards/index.asp.
3. Bátiz-Lazlo, B. and Reid, R.J.K. (2008) Evidence from the Patent Record on the development of cash dispensing technology, History of Telecommunications Conference, 2008. HISTELCON 2008. IEEE, also available at: https://mpra.ub.uni-muenchen.de/9461/.
4. Bátiz-Lazo, B. (2009) Emergence and evolution of ATM networks in the UK, 1967–2000. *Business History*, 51:1, 1–27.
5. Goodfellow's patent was filed on May 2, 1966 (GB1197183).
6. Booth, A. Technical change in branch banking at the Midland Bank 1945–75. *Accounting, Business & Financial History*, 14:3, 277–300.
7. Bátiz-Lazo, B., Karlsson, T. and Thodenius, B. Building Bankomat: The development of on-line, real-time systems in British and Swedish savings banks, c.1965–1985, available at: http://ssrn.com/abstract=1734004.
8. Bátiz-Lazo, B. (2009) Emergence and evolution of ATM networks in the UK, 1967–2000. *Business History*, 51:1, 1–27.
9. Data from: Bank of International Settlements, Committee on Payments and Market Infrastructures, available at: http://www.bis.org/list/cpmi_all/sdt_1/page_1.htm, and following pages.
10. Hayashi, F., Sullivan, R., and Weiner, S.E. (2003) A guide to the ATM and debit card industry, available at: http://info.worldbank.org/etools/docs/library/83934/Hayashi_April2003.pdf.
11. Howells, J., and Hine, J. (1991) Competitive strategy and the implementation of a new network technology: The case of EFTPOS in the UK. *Technology Analysis & Strategic Management*, 3:4, 397–425.
12. A history of banking: from coins to pings, *The Telegraph*, June 1, 2015.
13. Horwood, R. From quill pen to rocket science. (1996) *Computing & Control Engineering*, 7:6, 264–266; Mondex, available at: https://en.wikipedia.org/wiki/Mondex.
14. Elliott, H. No accounting for Lloyds online. *The Times*, July 21, 1999: p.10.
15. UK spends record amount spent on plastic cards. *Card Technology Today*, July/August 2007.

25

Saving TV Programmes: Video Recording

Wouldn't it be nice if our lives were like VCRs, and we could fast forward through the crummy times?

Charles M Schulz, 'Peanuts'

Film provided a convenient way of recording moving pictures. However, with the rise of television it was not very satisfactory to convert the TV images to film, which required the time-consuming business of processing it. To regenerate television pictures required a further conversion back from the film. This was not something that could be done quickly; it produced poor quality pictures, and was inconvenient for regular use.

In America, with its different time zones, this was a particular problem for broadcasters wanting, for example, to transmit news at a convenient time in that region. In practice, it meant that the newsreader had to repeat the program, sometimes several times. There was a great need for a simple machine that could record the television picture and play it back at any time that it was required.

The development of tape recorders for audio after the Second World War, as described in Chap. 7, seemed an obvious place to start to try to make a video recorder. The problem is that the range of frequencies it must record is at least a hundred times greater.

The most obvious way to achieve this is simply to send the tape past the recording and playback heads at high speed. This was the approach taken initially by Bing Crosby Laboratories, RCA, and the BBC in Britain. The BBC developed their Vision Electronic Recording Apparatus (VERA) machine and demonstrated it in 1956.[1] The tape traveled at nearly 21 m/s and was wound on to 53 cm-diameter reels. Despite their huge size, these could only hold 15 min of program of the 405-line black-and-white standard then in use.

Unbeknown to them, a small company in America called Ampex had a better idea. If the signal could be scanned across the tape as well as it moving along, then the required frequency range could be achieved at a reasonable tape speed. The result, the Mark IV, was a machine that only used 25 cm reels for a one hour program with good picture quality. When it was launched in 1956, despite selling at $50,000, it immediately became the standard, and most other developers gave up their efforts.[2]

J.B. Williams, *The Electronics Revolution*, Springer Praxis Books,
DOI 10.1007/978-3-319-49088-5_25

The Ampex system, the 'Quadruplex' transverse scanning method, used four heads mounted on a disc which spun at right angles to the direction the tape was moving.[3] The result was strips of magnetization which ran across the tape but tilted at a slight angle. With their use of 50 mm-wide tape, this gave the required bandwidth with a tape speed slow enough to give an adequate playing time on reasonably size reels.

Once the basic idea of scanning across the tape was established, a large number of variants were tried, and development work continued particularly on the helical scan system which produced diagonal tracks on the tape. Though it had been invented first, it wasn't until into the 1960s that reliable systems were achieved. In 1965, Ampex produced an improved machine using this, which only required tape 25 mm wide. The next step was in 1976, when Bosch introduced a machine also using 25 mm tape which found wide use among European broadcasters, while Sony and Ampex collaborated to produce a machine for American and Japanese TV stations.

Without these recorders it would have been difficult for television to develop in the way it did. Otherwise, programs either had to be 'live' or to be on film and broadcast via a telecine machine which converted film to a local TV standard. With machines such as the Ampex, recordings could be made and simply edited to produce the final program. This gave far greater creative freedom to producers.

In the early 1960s, Sony in Japan started to look at video recorders with a view to making one for the home market, which was where their interest mainly lay. By 1963, they had produced their first machine. It was still reel-to-reel, but though much cheaper than the broadcast machines it was still out of reach for most households.[4] As a result, these video recorders were mostly used by businesses, medical officials, airlines and the education industry.

Within the next 2 years Sony produced the next model, the CV-2000; it was still reel-to-reel and, though smaller and cheaper, still didn't quite reach the intended market.[5] The dream of producing a machine for use in the home still eluded the manufacturers. Partly this was to do with price, but the complexities of threading the tape from the reel was also a drawback. What was needed was something simpler.

The answer was to follow the lead of the audio cassette where all the tape handling took place inside a plastic unit which could simply be dropped into the player and recording or playback started. The question was: could something similar be produced for a video recorder and produced cheaply enough for households to afford?

By 1970, Sony had produced their U-Matic machine which used a special cassette, and could handle color television signals. It was the first cassette machine on the market and the cassettes could record 90 min of program. However, despite all their efforts, the price was still a little too high for their real objective of the home market and it required a special monitor rather than a television set to view the pictures. It still wasn't the ideal product.

The following year, Philips got into the game with their N1500 machine, which used a different cassette.[6] It was a little closer to the requirements of a home machine as it had a built-in tuner which meant that it could record programs straight from air. It also had a timer which was valuable for presetting the machine to record a program. Despite this, few homes had them and they were still mainly used in schools and businesses.

Back at Sony, Masaru Ibuka, the technical driving force, could see the next step to get closer to his dream of a truly home VCR. The cassette should be the size of a paperback book and should be able to record a full television program in color. To achieve this, a more sophisticated recording system was needed which packed more information on to the tape, meaning that the smaller size would still last one hour. The machines could be connected to a TV set to watch the recorded program.

The resulting product, launched in 1975, was called Betamax; priced at about the same level as a large color television it was thus affordable for home use. The price of the cassettes was quite low and Sony thus appeared to have reached their goal.

The Betamax was marketed as a 'timeshift' device. Off-air programs could be recorded at one time and watched at another. The machines started to sell steadily but the number sold in the first year was less than 100,000 which, while considerable, was not the real level of mass production that Sony wanted to reach. Just as they thought that the product was ready to take off they had two nasty surprises.

The first was that Universal Pictures took Sony to court in California on the grounds that the machines would cause copyright infringement by the users and Sony were liable for this.[7] Sony won the case but Universal appealed to the Circuit Court which overturned the decision. Sony then took the case to the American Supreme Court, which eventually found in their favor by a close five to four majority. The whole process took 8 years but eventually the threat, which could have stopped the sale of video recorders in the USA, was removed.

The other surprise was that, in September 1976, Japan Victor Company (JVC) introduced a video recorder using their Video Home System (VHS).[8] It used a cassette that was a little larger than the Betamax and was not compatible (Fig. 25.1).[9] What offended Sony was that they had disclosed some of their technology to other companies in an attempt to get them to use the Betamax as the standard. When they opened the VHS recorder it was clear that some of this had been incorporated.

The industry in Japan divided into two camps. Sony, Toshiba, Sanyo Electric, NEC, Aiwa, and Pioneer supported Betamax, while JVC, Matsushita, Hitachi, Mitsubishi

Fig. 25.1 Betamax cassette (*top*) and VHS (*below*) and a Philips Video 2000 cassette (*right*). *Source:* https://en.wikipedia.org/wiki/VHS#/media/File:Betavhs2.jpg; https://en.wikipedia.org/wiki/Video_2000#/media/File:Grundig-Video2000-VCC-Kassette-1983-Rotated.jpg

Electric, Sharp, and Akai Electric went for VHS. Battle lines were drawn and a destructive competition ensued. The failure to agree on a common standard, despite Sony's efforts, was a drawback in the marketplace as many users held back from purchasing machines until they could see the likely winner.

The problem was that, though the picture quality of the VHS was probably not quite as good, it was adequate. The machine was simpler and therefore cheaper to make. In addition, JVC had a different conception of the market. Rather than just pushing the machines for timeshift where users recorded their own material, they concentrated on prerecorded tapes of films or television programs. The fact that the VHS tapes held 2 h of material rather than Betamax's one was a distinct advantage.

Suddenly, users could watch their favorite old films from the comfort of their own armchairs, or use a popular program to babysit the children. A whole industry grew up to supply the tapes which were either sold outright or rented. Rental companies such as Blockbuster grew rapidly, providing a wide range of titles that could be hired for a period and then returned. Of course, the idea was that in doing so the user would see another film they wanted to watch, and so on. It was very successful.

In 1979, Philips launched their Video 2000 system which used a cassette very nearly the same size as the VHS but was incompatible. Though this had a number of features that made it superior to VHS, such as 4-h tapes, the machines were expensive, they were too late into the market and the rivals were too entrenched for yet another standard.[10] It was a further example where grip on the market was more important than the technical performance.

It was the domination of the rental market that was decisive in the ensuing standards war. At first, it was unclear who would win but steadily VHS gained ground and, despite various attempts by Sony and their allies, by the mid 1980s it was all over. Philips gave up in 1985 and by 1988 Sony were also manufacturing VHS recorders. The market had in effect imposed a standard. There were many unhappy owners, particularly of Betamax machines, who found that they had backed the wrong horse.

The take-up of VCRs was quite rapid. In 1984, some six million American households had a machine—about 7.5% of them.[11] In 1985, VCRs were selling at the rate of almost a million a month.[12] The projection was that ownership would reach 25% by 1990, but in the event it was more than double that. Growth was so fast that 85% of US households had at least one VCR by 1996 and the rental market was $1 billion a year from a standing start in 1977 (Fig. 25.2).[13]

What was more remarkable was that the rise was even faster in the UK. Despite a slightly later start, the take-up almost immediately overtook the US (though later America was to catch up). Two events drove this: first was the royal wedding of Prince Charles and Lady Diana in 1981, and then the football World Cup the following year.[14] Even taking these into account, it is extraordinary that Britain should be ahead of most counties in the percentage of households with VCRs. At the beginning, renting the machines was popular, but as prices fell this steadily changed over to nearly all being purchased.

The cassettes were sold as blanks for the users to record their own programs, or already prerecorded. These prerecorded tapes were also available for hire. At first rental was very popular; after they had first been shown in cinemas, many films were made available on cassettes for hire. The large film distributors thought initially that if they sold the films

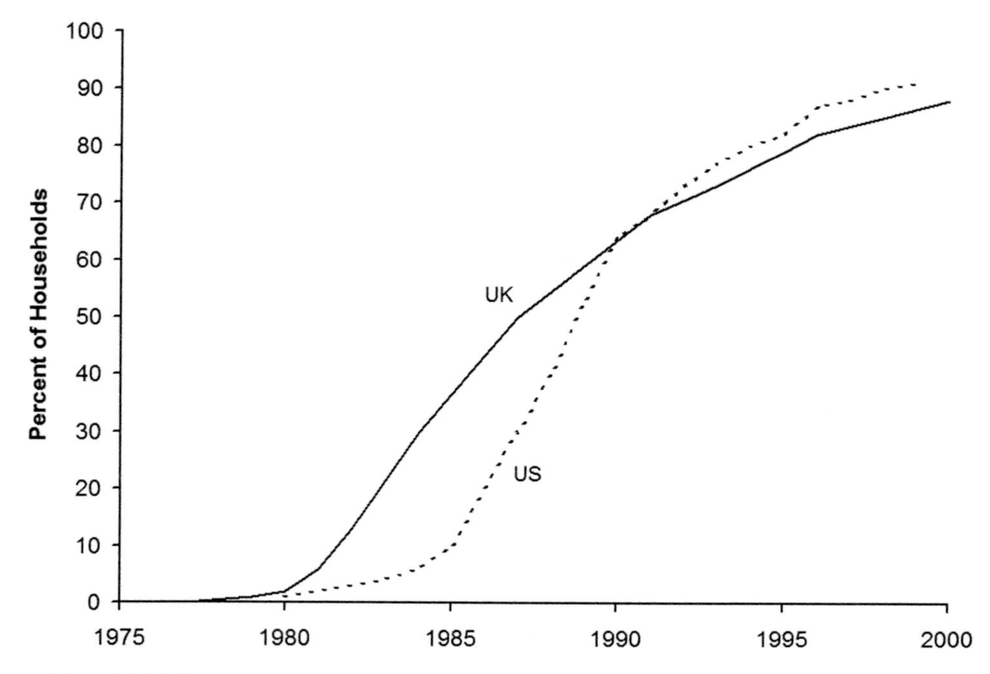

Fig. 25.2 Rise of video cassette recorders in UK and US households, 1975–2000. *Source:* Author[16]

outright this would affect the overall earnings that they could make from the film. Later, they changed their minds and sales started to take off.

Feature films dominated the prerecorded market for hire and sales.[15] Most popular were the 'blockbuster' films, particularly thrillers or drama, though comedy and adventures came close behind. Contrary to what the industry thought, the video market rather increased the attendance at cinemas than detracted from it. Advertising created a greater 'buzz' and this generated larger revenues as some people preferred to watch the film at home . It was much cheaper for a family to hire the film and all watch it rather than having to buy cinema tickets for everyone.

The rental market tended to be dominated by the younger age groups looking for an evening's entertainment.[17] The sudden reversal of the rental market in around 1990 was probably due to this group finding other things to do, such as play computer games (Fig. 25.3). There is a tendency for users to hire more programs when they first have the machine and gradually decrease as time goes on.[18] Also the rise of sales of cassettes bit into rental. If you could buy the same thing and wanted to watch it more than once, it soon paid to purchase it. The suppliers were happy as there was more money to be made from direct sales than selling to the rental trade.

Just as the video cassette market really thought that it had it made, there was a cloud on the horizon—digital. In the early 1980s, the Compact Disk (CD) had been introduced and steadily replaced the audio cassettes, as described in Chap. 7. This was partly due to the improved quality of the reproduction, but they were also more convenient and took up less

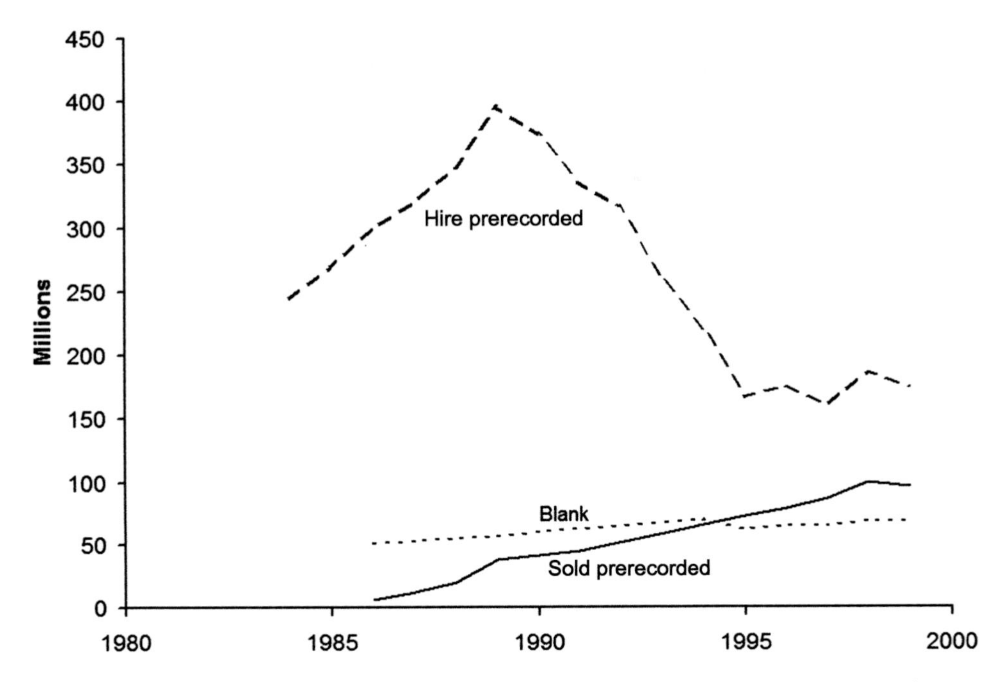

Fig. 25.3 Rental and sales of video cassettes in the UK, 1980–2000. *Source:* Author[19]

space. 'Rewind' was more-or-less instantaneous, and on a disk with multiple tracks the user could easily jump from track to track without having to wind through a lot of tape to find the right place.

The question that the manufacturers were asking themselves was: could something similar be produced for video recording? It would need to have five to ten times the capacity of the audio disk, but perhaps that might be possible. By the early 1990s, Philips and Sony were working together to produce their MultiMedia Compact Disc (MMCD) system.[20] The trouble was that Toshiba, in league with the movie company Time Warner, and supported by other companies, was also working on an incompatible system of their own which they called Super Density Disc (SD). Format wars were about to break out again.

Fortunately, this time there was a plan to use these disks in computers to store programs. When IBM heard about the two rival systems they called a meeting of the representatives of computer manufacturers who voted to boycott both systems until a common standard was reached.[21] This was a huge potential market and the warring parties had no option but to cooperate. After many months of haggling, in 1995 they finally agreed on a disk exactly the same size as a CD but using Toshiba's SD disk arrangement and Sony's coding system. Peace had broken out.

The following year, players started to appear in Japan. In 1997, they spread to America and subsequently to Europe and other parts of the world.[22] The disks could store 4.7 GB of information which was a huge jump from the around 700 MB of the CDs in the same size disk. It meant that several television programs or a complete film would fit on to a single disk.

Of course, the television signal really requires more data than this but fortunately the pictures contain large amounts of information that doesn't change from one frame to the next. It is thus possible to compress the information without significantly affecting the picture. The Motion Pictures Expert Group (MPEG) was set up in 1988 to produce standards for use in digital television signals.[23] Their MPEG-2 was defined in 1994 and so conveniently formed the basis of the encoding used on DVDs.

At the beginning, DVDs had one big drawback over tape: they could only play back a recording that had been manufactured into the disk. This was fine for the films and prerecorded programs, but didn't address the home recording and 'timeshift' uses of the video cassette recorder. The search was on for a solution to this with some form of 'writable' disk.

Several companies worked on this problem, but the first to announce a solution were Pioneer. Their disks could be written only once and then read as many times as required and became known by the description DVD-R.[24] The disks use a dye which absorbs the laser when the pattern is written on to the disc, hence producing the 'pits' which is how the information is recorded.

Quite soon these disks were followed by a DVD-RW version where they could not only be written but also erased and rewritten, meaning that they could now fulfil all the capabilities of a VCR. These were followed by other formats that used a slightly different dye system and were not completely compatible. The write-once versions were known as DVD+R with the read/writable version being designated DVD+RW.

At first, it looked as though the usual incompatible format wars were to break out again. Fortunately, the differences were not too great and the disk drive manufacturers soon overcame the difficulties and produced drives that could handle any of the various forms of disk. They were all physically the same size (though the thickness varied a little, but that didn't matter) so they would fit into the same drive. There was also a smaller version at 80 mm diameter instead of the usual 120 mm, but it was only used for special requirements and in any case the drives could also cope with it. So despite the range of slightly incompatible disk types the drive manufacturers managed to please everybody. The result was that all the disks remained in production without one dominating the others and the user could buy whichever one seemed suitable.

The take up of DVDs was even more rapid than VCRs, despite those still being available. One reason was the unity of the industry in offering only one standard which made it simpler for the users to choose.[25] The industry was also finding that the number of DVDs sold compared to the number of machines was far greater than for the VCRs. By 2002, the annual world production of DVDs had passed that for the falling numbers of video cassettes with both beyond a billion units.

The DVD market hadn't even settled down before the hunt was on for even greater capacities on the same size disk. The clever idea behind this was to change the laser wavelength from 650 nm (red), which these systems used, to the shorter one of 405 nm which is in the blue/violet region—hence the name BluRay.[26] This allowed smaller feature sizes on the disk enabling more to be packed on. By this and other changes the capacity was increased to around 22 GB which was needed for the high definition (HD) television signals that were starting to appear.

There was also an incompatible system called HD DVD but BluRay soon became the dominant system. For most purposes, the standard DVD was quite adequate and dominated the marketplace. Players which could also record appeared to use the writable disks, signing the death sentence for the VCRs which were so cumbersome by comparison.

Of course there was always something else on the horizon. Once the signals had been made digital then anything that handled digital signals could make use of them. That, of course, included computers. Once the power and speed of the computers reached the required level then there was no reason why, with some suitable software, they could not display a DVD. For convenience of supplying the large software packages, computer manufacturers started supplying their machines with DVD drives.

With drives that could also write, or 'burn', the disks then the computer could function as a DVD recorder. It could do more than that, as once it had a fast enough broadband internet connection it could download and store films or TV programs. They could be watched or even written to DVDs for more permanent storage or for use in DVD players. A whole new world of flexible content was opening up.

The net result of all this was that films or television programs that once had to be watched at the time of showing could be available exactly when the user wanted. Either stored on the VCR tape, or later the DVD or computer, they could be played when required, whether they had been recorded by the user or come in the form of a prerecorded cassette or DVD, perhaps as a 'box set' of their favourite programs or films.

What had once been a relatively rare resource could now provide information or entertainment as required. It was another step away from the 'mass entertainment' of music halls, theaters, and films that had begun the twentieth century, to a more personal form of entertainment where each user could choose exactly what they wanted to watch and when to view it. This was another move in the progression to greater individualism.

NOTES

1. The rise and rise of video, available at: http://news.bbc.co.uk/1/hi/entertainment/tv_and_radio/1182165.stm
2. CED in the history of media technology, available at: http://www.cedmagic.com/history/ampex-commercial-vtr-1956.html
3. Okuda, H. (2010) The dawn of video tape recording and development of the helical scanning system. Second IEEE Region 8 Conference on the History of Telecommunications (HISTELCON),
4. The changes to video recorders and VCR technology over the last 50 years, available at: http://www.thepeoplehistory.com/vcr.html
5. The video cassette tape, available at: http://www.sony.net/SonyInfo/CorporateInfo/History/SonyHistory/2-01.html
6. Philips N1500, N1700 and V2000 systems, available at: http://www.rewindmuseum.com/philips.htm
7. Sony Corp. of America v. Universal City Studios, Inc., available at: https://en.wikipedia.org/wiki/Sony_Corp._of_America_v._Universal_City_Studios,_Inc.
8. Sony goes to battle for its favorite child, available at: http://www.sony.net/SonyInfo/CorporateInfo/History/SonyHistory/2-02.html.
9. Shiraishi, Y., and Hirota, A. (1978) Video cassette recorder development for consumers. *IEEE Transactions on Consumer Electronics*, CE-24:3, 468–472.

10. Video 2000, available at: http://www.totalrewind.org/v2000.htm.
11. Levy, M. and Fink, E.L. (2006) Home video recorders and the transience of television broadcasts. *Journal of Communication*, 34:2, 56–71.
12. Harmetz, A., Video cassette revenues approach movie box office totals. *The Journal Record*, available at: https://www.questia.com/newspaper/1P2-33468551/video-cassette-revenues-approach-movie-box-office.
13. Bandebo, R. Video rental industry, available at: http://historybusiness.org/2790-video-rental-industry.html.
14. *Video Recorders: An Industry Overview*. Key Note Publications, 1984.
15. *Video Retail and Hire, 1995 Market Report*, Key Note Publications.
16. UK data from: Video recorders: An industry overview, Key Note Publications 1984; calculated from O'Hare, G.A. Towards the 1990s: Developments in the British video market, Goodall Alexander O'Hare; General Household Survey; US data from *The Wall Street Journal*. Tuning in: Communications technologies historically have broad appeal for consumers, available at: http://www.karlhartig.com/chart/techhouse.pdf.
17. *Video Retail and Hire, 3rd ed.*, 1991. Key Note Publications.
18. O'Hare, G.A. Towards the 1990s: Developments in the British video market. Goodall Alexander O'Hare.
19. Data from: *Video Retail and Hire, 1995 Market Report*. Key Note Publications; *Video Retail and Hire, 3rd ed.*, 1991, Key Note Publications; *Video Retail and Hire, 2000 Market Report*, Key Note Publications; O'Hare, G.A. Towards the 1990s: Developments in the British video market. Goodall Alexander O'Hare.
20. Chapin, R. History of DVD, available at: http://www.miqrogroove.com/writing/history-of-dvd/: History of DVD, available at: http://didyouknow.org/dvdhistory/.
21. DVD, available at: https://en.wikipedia.org/wiki/DVD.
22. Schoenherr, S. Recording technology history, available at: http://www.aes.org/aeshc/docs/recording.technology.history/notes.html#digital.
23. Fairhurst, G. MPEG-2. University of Aberdeen School of Engineering, available at: http://www.erg.abdn.ac.uk/future-net/digital-video/mpeg2.html.
24. DVD-R, available at: https://en.wikipedia.org/wiki/DVD-R.
25. Whiting, V. (1998) DVD video: European Market Assessment and Forecast 1998–2000. Screen Digest Games.
26. BluRay disc format, available at: http://www.blu-raydisc.com/Assets/Downloadablefile/White_Paper_General_3rd_Dec%202012_20121210.pdf.

26

Electronics Invades Photography: Digital Cameras

The digital camera is a great invention because it allows us to reminisce. Instantly.

Demetri Martin

Originally, of course, photography was a matter of chemicals. Electricity, in any form, wasn't involved but, in low light levels, something was needed to brighten the scene and that was the flash. Initially, a number of evil substances were used that, when triggered, would burn or explode producing a short but brilliant light. It wasn't long into the twentieth century before a battery and some fine wire were being used to trigger the flash.[1]

This was dangerous—a small explosion—and the hunt was on for a better method. In the late 1920s, flash bulbs started to appear which used aluminum foil in a glass envelope filled with oxygen.[2] The first was the wonderfully named 'Vacublitz' from the German company Hauser, followed shortly after by a similar product from General Electric in America. With both of these, when an electric current from a battery flowed through the foil, it burnt in the oxygen producing a burst of brilliant white light.

At the same time, Harold Edgerton, working at MIT, was experimenting with arc discharge tubes and chose the xenon tube as it gave a white light.[3] Though he was interested in freezing motion with fast repeated flashes, a single flash for use with a camera was also possible. Interest in this technique began to grow after he published spectacular pictures of frozen action, such as a drop falling into a bowl of milk.

By charging a capacitor, and then suddenly discharging it into the xenon tube, a very fast pulse of light is produced. This was ideal as a flash for use with cameras and before long such devices found their way into photographic studios. However, it needed the introduction of semiconductors before small 'electronic' flash units which could be mounted on the camera became available. These required an electronic circuit to convert the low voltage of the battery to the high one needed by the discharge tube.

On a simple camera everything was fixed, and all the user had to do was press the button to fire the shutter. However, for more sophisticated use the shutter speed, the aperture of the diaphragm, the film speed and the focus distance were all variables. While the film speed could be set from the manufacturer's information and the focus achieved by ensuring that the relevant part of the picture was sharp, there were considerable difficulties in

J.B. Williams, *The Electronics Revolution*, Springer Praxis Books,
DOI 10.1007/978-3-319-49088-5_26

deciding what combination of aperture and shutter speed was required. This required skill and experience.

Some means of measuring the light level was desirable and though many schemes were tried it wasn't until 1931 that the Rhamstine Electrophot exposure meter appeared.[4] This used a selenium cell which was sensitive to light and showed the result on a moving coil meter. Within a year, a number of companies had entered the market, in particular Weston with their 617 model which became popular.[5] These gave a value for the light level, but it was still necessary to work out the correct combination of aperture and shutter speed.

To achieve this, calculators were used; sometimes the meter had recommended settings against the light level indication, or disk calculators built in. These were all very well but the photographer might have special requirements, such as wanting to take something moving where he would need the fastest speed he could and would have to open up the aperture to compensate. However, if greater depth of field was required the aperture needed to be as small as possible and the speed reduced to compensate.

As the 1930s wore on the sensors became smaller and it was possible to incorporate them into the camera rather than have them as separate items. The advantage of the selenium cells was that they were photovoltaic, in other words they generated a voltage which could be measured by the meter. The disadvantage was that they didn't work at low light levels, and as films became faster this became a problem.

The solution was to use Cadmium Sulphide (CdS) sensors. These were photo resistive, meaning that their resistance changed with light level, but a battery was needed for them to operate. However, a battery in a camera could be used for other purposes such as adjusting settings automatically, though the metering was still rather crude.

It took until 1960 for the metering to be through the lens, which was a great improvement.[6] Around this time, with semiconductors now being available, various degrees of automation appeared. Some were semi-automatic where the light meter would set either the speed or the aperture leaving the other to be set manually. Then came 'fully automatic' cameras, where both aperture and speed were set without the user being involved. The camera could only set these to a predetermined pair of settings depending on the light level, but for most purposes this was quite satisfactory.

What the automatic settings meant was that a novice user could achieve good pictures most of the time with a sophisticated camera and all they had to do was focus the shot. (Of course, they had to set the film speed when it was inserted, though later this became automatic as it could be read by the camera from bars on the cassette.) This was getting near the 'point and shoot' simplicity of the fixed lens 'box' cameras.

Of course, to achieve this level of automation electronics had crept into the cameras. Not only was this in the light sensor, which by now was usually silicon, but it required power to operate the adjustments to the shutter speed and the aperture. With 35 mm Single Lens Reflex SLR cameras where the view finding was also through the main lens, power was also necessary to flip the mirror out of the way after the scene had been viewed ready to take the picture.

There was one further step to take and that was to automate the focus. Cameras that achieved this were starting to appear by the late 1970s and early 1980s.[7] These were often compact devices ideal for the 'point and shoot' amateurs who wanted to capture their holiday on films, particularly now that many were going to more adventurous places. Some

SLR cameras even had autofocus built into their detachable lenses which coupled with auto exposure modes on the body, gave a device that was fully automatic for ease of use. The exposures could also be set manually if something more complex was needed.

In 1969, Bell Labs were on a quest to produce more efficient memories for computers so that they could automate more of the exchange operations at their parent company AT&T.[8] One area of research was magnetic bubble memories where a small magnetic 'bubble' could be moved along between minute sites. Thus, by inserting a signal at the beginning of a row of these bubbles and 'clocking' it along and then feeding the output back to the input a recirculating memory could be achieved.

Jack Morton, the legendary head of the electronics division at AT&T, challenged Willard S. Boyle, the semiconductor section director, to come up with a semiconductor device to compete with the bubbles. Bill Boyle called in one of his department leaders, George E. Smith, and they put their heads together. It didn't take them long to realize that an electric charge was the analogy of the magnetic bubble, and that this could be stored on a tiny capacitor formed as part of the Metal Oxide Semiconductor (MOS) process, something they were also working on (see Chap. 11).

By this time, it was reasonably straightforward to fabricate a series of MOS capacitors in a row and connect a suitable voltage to them so that the charge could be stepped along from one to the next. Quickly, Morton and Boyle designed and had made a suitable test circuit. It worked first time. Though it took a while to perfect, they had the device that they had been challenged to make. It was only subsequently that it was found to be too susceptible to stray radiation and would not make an effective memory.

This disadvantage could be turned to an advantage as the capacitors could take up a charge corresponding to the light falling on that spot for a short time. The set of signals could then be clocked out and it thus could form a linear camera. By mechanically scanning it in the other direction a full camera could be produced, or it was possible to have a two-dimensional Charge Coupled Device (CCD).

By 1971, they had built a two-dimensional CCD, and then a camera using this chip. One of the great advantages was that they were using variants of semiconductor processing that was already in place and reasonably well understood. It was thus fairly straightforward to make the devices. The camera produced was much smaller and more robust than conventional television cameras.

Seeing that another pet project at Bell labs was the 'Picturephone', something that could make a simpler camera than the usual television camera tubes was of great interest. This was particularly so as the normal tubes were large, fragile and didn't last very long. The CCD was far more promising than the television tube with a silicon array that was being investigated. As it turned out the Picturephone never caught on and so this use never materialized.

With the publication of the basic information about the devices in 1970, other semiconductor companies, particularly Fairchild, immediately became interested in the possibilities.[9] Many of these organizations had the technology to make this sort of device, and it was reasonably straightforward as it contained a regular pattern much like a memory device. However, to achieve a good picture a large number of capacitors, later to be known as pixels, were needed and the technology to produce large-scale integrated circuits was still in its infancy.

As time progressed, large chips could be made, and by 1975 Bell Labs had demonstrated a CCD camera with image quality good enough for use in a broadcast television camera.[10] Faichild produced a commercial chip with 100 × 100 pixels which could be used to make simple cameras.[11] It was really too small for most purposes, but was useful to explore the possibilities.

Keen to protect their huge film business, Kodak were watching these developments. In 1975, one of their engineers, Steve Sasson, cobbled together a functioning digital camera around the CCD sensor.[12] It only had the 10,000 pixels (0.01 megapixels) of the Fairchild chip, and produced black and white images. These were stored on an audio cassette tape and took 23 s to process. It was the size of a large toaster and weighed 4 kg.[13] Clearly, the technology still had some way to go, but the potential was evident.

The following year, Fairchild produced a commercial camera, the MV-101, for industrial uses such as inspecting products at Procter and Gamble. Progress was quite slow while waiting for larger sensors to appear. In 1981, Sony produced their first Mavica camera.[14] While this was not strictly a digital camera it was based around a 570 ×y 490 pixel (more than a quarter megapixel) CCD. It was really a still video camera that recorded its images on a small floppy disk.

By 1983, though the sensors were expensive, the astronomy community took to them as the sensitivity was much better than film and greatly increased the images they could obtain from their telescopes. Using multiple 1 megapixel sensors they could get extremely good pictures in seconds that had taken many hours before.

By 1986, Kodak passed the 1 megapixel barrier for a sensor with their 1.4 megapixel unit, which was good enough for a 5 × 7 in. photo quality print.[15] This, of course, was only part of what was needed to have a true digital camera. There had to be some way of storing the images and then a means to render the image visible again. One way was by downloading to a computer, or possibly printing it on a suitable printer. Also needed were standards for how the digital image was to be stored. All these were required before digital cameras became a practical proposition.

In 1987, Kodak set out to tackle some of these issues. They released seven products for recording, storing, manipulating, transmitting and printing electronic still video images. The following year, the Joint Photographic Expert Group issued their JPEG format which compressed digital images and was particularly good with color.[16] Progress was being made but there was still a lack of a really practical storage medium.

The first true digital camera appeared in 1990. There were two versions, the Dycam Model 1 and the Logitech Fotoman, and they seemed to be only differentiated by the color of the case.[17] It used a very small CCD sensor of 284 × 376 pixels (around 0.1 megapixels), presumably for cheapness as the technology was well past this size. The images were stored on a Random Access Memory (RAM) and survived while the battery charge remained which it only did for 24 h. The pictures could be downloaded to a computer and stored more permanently there. It wasn't very good, but it pointed the way to the future.

Meanwhile, Kodak had been busy. They had their CCD sensor and it was time to do something with it. Their approach was to use professional camera bodies from the likes of Nikon and Canon and put on a 'digital back'.[18] The storage and processing unit was in a separate box which had to be carried around. At first they concentrated on the military uses, but by 1991 they used a Nikon F3 camera and added their back to produce the first

DCS digital camera for use by photojournalists. The camera was handy but it still had a large box connected to it which housed batteries and a hard drive for storing the pictures.

The next year, they surprised the market with the DCS200. Though this had a sizable lump hanging below the camera body and stretching up the back, it was all integrated into the one unit. It used a slightly upgraded sensor at 1.5 megapixels and incorporated a small hard drive within the unit. Though still in the realms of the professional, the price was much reduced. Kodak went on to produce a whole range of similar cameras, gradually increasing the performance as the 1990s progressed.

Other matters were being addressed with the arrival of Adobe's 'Photoshop' program which allowed the digital photographs to be manipulated on the computer.[19] Of course, the pictures still needed to be committed to paper, and in 1994 Epson produced their first 'photo quality' ink jet printer, the MJ-700V2C, which achieved 720 × 720 dots per inch (dpi).[20] This was better than the cameras at the time. The 1.5 megapixel CCD was roughly 1000 × 1500 pixels which, when printed as a normal photo, meant that there were only 250 dots to each inch. Even the newer devices appearing that year at 6 megapixels, with twice the number of pixels in each direction, only needed 500 dpi.

Now most of the pieces were in place and what was needed was for Moore's Law to take its course. The sensors needed to have more pixels, while effective semiconductor memory would mean that photo storage on the camera could be shrunk. More powerful microprocessors would lead to more complex image processing and faster response times. The progress of the technology would inevitably enable the prices to be brought down to an affordable region.

One of the next cameras to appear was from the computer company Apple, with their Quicktake 100, made by Kodak.[21] This was a curious device shaped rather like a pair of binoculars. At first sight it was a retrograde step, with a sensor of only 640 × 480 pixels (0.3 megapixels) but it was designed to match the resolution of the computer screen, which for Apple was of course its Macintosh (though later a Windows version was made available). Its real breakthrough was price at less than $1000 and about £535.

It was inevitable that the Japanese electronics and camera firms would get in on the act once they could produce something at consumer prices that looked like a camera. The first was Casio in 1995 with their QV-10 which only had a CCD of 460 ×y 280 pixels (0.13 megapixels).[22] It introduced two important steps forward. The first was that it had a small LCD screen on the rear so that the photo could be viewed as soon as it was taken. This was an enormous advantage over film cameras where the result could only be seen after the film was processed.

The second innovation was the use of 'flash' memories to store the pictures. These were semiconductor devices that could have data written to them but could also be erased electrically. This made them the ideal store for the pictures. They were small in size but the pictures were not lost when the power was switched off. Once downloaded to a computer, the store could be erased and reused. The combination of this and the display meant that if a picture was taken and it was not satisfactory it could immediately be erased and another taken.

Unfortunately, the camera had a number of faults, not least of which was that it ate batteries. The picture resolution was very poor which presumably had to be sacrificed to keep the cost down and allow the memory to store 96 images. The company rapidly produced revised versions which addressed some of the problems. They had, however, shown the way forward and the form that digital cameras needed to take.

Now other companies started to pile in, with Ricoh, Olympus, Canon and many others, producing ever improving cameras based on this formula. As the semiconductor technology improved the megapixels and the number of pictures stored went up, while the power consumption and price went down. By the turn of the century, compact cameras usually offered 2 or 3 megapixel sensors and were starting to make serious inroads into the camera market.

For a considerable time another type of sensor had been available: Complementary Metal Oxide Semiconductor or CMOS. Here the basic light-sensitive cell was fairly similar to the CCD but each one had some amplification immediately connected to it rather than waiting until the signal had been clocked out of the complete row.[23] The advantage of this was that the chip was closer to normal integrated circuits and so could be more easily processed on the production lines. The downside was that it was more complex.

In the early days of digital cameras, the advantage lay with the CCD as these could be made more easily but at the cost of all the associated circuitry, such as the analogue to digital converter, being on separate devices. Once the semiconductor processes reached the stage where the CMOS chips could easily be made, then the advantage swung to them as the rest of the circuitry could be fabricated on the same chip.

The CMOS chips thus became cheaper in use and drew far less power. They also read out the picture more quickly as they could do more operations at the same time whereas with the CCD essentially everything had to go through a single connection at the end of the row. One key point was that with the 'on chip' processing the camera could be made very small. This meant that it could have many uses outside the conventional 'camera' format, such as a video camera for security surveillance.

What it also meant was that a camera was now small enough to incorporate into a mobile phone. In 1997, inventor Philippe Kahn took a picture of his newborn child with an experimental mobile phone and sent it to 2000 friends and relations. This galvanized the mobile phonemakers and in 2000 Sharp introduced their J-SH04 in Japan as the world's first camera phone.[24] Quite soon, all new mobile phones came with a camera so that images could be taken and sent without a separate camera being needed.

The rise of digital cameras was so rapid that in 1999, in Japan, film cameras sold at the rate of more than three times that for digital ones, but by 2001 this had crossed over and digital camera sales were greater.[25] In the Americas and Europe this growth took until 2003 or 2004, but the rest of the world hung on to their conventional devices for longer. By 2003, sales of digital cameras had reached 50 million units worldwide which was only just behind that for film ones.

It was hardly surprising that the following year Kodak stopped selling film cameras in order to concentrate on digital, where it soon became the largest supplier.[26] However, within 5 years it gave up making film altogether, which led to its demise not long afterwards, ending its 130-year history in the business. Surprisingly, it took camera phones some 10 years before they had any significant effect on the sales of digital cameras.[27]

In 2009, just when digital cameras had won the war, and electronics had conquered chemical film, one half of the Nobel Prize for physics was awarded to Willard Boyle and George Smith for the invention of the CCD. (The other half going to Charles Kao for his work on glass fiber as described in Chap. 23.) It was a fitting tribute to all those who had made this triumph possible.

Thus in the twentieth and twenty-first centuries, pictures had changed from something unusual and special to the everyday. They could now be taken and disposed of at will. It was all part of a world with images everywhere, and where everyone could join in, as taking and distributing them was so easy. Once again, electronics had changed the way people related to the world.

NOTES

1. Flash (photography), available at: https://en.wikipedia.org/wiki/Flash_(photography).
2. Carter, R. Digicamhistory 1920s, available at: http://digicamhistory.com/1920s.html; Tolmachev, I. A brief history of photographic flash, available at: http://photography.tutsplus.com/articles/a-brief-history-of-photographic-flash--photo-4249.
3. H.E. Edgerton, 86, dies: Invented electronic flash, *New York Times*, available at: http://www.nytimes.com/1990/01/05/obituaries/h-e-edgerton-86-dies-invented-electronic-flash.html.
4. Fisher, M. Wellcome & Johnson Exposure Calculators, available at: http://www.photomemorabilia.co.uk/Johnsons_of_Hendon/JoH_Exposure_Calculators.html.
5. Bilotta, S. Scott's Photographica Collection: Weston Electrical Instrument Corp. Model 617 Exposure Meter, available at: http://www.vintagephoto.tv/weston617.shtml.
6. Exposure meters, available at: http://www.earlyphotography.co.uk/site/meters.html.
7. Masoner, L. A brief history of photography, available at: http://photography.about.com/od/historyofphotography/a/photohistory.htm.
8. Smith, G.E. The invention and early history of the CCD. Nobel Lecture December 8, 2009, available at: http://www.nobelprize.org/nobel_prizes/physics/laureates/2009/smith_lecture.pdf.
9. Boyle, W.S. and Smith, G.E. (1970) Charge coupled semiconductor devices. *The Bell System Technical Journal*, 49:4587–593.
10. 2009 Nobel Prize in Physics to former Bell Labs researchers Boyle and Smith, available at: https://www.alcatel-lucent.com/blog/corporate/2009/10/2009-nobel-prize-physics-former-bell-labs-researchers-boyle-and-smith.
11. The history of the digital camera, available at: http://www.cnet.com/news/photos-the-history-of-the-digital-camera/4/.
12. Skipworth, H. World Photography Day 2014: The history of digital cameras, available at: http://www.digitalspy.co.uk/tech/feature/a591251/world-photography-day-2014-the-history-of-digital-cameras.html#~pseqfOZ6v9CAsI.
13. The history of the digital camera, available at: http://www.cnet.com/news/photos-the-history-of-the-digital-camera/.
14. The history of the digital camera, available at: http://www.cnet.com/news/photos-the-history-of-the-digital-camera/2/.
15. Bellis, M. History of the digital camera, available at: http://inventors.about.com/library/inventors/bldigitalcamera.htm.
16. The history of the digital camera, available at: http://www.cnet.com/news/photos-the-history-of-the-digital-camera/6/.
17. Henshall, J. Logitech Fotoman Digital Camera, available at: http://www.epi-centre.com/reports/9301cs.html.
18. McGarvey, J. The DCS story, available at: http://www.nikonweb.com/files/DCS_Story.pdf.
19. The history of the digital camera, available at: http://www.cnet.com/news/photos-the-history-of-the-digital-camera/7/.

20. The history of the digital camera, available at: http://www.cnet.com/news/photos-the-history-of-the-digital-camera/10/.
21. Henshall, J. Apple Quicktake 100, available at: http://epi-centre.com/reports/9403cdi.html.
22. The history of the digital camera, available at: http://www.cnet.com/news/photos-the-history-of-the-digital-camera/11/; Wherry, P. The Casio QV-10 Digital Camera, available at: http://www.wherry.com/gadgets/qv10/.
23. CCD and CMOS sensor technology, available at: http://www.axis.com/files/whitepaper/wp_ccd_cmos_40722_en_1010_lo.pdf.
24. Davlin, A. History of cameras: Illustrated timeline, available at: http://photodoto.com/camera-history-timeline/.
25. 50 million digital cameras sold in 2003, available at: http://www.dpreview.com/articles/5474101424/pmaresearch2003sales.
26. Sparkes, M. Kodak: 130 years of history, *The Telegraph*, available at: http://www.telegraph.co.uk/finance/newsbysector/retailandconsumer/9024539/Kodak-130-years-of-history.html.
27. Griffith, A. Crunching the numbers: Four insights we can glean from camera sales data, available at: http://petapixel.com/2013/12/18/crunching-numbers-4-insights-camera-sales-data/.

27

Seeing Inside the Body: Electronics Aids Medicine

Every year, more than 300 million X-rays, CT scans, MRIs and other medical imaging exams are performed in the United States, and seven out of ten people undergo some type of radiologic procedure.

Charles W. Pickering

The relationship between muscles, including those of the heart, and electricity had been known since Galvani's experiments in the 1780s, or possibly even before. However, in 1887, Augustus D. Waller at St Mary's Medical School in London showed that electrical signals of the heart could be monitored from outside the body.[1] He used a capillary electrometer, invented in 1872 by French physicist Gabriel Lippmann, to measure the signals.[2] This was a crude device where a column of mercury met sulfuric acid and the meniscus between the two would move when a current flowed from one liquid to the other. The result was recorded on a moving photographic plate.

Willem Einthoven was a Dutch physiologist who in 1889 saw one of Waller's demonstrations and became fascinated by the subject. He set to work to investigate further, and in 1893 he introduced the term 'electrocardiogram', though modestly he ascribed it to Waller. Although he managed to improve the results that could be obtained from the capillary electrometer he realized that a better measuring instrument was required.

The device he came up with was the string galvanometer. A very thin silver-coated quartz filament was suspended between two electromagnets and the current to be measured passed through the string.[3] By the normal laws of electromagnetism the string would move and this could be observed with a microscope. In order to produce electrocardiograms, the movement of the string was magnified optically and the light shadow left a trace on a moving photographic plate. In 1902, he published electrocardiograms produced this way.

Einthoven had laid the foundation of the whole subject and converted it from scientific curiosity to a useful medical tool. He clarified the various actions of the heart that could be measured and introduced the nomenclature that is still used. His contribution, both to the science of electrocardiograms and to their measurement, was such that he received a Nobel Prize for Physiology or Medicine in 1924.

J.B. Williams, *The Electronics Revolution*, Springer Praxis Books,
DOI 10.1007/978-3-319-49088-5_27

Fig. 27.1 An early ECG machine when connections were being made to hands and foot instead of the chest. *Source:* https://en.wikipedia.org/wiki/String_galvanometer#/media/File:Willem_Einthoven_ECG.jpg

Although it worked very well, the problem with Einthoven's string galvanometer was that it was so large it occupied two rooms and took five people to operate it.[4] It was only after a further paper was published in 1908 that the utility of the method became clear, and the Cambridge Scientific Instrument Company redesigned the instrument to make it a manufacturable device, leading to their supply later that year (Fig. 27.1).

One of the first people to have an electrocardiogram was Sir Thomas Lewis of the University College Hospital in London. From using it, he was able to publish his classic text book, *The Mechanism of the Heart Beat.* Now that the subject was beginning to be reasonably well understood, all that was needed was a simpler and more practical instrument so that electrocardiograms could become a standard diagnostic tool in hospitals.

After the First World War electronics had developed to a degree that vacuum tube amplifiers were commonplace. It was not too difficult to use them to amplify the weak signals coming from the body and drive a pen recorder. By producing a multichannel fast response recorder all the relevant signals from the heart could be displayed at once, making a convenient device for examining electrocardiograms. Their use in hospitals began to grow and over the years more compact versions appeared as the technology improved.

The idea that it might be possible to detect electrical activity in the brain went back further than that in the heart. In 1875, Richard Caton, a physician in Liverpool, presented a paper to the British Medical Association claiming that he had been able to detect electrical activity in the brains of animals by placing electrodes on the surface of the grey matter. The instruments in those days were simply not sensitive enough to detect the signals on the surface of the skin.

Over the remainder of the Nineteenth century a number of other researchers found similar results but to carry out experimental work on humans was very difficult. One researcher who took an interest in this was the German psychiatrist Hans Berger, who eventually published his findings in 1929.[5] Despite a certain amount of interest the subject didn't really take off.

What made people take notice was when the British neurophysiologist Edgar Douglas Adrian demonstrated the scientific validity of electroencephalography (EEG) to an astonished international audience at the Cambridge meeting of the Physiological Society in May 1934. It was, of course, the arrival of electronic amplifiers that had made this subject possible. Also some of the signals were too rapid for the pen recorders and required the electronic oscilloscope to display them.

By 1936, EEGs were becoming useful to understand the patterns of 'brain waves' and so help diagnose conditions such as epilepsy.[6] In the same year, British scientist W. Gray Walter proved that, by using a larger number of small electrodes pasted to the scalp, it was possible to identify abnormal electrical activity in the brain areas around a tumor.[7] A few years later, in the 1940s, EEG use was being pushed further and further with attempts to use them to diagnose criminal behavior and even some mental illnesses. Another use was as part of lie detector machines. It took until the 1960s before there was some sort of consensus that they could not reliably detect criminality, mental illness or dishonesty. However, EEGs remain useful as a tool to discover brain problems and in some areas of research such as into sleep.

The genesis of ultrasonics can be traced back to the sinking of the 'Titanic' in 1912.[8] This tragedy motivated a number of researchers to seek ways of detecting icebergs and other hazards at sea. Amongst them were Alexander Belm in Vienna, Lewis Richardson in England, Canadian wireless pioneer Reginald Fessenden in the US, and French physicist Paul Langévin and Russian scientist Constantin Chilowsky in France.[9]

The principle was to use piezoelectric crystals bonded to steel plates to couple the high-frequency vibration into the water. While they were originally intended to help shipping avoid icebergs, with the coming of the First World War these 'Hydrophones' rapidly became refined for the detection of submarines. This generated a wider interest in the possibilities of 'ultrasonics'.

It took the introduction of electronics and the development of radar to move ultrasonics on to the detection of cracks in metals, particularly flaws in armor plate or the hulls of large ships. By the 1940s, and with another war raging, this became a valuable tool to quickly and cheaply ensure that the materials were of sufficient quality.

Quite soon, medical researchers started to use these devices to see what information they could gain about soft tissues in the human body. While X-rays were useful for bones they were not good at distinguishing between other structures inside bodies. Results were obtained using ultrasonics but there were considerable doubts about some of them, and there were problems with exactly how to couple the ultrasonic signals to the body.

By the 1950s, progress was beginning to be made in the US and Japan. One of those pushing forward was John Julian Wild, an Englishman working at the Medico Technological Research Institute of Minnesota. He found that tumors produced a different echo than normal tissue and hence the method could be useful for diagnosis. He and his collaborator, electronics engineer John Reid, built a number of ultrasonic devices and were able to detect cancerous growths in the breast and tumors in the colon.

While Wild was lecturing in the UK in 1954, among his audience was Ian Donald who was about to become Professor of Midwifery at Glasgow University. He immediately became interested in the possibilities of ultrasonics for medical investigations.

Though trained as a doctor, he had been in the RAF during the war and gained knowledge of radar and sonar. Being rather gadget minded, ultrasonic diagnosis naturally fitted his varied skills.

Once again he reached towards the industrial metal flaw detectors and tried these on various pieces of tissue. He was lucky to team up with an engineer, Tom Brown, and with Dr. John MacVicar. They set to work to investigate the value of ultrasound in differentiating between cysts, fibroids and any other intra-abdominal tumors. However, the early results were disappointing and many people thought that they were wasting their time.

What changed everything was when they met a woman who had been diagnosed with an inoperable cancer of the stomach. They undertook an ultrasound investigation that showed that she had a huge ovarian cyst which was easily removed and her life saved. The team wrote this up in an article which eventually appeared in *The Lancet* in 1958 under the uninspiring title,' Investigation of Abdominal Masses by Pulsed Ultrasound'.[10] It has been claimed that this is the most important paper ever published on medical diagnostic ultrasound. People started to take the technique seriously.

As Ian Donald was to say later, 'As soon as we got rid of the backroom attitude and brought our apparatus fully into the Department, with an inexhaustible supply of living patients with fascinating clinical problems, we were able to get ahead really fast. Any new technique becomes more attractive if its clinical usefulness can be demonstrated without harm, indignity or discomfort to the patient.'

In 1959, Wild and his colleagues were able to show the value of the technique for looking at the heads of babies while still in the womb. In the next few years the technique was developed to examine the development of the fetus during the whole pregnancy. The deciding factor that pushed the method into the mainstream was the work of the British epidemiologist Alice Stewart, who showed that the standard method of examining the fetus using X-rays was harmful.[11] Ultrasound was a better and safer alternative.

Once ultrasonics had been accepted, and small portable machines became available, its use spread to other medical specialties such as cardiology and critical care.[12] It was useful for checking for gallstones, and other lumps and bumps as well as the obvious cancers. It provided a very convenient way of seeing the softer tissues inside the body and provided a useful complement to X-rays.

There was still a desire to have a more powerful tool to look inside the body. The man who thought he could see a way to do this was Godfrey Hounsfield who worked at EMI Central Research Laboratories in Hayes, west London.[13] From an unpromising start he had worked his way up via the RAF during the Second Wold War, and been heavily involved in the development of computers.[14] He was transferred to the EMI's research labs where he made an unsuccessful attempt to build a large computer memory store; luckily he was then given time to think of a project he could undertake.

Hounsfield began to examine the possibilities of pattern recognition and in 1967 gradually the idea for a scanner took shape. The basic problem with X-rays is that, when a beam is passed through a human body, most of the information is lost. It was hoped that a scanner could slowly rotate the source and sensor around the object of interest, take a huge number of samples and then, using computing power, to build an image of that 'slice' through the object.

The mathematics to do this is quite formidable but fortunately the South African, Allan Cormack, had already undertaken a considerable amount of theoretical work on this subject. Hounsfield's early experiments involved a lathe bed to rotate the gamma-ray source and sensors on the other side picked up the signals.[15] Because of the low power source it was painfully slow, taking 9 days to do one scan.

Changing to an X-ray source brought the scan time down to 9 h but it still required two-and-a-half hours of computer time to process the result. The output was a paper tape which was then used to modulate a spot on a cathode ray tube in front of a camera. This too was a slow process; it took 2 h to produce a photograph. It was tedious, but it worked.

Gradually, the speed increased and results were sufficiently encouraging that the team built a much more sophisticated machine that could scan the brains of living patients. What was essential was that tumors would show up, otherwise it was useless. In 1971, they were ready and the first patient, a woman with a suspected brain lesion, was scanned at Atkinson Morley's Hospital in London. The picture showed the clear detail of a large circular cyst in the brain. When the surgeon operated, he said that the tumor looked exactly like the picture. Clearly, the machine had a future (Fig. 27.2).

The next steps were obvious: more speed—which was largely achieved with a larger number of sensors and greater computing power—and to move on to a whole body scanner. This machine was able to take a high resolution picture in 18 s compared with the 5 min needed by the previous device. By 1975, EMI were marketing a full body scanner, the CT5000, and the first one was installed in the Northwick Park Hospital in London. From then on it was a matter of increasing speed and sophistication as other companies entered the field.

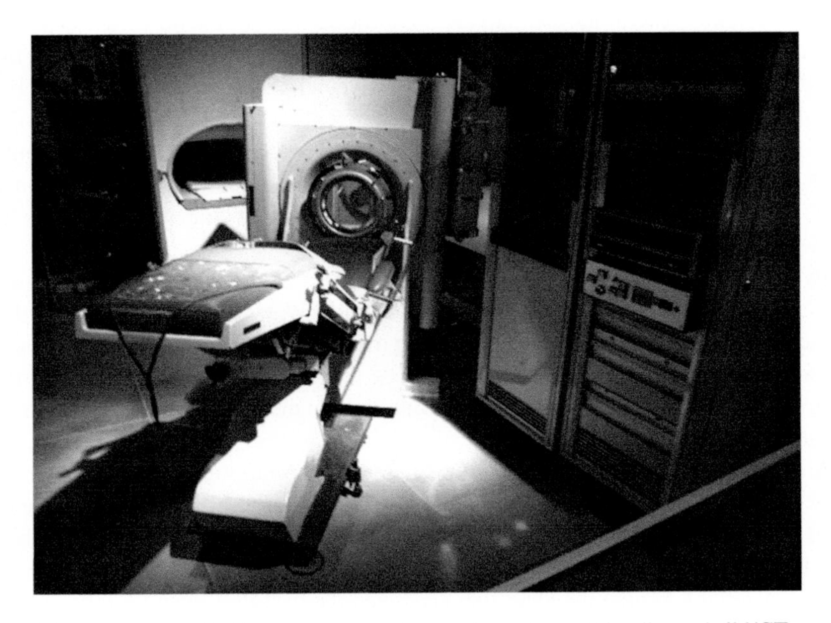

Fig. 27.2 The original EMI head scanner. *Source:* https://en.wikipedia.org/wiki/CT_scan#/media/File:Emi1010.jpg

In 1979, Godfrey Hounsfield and Allan Cormack were awarded the Nobel Prize for medicine. This was surprising in a couple of ways. Normally, the committee likes to wait a considerable length of time to be sure that the development is of lasting benefit. The advantages of the scanner and the take-up in hospitals were already clear by that time and they had no difficulty in awarding the prize. The other odddity was that, as Allan Cormack said, 'There is irony in this award, since neither Hounsfield nor I is a physician. In fact, it is not much of an exaggeration to say that what Hounsfield and I know about medicine and physiology could be written on a small prescription form!'[16]

Just after the Second World War two men, Felix Bloch and Edward Purcell, independently made the same discovery.[17] The atomic nuclei in solids and liquids placed in a magnetic field and subjected to radio waves will 'flip', and in doing so will give off another radio signal which can be detected. This phenomenon, known as Nuclear Magnetic Resonance (NMR), gave different frequencies for different atoms and so could be used to determine exactly what chemicals were present. In 1952, the two men were jointly awarded the Nobel Prize for physics for their discovery.

In the early 1970s, Paul Lauterbur, working in New York, found that this method could be extended to the cells in a human body.[18] He found that by introducing a gradient into the magnetic field he could distinguish hydrogen nuclei in different parts of a sample. By applying these gradients at different directions he could build up a three-dimensional image of the positions of the hydrogen nuclei and hence a picture of the tissue. Peter Mansfield, in Nottingham, England, went on to develop these methods which allowed the images to be collected in a matter of seconds. They were jointly awarded the Nobel Prize for Physiology or Medicine in 2003 for their work.

The man who turned these ideas into a machine for examining patients was American doctor and scientist Raymond Damadian.[19] He found that Magnetic Resonance Imaging (MRI), as it was now called to remove the stigma of 'nuclear', was useful for medical diagnosis and in particular it could distinguish between healthy tissue and tumors. In 1977, he built the first whole body MRI scanner which he called the 'Indomitable'.

After that, with the success of the CT scanner, it was natural that MRI should also develop rapidly. By the early 1980s, MRI equipment was starting to appear in health care.[20] In 2002, some 22,000 MRI scanners were in use worldwide, and more than 60 million MRI examinations were being performed.

Both CT and MRI scanners could image the tissues of the body, but not look at body functions such as blood flow. Before the Second World War, Georg von Hevesy had undertaken a considerable amount of work exploring the use of radioactive isotopes for tracking chemical processes.[21] He used the same methods for tracing fluid flows in biological systems, first in plants and later in animals and finally humans. In 1943 he received the Nobel Prize for Chemistry for his work.[22]

In America, Michel Ter-Pogossian, of Armenian extract though born in Berlin, became interested in the possibility of using these isotopes for examining functions of the body.[23] He used the method to track the progress of chemicals, such as oxygen, around the body. In the 1970s, with the coming of scanners, he was a pioneer of Positive Emission Tomography (PET). He led the team that produced the first PET scanner for use in hospitals. Though not as widely used as the other types of scanners, PET does have a value when dynamic processes need to be studied.

All these scanners depended on electrical systems to operate and, crucially, on computing power to process the results which were displayed on screens. Without these advanced technologies they could not exist. It is for this reason that, though the basic processes on which they depended were known for many years, they didn't start to appear until the 1970s and 1980s when adequate computing power was available sufficiently cheaply and of practical size.

In the 1960s, attempts began to be made to bring together a number of measurements of vital signs in a patient so that they could be monitored constantly by a machine that would sound an alarm if all was not well.[24] It required that basic sensors for heart rate, blood pressure and respiratory (breathing) rate were all made automatic and these were then coupled with an ECG. Alarms were registered if each of these strayed outside acceptable limits, and also for combinations of these which were symptomatic of various medical conditions such as shock.

The potential alarm conditions were quite complex and this required electronic logic or, preferably, computing power. Once this became available in convenient sizes, and with suitable displays to show the heart traces, then patient monitors became practical devices. They were particularly useful for critical or intensive care where they could lighten the load on the nursing staff who knew that the patient was being constantly monitored; alhough they needed to keep an eye on the situation they could largely rely on the alarms to alert them to problems.

A further sensor that would add to the patient monitor was to measure the oxygen saturation of the blood. Numerous attempts had been made at this, comparing two signals passing through a thin section of the body such as the ear lobe. The principle was that by using a red light and one in the infra red only the first was affected by the oxygen in the blood. The problem was that it was difficult to get an absolute reading.

This difficulty was overcome in the 1970s in Japan by the invention of the pulse oximeter by Takuo Aoyagi of the Nihon Kohden Corporation as a lucky accident when he was actually testing something quite different.[25] It was trialled in Japan in 1975 and spread to the US. However, the company didn't progress the device very well and failed to patent it in America. Others took up the banner and the whole thing ended in patent battles.

Nevertheless, by the mid 1980s small portable oximeters were available from numerous suppliers. The oximeter usually had a clip that went over a finger with the two wavelengths of light passing through that to make the measurement. It was then natural that oxygen measurement should be added to the patient monitors making a more complete overview of their condition.

By the end of the century electronics, and particularly computing power, had made a very significant contribution to health care. It was heavily involved in giving doctors ways to see inside the body to discover abnormalities. It was also providing the watchful eye on very ill patients, monitoring their vital signs for any significant deterioration so that that the nurses and doctors could intervene in good time.

This subject has had more than its fair share of Nobel Prizes. Maybe the Nobel committees like medical advances particularly, but even so it has collected an extraordinary number of them. Medicine is normally regarded as a rather conservative field, and often it takes time for the advantages of some new process to be appreciated. Once they are, however, progress has often been rapid.

NOTES

1. Waller, A.D. (1887) A demonstration on man of electromotive changes accompanying the heart's beat, *Journal Physiology*, 8, 229–234, also available at: http://www.ncbi.nlm.nih.gov/pmc/articles/PMC1485094/pdf/jphysiol02445-0001.pdf
2. A (not so) brief history of electrocardiography, available at: http://www.ecglibrary.com/ecghist.html
3. Rivera-Ruiz, M., Cajavilca, C., and Varon, J. (2008) Einthoven's string galvanometer: The first electrocardiograph. *Texas Heart Instution Journal*, 35:2, pp174–178
4. Burch, G.E. and DePasquale, N.P. (1964) *A History of Electrocardiography*. Chicago: Year Book Medical Publishers, p.115
5. Borck, C. (2006) Between local cultures and national styles: Units of analysis in the history of electroencephalography. *Comptes Rendus Biologies*, 329:5–6, 450–459
6. Brought to life. Exploring the history of medicine, EEG, available at: http://www.sciencemuseum.org.uk/broughttolife
7. Sabbatini, R.M.E. The history of the electroencephalogram, available at: http://www.cerebromente.org.br/n03/tecnologia/historia.htm
8. Eik-Nes, S.H. (2013) Presentation of the Ian Donald Medal for Technical Development to Mathias Fink. *Ultrasound Obstetrics and Gynaecology*, 41, 114–120, also available at: http://www.isuog.org/NR/rdonlyres/632FE947-9766-4773-999E-33DA0180D7FE/0/2012_UOG_Congress_Report.pdf
9. Woo, J. A short history of the development of ultrasound in obstetrics and gynecology, available at: http://www.ob-ultrasound.net/history1.html
10. Donald, I., Macvicar, J., and Brown, T.G. (1958) Investigation of abdominal masses by pulsed ultrasound. *The Lancet*, 271:7032, 1188–1195
11. Brought to life, Exploring the history of medicine, Alice Stewart (1906–2002), available at: http://www.sciencemuseum.org.uk/broughttolife/techniques/~/link.aspx?_id=694B6EF236C34F29B07EA0236636222B&_z=z
12. Tsung, J. History of ultrasound and technological advances, available at: http://www.wcume.org/wp-content/uploads/2011/05/Tsung.pdf
13. A brief history of CT, available at: http://www.impactscan.org/CThistory.htm
14. Hounsfield, G.N. Biographical, Nobelprize.org. Nobel Media AB 2014, available at: http://www.nobelprize.org/nobel_prizes/medicine/laureates/1979/hounsfield-bio.html
15. Hounsfield, G.N. Computed medical imaging. Nobel Lecture, December 8, 1979, available at: http://www.nobelprize.org/nobel_prizes/medicine/laureates/1979/hounsfield-lecture.pdf
16. Cormack, A.M. Banquet speech, Nobel Banquet, December 10, 1979, available at: http://www.nobelprize.org/nobel_prizes/medicine/laureates/1979/cormack-speech.html
17. Pietzsch, J. The attraction of spin. The Nobel Prize in Physics 1952, available at: http://www.nobelprize.org/nobel_prizes/physics/laureates/1952/speedread.html
18. Pietzsch, J. Glimpse the life magnetic. The Nobel Prize in Physiology or Medicine 2003, available at: http://www.nobelprize.org/nobel_prizes/medicine/laureates/2003/speedread.html
19. Brought to life. Exploring the history of medicine, MRI, available at: http://www.sciencemuseum.org.uk/broughttolife/themes/~/link.aspx?_id=9CADB5F24D3249E7A95091E4A1984440&_z=z
20. Bellis, M. Magnetic Resonance Imaging, MRI, available at: http://inventors.about.com/od/mstartinventions/a/MRI.htm
21. Westgren, A. Award Ceremony Speech, The Nobel Prize in Chemistry 1943, available at: http://www.nobelprize.org/nobel_prizes/chemistry/laureates/1943/press.html

22. Though the prize was for 1943 it was not actually presented until the following year. See: http://www.nobelprize.org/nobel_prizes/chemistry/laureates/1943/
23. Brought to life. Exploring the history of medicine, Michel Ter-Pogossian (1925–96), available at: http://www.sciencemuseum.org.uk/broughttolife/people/michelterpogossian.aspx
24. Stewart, J.S.S. (1970) The aim and philosophy of patient monitoring. *Postgraduate Medical Journal*, (46, 339–343, also available at: http://www.ncbi.nlm.nih.gov/pmc/articles/PMC2467176/pdf/postmedj00354-0010.pdf
25. Severinghaus, J.W. and Honda, Y. (1987) History of blood gas analysis, VII Pulse oximetry. *Journal of Clinicai Monitoring*. 3:2, also available at: http://www.masimo.com/Nellcorfiction/PDF_FF/History%20of%20Pulse%20Oximetry.pdf

28

Knowing Where You Are: GPS

Right now there are thirty-one satellites zipping around the world with nothing better to do than help you find your way to the grocery store.

Ed Burnette

On October 22, 1707 the wonderfully named Admiral Sir Cloudesley Shovell was bringing the British Mediterranean fleet back to home waters.[1] He called his navigators together and the consensus was that they were off the French coast near Ushant, though one thought they were further north. A few hours later the leading ships crashed into the rocks of the Scilly Isles. Shovell's flagship, the 'Association', and two others, the 'Romney' and the 'Eagle', sank and there was only one survivor out of the 1300 crew.

It was a disastrous navigation error, and pointed up the failings of the methods of estimating position at the time. The normal method was called 'Dead Reckoning' and many wags said that was appropriate due to its inaccuracy. The principle was to estimate the speed of the ship with the so-called 'log'; then, knowing the direction from the compass, together with the time taken in that direction, the present position could be calculated as the distance from the previous known point.

The problem was that wind, tides and currents caused deviations from this course, though to some extent these could be compensated if they were known. Unfortunately, in the 1700s most were not, so large errors could occur. That was on top of the uncertainties in the measurement of both speed and time with the poor hourglasses available.

Sir Cloudesley Shovell's error was mostly one of latitude: they were further north than they thought they were. Their difficulties had been compounded by overcast weather; they couldn't measure the angle of the sun at midday which would have allowed them to calculate their latitude. However, they were also further to the west than they thought and had no way of checking their drift in that direction—their longitude—as no satisfactory method existed.

In 1714, the Longitude Act proposed a scheme echoing the existing Spanish and Dutch ones.[2] It provided for 'Publick Reward for such persons as shall discover the Longitude at sea'. There were three levels of reward for the inventor: £10,000 for someone who could find longitude 'to One Degree of a great Circle, or Sixty Geographical Miles'; £15,000 if

J.B. Williams, *The Electronics Revolution*, Springer Praxis Books,
DOI 10.1007/978-3-319-49088-5_28

the method could find the longitude to two-thirds of that distance; and £20,000 if it found the longitude to half of the same distance.[3] These were fabulous sums at the time and showed the importance of the subject.

There were two contending methods. The first was championed by the astronomers who were producing tables of the positions of celestial bodies, so that measurements of their angles to the horizon, or between them, could be used to calculate longitude. It was a tremendous task which depended on many accurate measurements, and then the vast labor of producing the tables. Even then, it required accurate instruments for use on the ships and men who knew how to use them. Inevitably, progress was very slow.

The other approach was to determine the local midday by finding the highest point of the sun and then comparing that with the time at Greenwich. Finding the local midday was not too difficult as this was part of finding the latitude and the quadrants and sextants used to measure this were being steadily improved. The real difficulty lay in having an accurate clock which could maintain Greenwich Time.

The man who set himself the task of solving this was John Harrison, a carpenter and clockmaker (Fig. 28.1). Between 1735 and 1772 he built a series of clocks, greatly improving on current clocks which were only accurate to a few minutes a day, which was nowhere near good enough. Eventually, he realized that the way to overcome the difficulties of motion at sea affecting the clock was to make it as small as possible.

The final device was like a large pocket watch and was accurate to around 1/5 of a second a day.[4] Trials were undertaken in 1764, including the formal one to Barbados where its performance was compared to astronomical measurements taken by the Astronomer Royal. The watch's error was found to be just 39.2 s or 9.8 mi (15.8 km) and thus easily should have been awarded the full prize under the 1714 Act.[5]

Fig. 28.1 John Harrison—undoubtedly a difficult man. *Source:* https://en.wikipedia.org/wiki/John_Harrison#/media/File:John_Harrison_Uhrmacher.jpg

That was when the trouble started, as the Board of Longitude refused to pay out. Not only that, but they recommended to Parliament that Harrison should be awarded only £10,000 when he demonstrated the principles of the watch. The second £10,000, less what had already been paid, he should only get when other watchmakers had shown that they could make similar devices. The Board felt that they were protecting their interests and this was the spirit of the 1714 Act. Harrison felt cheated and that the rules had been changed on him. There was a suspicion that behind this were the machinations of the astronomers.

Eventually copies were made and proved how good his design was. He was rewarded eventually by Parliament, probably with the King's urging, but the Board of Longitude never really relented. Undoubtedly, Harrison was difficult and didn't want to let go of the secrets that he had worked a lifetime to perfect, but it was badly handled by the Board who wanted watches that every ship could use.

Quite soon chronometers, as these accurate watches were known, were an essential on ocean-going ships and, together with more accurate sextants, provided the basis for much more satisfactory navigation on the long-distance voyages which mapped so much of the world. Navigation was a matter of time or angles and often both. Apart from steady improvements in the equipment, little changed for more than a century.

In 1886, Heinrich Hertz was able to show that radio waves radiated from an open loop aerial were directional. Once the technology of making and receiving them was under some sort of control in the early twentieth century it was natural for investigators to see if the waves could be used for navigational purposes.

As early as 1906 the US Navy was experimenting with a radio direction finder, which they called the Radiocompass.[6] This consisted of a radio transmitter on shore and a directional antenna on a ship. The problem was that this was fixed so the ship had to be turned to try to find the maximum signal. With no form of amplification this was very difficult and the Navy lost interest.

The following year in Germany two companies, Lorenz and Telefunken were testing their 'Funkbaken' or radio beacons.[7] Otto Scheller of the Lorenz company patented a system with crossed antennae in which he proposed to send two separate signals on the two sides so that along the line between them both were of equal strength. By having complementary signals on the two sides a continuous signal would be received on a narrow center line and the different signals found if the user strayed to one side or the other.

He proposed to use Morse A (dot dash) on one side and Morse N (dash dot) on the other.[8] By receiving the A or the N signal, the navigator would know their position relative to the center line and could correct their course until they received a continuous tone when they knew they were on the center line. The system did have the advantage that the user only required a nondirectional antenna on the ship but it did give a navigable line either towards or away from the beacon. However, it was only really useful as a homing device.

In 1908, Telefunken unveiled an even more elaborate system, their Kompass-Sender. It consisted of a whole circle of antenna pairs, like the spokes of an umbrella. The transmitted signal was radiated briefly from all the spokes to provide a starting (or north) point and then each pair was connected separately to the transmitter in sequence with a fixed time at each. The user could pick up the north signal and start a watch, keeping track of the signals until he found the maximum. He then stopped the watch and, knowing the sequence direction of the signals would know the direction of the beacon. Over the next few years the system was developed further with several beacons and it was used in the First World War by German Zeppelins bombing targets in England.

When using a simple fixed transmission on shore, the problems with having to rotate the ship could easily be solved in one of two ways. The first was to go back to Hertz's original loop antenna and simply rotate that. The American Navy used a variant of this, the Kolster system, which used a coil instead of the loop, but at first the conservatism of the service meant that it was barely used. This changed when America entered the war in 1917 and then ships' crews rapidly discovered its usefulness.

The other approach was the Bellini–Tosi system, invented by two Italians. This used a fixed pair of antennae coupled them to a goniometer. This device, later to find use in radar systems (see Chap. 5), used three coils, two of which were connected to the antennae. The third could be rotated to find either the maximum or minimum signal and hence the bearing of the transmitter. This was easier than rotating the antenna and could be undertaken in the radio room.

What made all the difference to these systems was the introduction of vacuum tubes which meant that the signals could be amplified and processed much more easily. It was arguable which system was best; they all had their advantages and disadvantages and, hence, champions and detractors. Nevertheless, they had changed the face of navigation as it was now possible to get a more accurate fix on one's position by finding the bearings of two or more radio beacons even in bad weather.

Between the wars, many organizations, including the Marconi Company in Britain, set up radio beacons which relied on some variant of these basic systems. Their use soon extended to aircraft where methods that could give a quick answer were essential. This was particularly so in the US, where the aircraft industry took off more quickly than in other countries, due to the greater distances.

The Lorenz system was extended for use as a blind approach or instrument landing system to guide aircraft on to the runway in bad weather. The signals on each side were simplified to a dot on one and a dash on the other. Though at first an audible arrangement was used, this was modified to drive a meter which gave an indication of left or right and centred itself when on the correct line of approach. Coupled with markers to check that the approach height was correct at appropriate points it made an effective system.

Later in the war, the blind landing system was extended by the Germans, with superior aerial systems and higher powers, as a navigation aid to their bombers over Britain. The planes flew along one 'beam' and a second cutting across it was used as a marker to warn that they were approaching the target. Unfortunately for them, it proved quite easy to jam so they moved on to more complex arrangements. Though these still used a beam they also depended on sending a signal to the plane which contained a 'transponder' that sent the signal back and the time of flight was used to measure the distance knowing the speed of light, or in this case electromagnetic waves.

These systems were really guidance rather than navigation systems. The Germans found that they were more economical than teaching all flight crews how to navigate by conventional methods including taking 'sights' of the angles of stars at night. British crews used these methods and it was well into the war before it was proved how inaccurate these methods were.

The British finally used a system called 'Oboe' which used the 'time of flight' method but in a different way.[9] The pulses in the signal from the prime station were carefully changed depending on whether the plane was too near or too far from the target. In practice, this meant that, in order to keep at the correct distance, the plane flew along an arc of a circle centred on the transmitter. A second station, spaced well away from the first, used their signal to indicate that the target had been reached.

It was a very accurate system, but suffered from operational weaknesses. The bombers could be seen following the curved path and that meant that their track could be followed by the enemy. This was overcome by using fast Mosquito planes, flying at high altitude as path finders, to drop flares on the target which could be used by the main bomber force. The Germans tried to jam the system but the British moved to the 10 cm 'radar' wavelengths, leaving the old ones in operation as a blind.

The British also needed a more general navigation system. This was provided by an elegant system called 'Gee', standing for 'Grid', which was the first 'hyperbolic' system. A master station sends out a pulse and a millisecond later the slave 80–160 km away sends out a second pulse. The receiver in the aircraft looks at the timing of these signals and determines the difference in the time for each signal to reach them and hence the difference in the distance between the stations.

At first sight this isn't very useful as it merely defines a curve of equal difference between the stations. This curve is called a hyperbola and is only straight for the condition where the delays are equal, i.e. when it is exactly the same distance from the two stations. However, if the master has a second slave also some 80–160 km distant, then timing the difference of this signal gives a second hyperbola. Where these two cross is the position of the aircraft.

The user of the system needs a set of hyperbola Lattice charts to work out their real position once the receiver has obtained the basic information.[10] Because it used frequencies in the 15–3.5 m (20–85 MHz) region it was a 'line of sight' system with a range limited to about 400 km at normal flying altitudes. It was not really accurate enough to be used as a bombing aid, but for general navigation it worked well.

As part of the Tizard Mission (see Chap. 5), information about Gee was taken to the US where it attracted immediate interest. They decided to develop their own version that could have greater range. To achieve this, the frequency was lowered to 171–154 m (1.75–1.95 MHz) so that it would propagate farther through the atmosphere, particularly through ionospheric bounce.

The system, known as LORAN, for Long Range Navigation, used faster pulses and more slave stations than Gee, but otherwise was fairly similar. It did, however, have a range of about 1100 km with an accuracy of around 1% of the distance from the transmitter. This was not good enough for bombing but as good as celestial navigation and easier to use.

The first stations appeared along the East coast of America, where LORAN could assist shipping in the Atlantic. Later, they were extended to Britain so that the system could be used for guiding bombers deeper into Germany than Gee could reach. By the end of the war, stations appeared all over the world to aid operations in many different areas of combat.

In Britain, another hyperbolic navigation system was developed by the Decca Radio Company and was known as the Decca Navigator. This used even lower frequencies in the 4285–2307 m (70–130 kHz) range and so could propagate over very long distances, but at the cost of accuracy. As a result, it was mostly used for maritime navigation. It used a slightly different arrangement of continuous signals rather than pulses and obtained its information from the phase shift between the signals rather than the time difference between pulses.

Decca Navigator first saw service guiding Royal Navy minesweepers during the Normandy invasion in June 1944, but it continued in use for the rest of the century. Like most of these systems it was modified a number of times but the basic systems stayed the same. Variants of hyperbolic systems remained the basic navigation arrangements for many years after the war.

In 1957, the launching of the Sputnik satellite by the USSR caused a considerable stir in America (see Chap. 20). At the Advanced Physics Laboratory (APL) of the John Hopkins University in Maryland two young men, William H. Guier and George C. Weiffenbach, became interested in trying to receive the signals that it transmitted.[11] They set up their receiver and soon could hear the bleeps as Sputnik passed by in its orbit.

Because they were radio experts they set their receiver precisely using the accurate 20 MHz signal from the nearby Bureau of Standard's radio station, WWV. The satellite conveniently transmitted at about 1 kHz higher than that, so they were able to obtain a signal. As the satellite moved around its orbit this changed, increasing in frequency by the Doppler Effect as the satellite moved towards them and decreasing when it was going away. This signal varied from about 1500 Hz to 500 Hz.

Then Gguier and Weiffenbach became interested in what they could do with the data. They looked at the range of the change of frequency and then roughly calculated the speed of the object they were following. It seemed to be about right for a satellite in a near-Earth orbit. Sputnik was definitely real. Over the next week or so, with help from their colleagues, they gradually obtained accurate data of the frequencies and time-stamped them with the WWV transmissions.

With data from nearly horizon to horizon they had enough information to try to calculate the orbit. They succeeded in doing this and after a while could even predict when the satellite would next appear, which showed that their estimates were quite good. Once the satellite stopped transmitting they were able to use the university's newly-acquired computer to try to determine the best accuracy that could be obtained in estimating the orbit using the Doppler method.

When the results were reported to the headquarters of Vanguard, the American program to launch satellites, they became very interested in the accuracy achieved which was rather better than other attempts. It led to an interesting interview with the young men's boss, Frank McClure, who asked them whether the process could be inverted. In other words, if the satellite position was known, could a point on the Earth be accurately determined. They did some calculations which showed that it could, and with surprisingly good accuracy.

The real interest came from the Navy. They were concerned about navigation on their Polaris missile submarines and needed to accurately know their position at the time of launch. Soon the essentials of what became known as the Transit or Navy Navigation Satellite System were designed. It would require multiple orbiting satellites radiating two ultrastable frequencies which would be encoded with their orbit information. This would be sent to them by fixed ground stations and then submarines with receivers and computers could determine their position about once an hour anywhere in the world.

That was the theory, but this was 1958 and the US had only just launched its first satellite, Explorer I. With the Cold War in full swing, the Navy was very keen and got APL to set up a program of three phases: experimental to determine all the data they needed, prototypes to test systems, and finally operational satellites. On September 17, 1959 they launched their first satellite, but it failed to achieve orbit.[12] Not a good start.

The next one did achieve orbit and worked for 89 days which was sufficient to gain useful data. Satellites were launched every few months with varied success, the best one working for 2 years. The results, however, were good enough to move on to the prototype phase. The first few launched were not fully functional, for one reason or another, but finally with the launch of 5BN-2 on December 5, 1963 the first operational satellite was in use.

From that point on the Navy always had at least one satellite that they could use for navigation. This was despite very considerable problems with the operational or Oscar satellites. The first three failed almost immediately and the next six remained operational only for a matter of months. Finally, with Oscar 12 and its successors the situation ws mastered, with the average satellite lifetime jumping to 14 years with some exceeding 20 years. The improvement was such that many of the satellites that had been ordered were never launched.

Though the first efforts had accuracies of only a kilometer or so, by the time the system was really operational at the end of 1963 this was down to around 100 m which was within the 0.1 nautical mile (185 m) target that had originally been set. As the years went by, and more and more allowances were introduced for problems such as the vagaries of the Earth's gravitational field, accuracy continued to improve reaching a few meters by the early 1980s. This could, in principle, be achieved anywhere in the world and in all weathers.

There was a downside. Depending on the number of satellites used it could be anything from an hour to many hours between 'fixes', and each one could take some time to achieve. While this was quite adequate for Navy ships, which had good inertial navigation systems and the satellite system was only being used for corrections, it was no use for aircraft or other fast-moving objects. While the system allowed many advances—for example, Hawaii was found to a 1 km away from where it had been thought to be—its limitations became increasingly obvious and the search was on for a better system.

In 1951 the Korean War in the balance and it was a time of great tension. When the US tested their first hydrogen bomb in 1952, the military were convinced that this was the way to deter the Russians from sweeping through Europe, but they were concerned about how they could deliver it. The concept was to use large Intercontinental ballistic missiles, but one of the problems was how these devices could be guided to their targets.

One of the people consulted about this was Dr. Ivan Getting, Rhodes Scholar and astrophysicist, who had just joined the Raytheon Corporation as vice president for engineering and research after a professorship at MIT and a short spell with the Air Force.[13] He proposed a system using satellites rather than ground-based transmitters where the 'time of flight' of radio waves from a number of satellites in known positions could be used to determine position. It was way beyond the technology at the time.

By 1960 things had moved on with the launch of the first satellites and the beginnings of the Transit system, and so it was time to re-examine Getting's earlier ideas for a navigational system. They were incorporated into a proposal called MOSAIC which was part of the Minuteman program, but it never went forward.[14] Later that year Getting left Raytheon and became president of Aerospace Corporation, an independent research organization sponsored by the Air Force to investigate advances in ballistic missiles and space systems.

The Navy were aware of the limitations of the Transit and started to explore another system called Timation. An essential part of this was the use of very accurate clocks in space. In 1967, they launched an atomic clock into space.[15] Despite this success, the other American armed forces were all pursuing their own incompatible navigation systems based on satellites.

In 1973, Col Dr. Brad Parkinson at the United States Air Force Space and Missile Systems Organization got tough with his colleagues and managed to produce one single program, the NAVSTAR Global Positioning system.[16] He managed to get support, in some cases lukewarm, from all the three services by cleverly incorporating aspects of all their systems and defining all the functions of the new system.

Because of continuing opposition in some quarters, the system was tested 'upside down' with transmitters on the ground and receivers in an aircraft in the air. This was sufficient to prove the basics of the system and so the first satellite was launched, but unfortunately it failed. However, 10 satellites were successfully launched in the following 7 years.

Like the Transit the system depended on the satellites being in precise known positions and this so-called 'Ephemeris' information was uploaded from ground stations. An important aspect was the very precise atomic clock (see Chap. 13) contained in each satellite which had been trialed in the Timation program. The satellite transmits the Ephemeris data and the receiver can calculate how far away it is from the time taken for the signal to reach it. This means that the receiver is somewhere on a sphere of that dimension centred on the satellite.[17]

The same information can be obtained by looking at further satellites, so the receiver is at the point where these spheres intersect. In theory, only three satellites need to be 'seen' by the receiver to get a fix, but as the receivers have no very accurate clock a fourth is needed to get full performance and this extra signal is used to bring the receiver into line with the atomic clocks in the satellites. The actual operation is complex and depends on a great deal of computation taking place in the receiving unit to calculate the exact position (Fig. 28.2).

In 1983, a South Korean airliner got lost over a remote area of Soviet territory and was shot down by Russian fighters.[18] President Reagan's reaction was that the GPS system, as it was generally known, should be opened up to civilian users. There was a catch in that it contained a deliberate arrangement to degrade its accuracy called Selective Availability. Despite that, for many purposes it was quite usable if the complex receiver and computing power could be obtained.

In 1989, the first of a new generation of satellites was launched and by 1990 there were enough for the system to become fully operational. By 1994, 24 were in use, and by 1997 there were a total of 28 which allowed for in orbit spares for any that might fail in service. In 1990, with the Gulf War, civilian use was curtailed and didn't return until 1993. Now with the increase of semiconductor technology it was possible to build small, relatively cheap GPS receivers with built-in computation. They started to be used for tracking fleets of vehicles and other such uses.

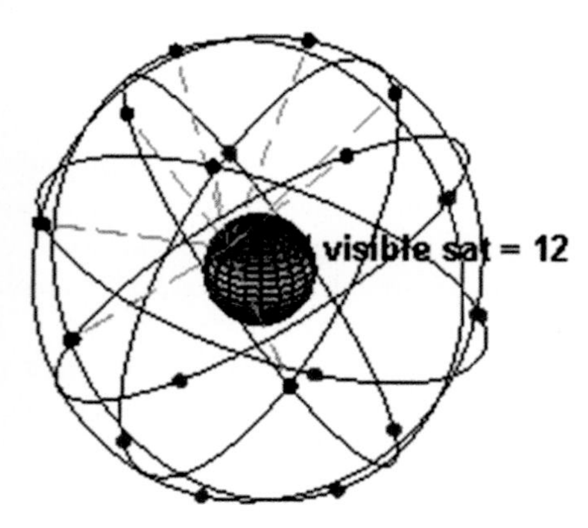

Fig. 28.2 Artist's impression of a GPS satellite over the Earth and the constellation of them in orbit. *Source:* https://en.wikipedia.org/wiki/Global_Positioning_System#/media/File:GPS_Satellite_NASA_art-iif.jpg; https://en.wikipedia.org/wiki/Global_Positioning_System#/media/File:ConstellationGPS.gif

It wasn't until 2000 that the full accuracy of the system, which had been gradually improving, was available to civilian users when President Clinton ordered that the Selective Availability should be switched off. The apparent accuracy improved from something like 100 m to 10–15 m overnight. Now it could find uses everywhere. In-car systems with built-in maps had started to appear a few years before, but they now really became useful. Within a few years it became the normal way for find one's way about.

With GPS becoming commonplace it was easy to forget the incredible technology involved. It required satellites to be launched, and this was only possible because of computing power to calculate orbits and electronics to control and track the launcher. Then the satellites had to be fitted with high frequency radio transmitters and receivers together with very precise atomic clocks. The users' units contained radio receivers, adjustable clocks, kept in line with the satellite's atomic clocks and which could be used as accurate timepieces. There was also a great deal of computing power together with a digital map and a display for the user to view the result.

It was unbelievable that such a complex arrangement could be made to work, but it did. The whole thing, as well as the space technology, required all the advances in electronics and computing to be possible. It was like the culmination of all the technologies that the century had been assembling.

NOTES

1. Navigation gone wrong. A British fleet runs aground, available at: http://timeandnavigation.si.edu/navigating-at-sea/challenges/british-fleet-runs-aground; Cavendish, R. (2007) Sir Cloudesley Shovell shipwrecked. *History Today*, 57:10, also available at: http://www.historytoday.com/richard-cavendish/sir-cloudesley-shovell-shipwrecked.

2. What made the search for a way to determine longitude so important?, available at: http://www.rmg.co.uk/discover/explore/what-made-search-way-determine-longitude-so-important.
3. Baker, A. Longitude Acts. Cambridge Digital Library, available at: http://cudl.lib.cam.ac.uk/view/ES-LON-00023/1.
4. Innovations in England, available at: http://timeandnavigation.si.edu/navigating-at-sea/longitude-problem/solving-longitude-problem/innovation-in-england.
5. Longitude found: John Harrison, available at: http://www.rmg.co.uk/discover/explore/longitude-found-john-harrison.
6. Howeth, L.S. (1963) *History of Communications-Electronics in the United States Navy*. Washington DC: US Government Printing Officer, p.261, available at: http://earlyradiohistory.us/1963hw22.htm.
7. Bauer, A.O. Some historical and technical aspects of radio navigation, in Germany, over the period 1907 to 1945, available at: http://www.cdvandt.org/Navigati.pdf.
8. Kendal, B. (1990) Air navigation systems, Chapter 3. The beginnings of directional radio techniques for air navigation, 1910—1940. *Journal of Navigation*, 43:03, 313–330.
9. Goebel, G. [10.0] Radio navigation systems, available at: http://www.vectorsite.net/ttwiz_10.html.
10. RAF navigation bits, available at: http://www.cairdpublications.com/scrap/navbit/navbit.htm.
11. Guier, W.H. and Weiffenbach, G.C. (1997) Genesis of satellite navigation. *Johns Hopkins APL Technical Digest*, 18:2, 178–181, also available at: http://techdigest.jhuapl.edu/TD/td1901/guier.pdf.
12. Danchik, R.J. (1988) An overview of transit development. *Johns Hopkins APL Technical Digest*, 19:1, 18–26, also available at: http://www.jhuapl.edu/techdigest/TD/td1901/danchik.pdf.
13. GPS Week looks back to 'Fathers of GPS', available at: http://www.schriever.af.mil/news/story.asp?id=123439399; Air Force Space Command, Air Force Space and Missile Pioneers, Dr. Ivan A. Getting, available at: http://www.afspc.af.mil/shared/media/document/AFD-100405-063.pdf.
14. Czopek, F.M. Pre GPS history of satellite navigation, available at: http://scpnt.stanford.edu/pnt/PNT08/Presentations/2_Czopek_PNT_2008.pdf.
15. GPS history, dates and timeline, available at: http://www.radio-electronics.com/info/satellite/gps/history-dates.php.
16. History of the GPS program, available at: http://www.aiaa.org/uploadedFiles/About_AIAA/Press_Room/Videos/IAF-60th-Anniv-GPS-Nomination.pdf.
17. Griffin, D. How does the Global Positioning System work?, available at: http://www.pocketgpsworld.com/howgpsworks.php.
18. Snively, J. GPS history - How it all started, Maps-GPS-info, available at: http://www.maps-gps-info.com/gps-history.html.

29

The Electronics Revolution

Study the past if you would define the future.

Confucius

On the stroke of midnight on New Year's Eve 1999, aircraft will fall out of the sky, lifts stop mid floor, power stations will cease generating, nuclear reactors will go critical and missiles will fire themselves at random.[1] Broadcasting will stop, vehicles will grind to a halt so food supplies will not get through, money will disappear as the banks' computers fail and civil disorder will break out. Those were only some of the disasters that were predicted.

Some people, particularly in America, took it all so seriously that they headed for the hills with enough equipment to make them self-sufficient.[2] Even in Britain families headed for farmhouses in remote Scotland burdened with supplies.[3] Some airlines decided that they would have no planes in the air over this critical time, just in case.[4] In Britain, the government took this sufficiently seriously to set up Taskforce 2000 as early as 1996 to warn of the threat of the Millennium Bug.[5]

These and many more were reported in *The Times*, hardly the most sensational of newspapers, so what was all the fuss about? The Millennium Bug, or Y2K, problem was real, but the difficulty was that no one really knew how serious—or otherwise—it was. In these circumstances there are always people who fear the worst and act accordingly. It wasn't helped by those, who perhaps should have known better, whipping things up and frightening susciptible people (Fig. 29.1).

The difficulty had been created many years before when computer resources were scarce and it seemed reasonable to represent the year by just two digits; assuming that the first two were 19, 1987 would just be stored as 87. This worked fine and nobody thought about the implications at the end of the century and what would happen when year 2000 arrived. In most cases, it was assumed the computer would compute the date as 1900.

For many applications this wouldn't really matter, but if the calculation was for the number of days between dates before and after the millennium then a completely wrong answer would be obtained.[6] If the calculation was to find the amount of interest on a loan, for example, then there was a real problem. In a stock control program, items entered after

J.B. Williams, *The Electronics Revolution*, Springer Praxis Books,
DOI 10.1007/978-3-319-49088-5_29

Fig. 29.1 The Millennium Bug logo as used by the campaign to deal with it. *Source:* http://www.bbc.co.uk/news/magazine-30576670

the end of the century would be deemed so old they would probably be up for scrapping and certainly wouldn't be seen as newer than items that were actually older.

There were dangers that critical information would be seen as very old rather than new, and so unimportant and hence deleted. On top of these were other more subtle problems in that many programers were unaware that the year 2000 was an exception to the normal rule that century years were not leap years. Thus date calculations beyond the end of February would be in error.

In practice, it took a great deal of time for programers to examine all this code and either fix the problems or declare the application free of them. It was a bonanza for computer consultants and there were those, usually wise after the event, that thought the whole thing had been a giant con.[7] They were just as wrong as the doomsayers because there were some actual problems that needed to be investigated.

In the event, the sky didn't fall in and nothing really catastrophic happened. Those who had real problems had largely dealt with them, life carried on as normal and the survivalists who had headed for the hills looked rather foolish. There were actual incidents that still got through, for example with documents with 1900 dates, but people are quite good at overcoming these difficulties when they know what it is all about.

What the whole thing demonstrated was how dependent everyone had become on computer systems, and how a simple technical error could threaten everyday life. Looking back to the beginning of the twentieth century, the whole thing would have baffled anyone living then. It was unimaginable that the world would become so dependent on something which wouldn't be invented for half a century that it could infect all aspects of people's lives. Even though many of the purported dangers were overblown, there were some real risks associated with the Millennium Bug.

Looking back, it can be seen how this castle was built; how one development enabled another and each took the technology a little further until the whole thing was constructed. This doesn't mean that progress is linear. One development enables others, but the route taken is often determined by other factors. We are dealing with human beings and their infinite variability. However, a general outline can be observed.

First came the vacuum tube, that child of the light bulb. Though wireless was invented first, it was in a blind alley until electronics enabled amplification, rectification and many other signal processing advantages. Then it could bring mass entertainment in the form of radio and ultimately television, but also radar and navigation systems. But these devices could only take the revolution so far.

It was the exploitation of the peculiar properties of, first, germanium and, later, silicon that brought so much to the second half of the century. They gave us the various forms of transistor which looked so puny and uninspiring at first, but gradually their potential began to be realized. It wasn't just the size of products that could be shrunk, but their lower power and greater reliability made things possible that were not before.

However, the simple idea of packing many of them on to a single piece of silicon—an integrated circuit or 'chip'—would have unforeseen consequences. Gordon Moore predicted his extraordinary 'law' suggesting that the densities of these devices would double every year or two. It didn't seem possible that this rule could go on holding for half a century or more.

Another child of electronics was the computer. While Charles Babbage had laid out the principles a century before, it was only with the coming of electronics that it was possible to sensibly build such machines. Only gradually did the enormous power of a fixed device, that could be programed to undertake different tasks, begin to be realized. These sets of instructions or 'software' could be varied in innumerable ways, but still run on the same physical device.

Developments in electronics brought down the size and cost of these machines, but it was the marriage of the computer and the integrated circuit, with its potential for huge numbers of transistors, that would transform the beast into something quite different. Development of a chip is a complex and expensive process so the manufacturer wants to sell as many as possible. The conversion of the computer into a microprocessor all on one chip was the ultimate device for the industry.

It was soon found that in anything that required sequencing or control or computation it was simpler to use a microprocessor and define its properties in software. Gradually, microprocessors crept in everywhere: into cars, washing machines, televisions, as well as the more obvious computers. They are in surprising places, for example even the humble computer 'mouse' contains a small microcontroller.

There was another strand in communications. It had begun with the telegraph, but that was starting to be superseded by the telephone at the start of the twentieth century. At first, this was purely an electrical device but electronics crept in and enabled interconnection and, ultimately, dialling to anywhere in the world. Electronics had brought 'instant' communication and the ability to get information across the world without the enormous delays of physically carrying the message.

When telecommunication and computing started to merge, then more things became possible. It became practicable to separate the telephone from its wire and produce the 'personal communicator' much beloved in science fiction. The mobile phone has become universal and most now regard it as a necessity and can't imagine life without it. The coming together of radio communications, computing, integrated circuits and display technology, to name only the main parts, has made this possible.

Largely the same bundle of technologies spun off another child—the Internet. It brought a further form of communication in electronic mail, without which so much of modern life could not function. Then it also spawned the World Wide Web. This was seen as a super library that everyone could access, but it has gone far beyond that basic idea. The simple, few clicks, accessing vast amounts of information, has changed so many fields that users cannot now imagine being without it.

Thus these various strands combine to produce new possibilities which in turn open up further generations. The myriad devices that inhabit modern homes are the result of all these intermingling lines of development. Each new generation produces opportunities for more. The greater number of strands, the larger number of possibilities there are to exploit, and so it has proved—progress has been exponential.

Obviously it has not been possible to include everything in one volume, and some subjects such as avionics, electron microscopes and computer games have been omitted. The selection has been of things felt to have the greater impact on everyday life. This is a set of personal choices and so will not please everybody, but hopefully it achieves its objective of tracing the main outlines and showing how these technologies have piled on each other and the impact they have had on people's lives.

It might be expected to include some look into the future at this point. This is, however, an area beset with difficulties as many futurologists have found to their cost. The factor that is often forgotten is the lifetime of the items that already exist. It is no good suggesting everyone will be living in plastic pod houses in 30 years when houses typically last much longer than that. Most of the housing stock will still be there then so only a small minority could be different.

With many more technological items, such as mobile phones, the physical, or fashion, lifetime is often much shorter, so some change can be expected. However, it is not always in a linear direction. Different factors come into play and progress heads off in unexpected ways. For example, the trend with mobile phones was that they became smaller and smaller. However, this reversed and they are becoming larger so that the bigger displays, necessary to accommodate the additional features, can be included.

Despite this, the trend is generally for smaller, lighter, and particularly lower power devices. This is partly due to economic forces to make items ever cheaper, which is an area where the electronics industry has an extraordinary record. Additionally, there is the pressure to minimize the impact on the Earth, either in the amount and type of raw materials consumed but also in the waste or by-products such as carbon dioxide.

We can also be confident that the convergence of computers, the Internet, mobile phones, cameras and televisions will continue. Now that these are all digital platforms it is easy for the information files to be swapped between devices. Because they can all share the same computing power and other resources such as displays, it is easy to pile even more features onto devices. The only restraint is that the complexity can become impossible for the user.

One thing we can be certain of is that electronics, and its offshoots of computing and telecommunications, will have an ever greater impact on how we live and work. As automation takes a greater role in manufacturing and computing in services, people will be left with the human interactions and attending to the machinery when it fails to perform correctly. The other area is the use of craft or skills for things too specialized or in too small quantities to make automation practical.

As the technology has clearly had such an impact, the question is often raised as to whether technology drives history. From Karl Marx onwards there have been proponents of the concept of technological determinism. If a particular technological development occurs, is the result a foregone conclusion? Marx is quoted as saying: 'The hand mill gives you society with a feudal lord; the steam mill, society with the industrial capitalist.'[8]

One does wonder with some of the commentators whether they have ever shaped metal with a file or wielded a soldering iron or, more particularly, were ever involved in the process of producing a new technological product. If so, they would know that nothing is certain about the result. Everything can be there that seems to be exactly what is required and then it is a complete flop. On the other hand, something that seems quite improbable can be a roaring success. Why did it take around a century for a quarter of homes to have a telephone, but the mobile phone take off much more rapidly? The answers to questions such as this don't lie in technological developments but in the way people respond to them.

A technological development does not dictate an outcome, it is an enabler; it sets up opportunities. There is no certainty which of these will be taken. Sometimes there is an immense length of time before interest grows, maybe because the products aren't quite right or simply the idea's time hasn't come. Microwave ovens, for example, took some 30 years or so from their invention to being a staple of the ordinary kitchen. There are now so many things that no one can imagine living without.

The idea of making your own entertainment has gone. Now it is dominated by television and other electronic devices. Amusements are now more individual and in the home. This has had the effect of separating people, but perversely they can also 'come together' in spirit for some great event either nationally or even across the world. On the other hand, there has been the rise of the music festival and huge pop concert where the young gather in vast numbers.

The nature of work has changed. Automation has increasingly replaced manual work, decreasing the number of repetitive jobs, and largely eliminating the dangerous ones. In the office, work is mostly based around computers which again have taken out much of the drudgery, leaving the people to deal with the human contact. For good or ill, this has spawned the call center which absolutely depends on communications and computing power. The type of jobs that people do has almost completely changed in the course of the twentieth century.

So much of this depends in some way or another on the development of electronics and its offshoots of telecommunications and computing. It is these that have shaped so many of the changes in our lives in the course of the twentieth century. This is the electronics revolution.

NOTES

1. Copps, A. Millennium mayhem or hype? *The Times*, 30 June, 1999.
2. Wheelwright, G. Doomsayers of Y2K head for the hills. *The Times*, 19 August, 1998.
3. Harris, G., Y2K: The end of the world as we know it. *The Times*, 16 January, 1999.
4. Keenan, S. Hoping to avoid disruption. *The Times*, 27 February, 1999.
5. Elliott, V. Doom merchant or a voice crying in the wilderness? *The Times*, 6 November, 1998.
6. Ainsworth, P. (1996) The millennium bug. *IEE Review*, 42:4, 140–142.
7. Kaletsky, A. The bug that never was. *The Times*, 6 January, 2000.
8. Smith, M. R. and Marx, L. (1994) *Does Technology Drive History?: The Dilemma of Technological Determinism*. Cambridge, MA, MIT Press, p.123.

Bibliography

Addison, P. (1995). *Now the war is over, a social history of Britain 1945–51*. London: Pimlico.

Addison, P. (2010). *No turning back: The peacetime revolutions of post-war Britain*. Oxford: Oxford University Press.

Anon. (1913). *Kinematograph year books*. London: Kinematograph.

Arthur, M. (2007). *Lost voices of the Edwardians*. London: Harper Perennial.

Atherton, W. A. (1984). *From compass to computer: A history of electrical and electronics engineering*. London: Macmillan.

Bennett, C. N. (1913). *Handbook of kinematography: The history, theory, and practice of motion photography and projection*. London: Kinematograph Weekly.

Bogdanis, D. (2005). *Electric universe: The shocking true story of electricity*. New York: Crown.

Bowen, H. G. (1951). *The Edison effect*. New Jersey: The Thomas Alva Edison Foundation.

Brown, L. (1999). *A radar history of World War II*. Bristol: Institute of Physics Publishing.

Buderi, R. (1996). *The invention that changed the world: How a small group of radar pioneers won the second world war and launched a technological revolution*. New York: Simon & Schuster.

Burch, G. E., & DePasquale, N. P. (1990). *A history of electrocardiography*. San Francisco: Norman.

Byatt, I. C. R. (1979). *The British electrical industry, 1875–1914: The economic returns to a new technology*. Oxford: Clarendon Press.

Clout, H. (2007). *The times history of London*. London: Times Books/Harper Collins.

Coursey, P. R. (1919). *Telephony without wires*. London: The Wireless Press.

Derry, T. K., & Williams, T. I. (1960). *A short history of technology*. Oxford: Oxford University Press.

Dimbleby, D. (2007). *How we built Britain*. London: Bloomsbury.

Elsevier. (1991). *Profile of the worldwide semiconductor industry—Market prospects to 1994*. Oxford: Elsevier Advanced Technology.

J.B. Williams, *The Electronics Revolution*, Springer Praxis Books,
DOI 10.1007/978-3-319-49088-5

Fessenden, H. M. (1940). *Fessenden builder of tomorrows*. New York: Coward-McCann.
Fleming, J. A. (1924). *The thermionic valve and its developments in radiotelegraphy and telephony*. London: Iliffe & Sons.
Garfield, S. (2004). *Our hidden lives*. London: Ebury Press.
Green, M. (2007). *The nearly men: A chronicle of scientific failure*. Stroud: Tempus.
Griffiths, J. (2000). *Video retail & hire, 2000 market report*. Hampton: Keynote Publications.
Harrison, B. (2011). *Seeking a role: The United Kingdom 1951–1970*. Oxford: Oxford University Press.
Hattersley, R. (2004). *The Edwardians*. London: Little, Brown.
Hennessey, R. A. S. (1972). *The electric revolution*. Newcastle upon Tyne: Oriel.
Hiley, N. (1998). At the picture palace: The British cinema audience, 1895–1920. In J. Fullerton (Ed.), *Celebrating 1895: The centenary of cinema*. Sydney: John Libbey.
Hiley, N. (2002). Nothing more than a craze: Cinema building in Britain from 1909 to 1914. In A. Higson (Ed.), *Young and innocent: The cinema in Britain* (pp. 1896–1930). Exeter: University of Exeter Press.
Howard, J. (2000). *Photocopiers and fax machines. Market report*. Hampton: Keynote Publications.
Howgrave-Graham, R. P. (1907). *Wireless telegraphy for amateurs*. London: Percival Marshall.
Howitt, S. (1995). *Video retail & hire, 1995 market report*. Hampton: Keynote Publications.
Hughes, T. P. (1983). *Networks of power: Electrification in Western Society, 1880–1930*. Baltimore: Johns Hopkins University Press.
Hunt, T. (2005). *Building Jerusalem*. London: Phoenix.
Keynote. (1984). *Video recorders: An industry overview*. London: Author.
Keynote. (1991). *Video retail & hire* (3rd ed.). Hampton: Author.
Kroll, M. W., Kroll, K., & Gilman, B. (2008). Idiot proofing the defibrillator. *IEEE Spectrum, 45*, 40e45.
Kynaston, D. (2007a). *A world to build*. London: Bloomsbury.
Kynaston, D. (2007b). *Smoke in the valley*. London: Bloomsbury.
Kynaston, D. (2009). *Family Britain 1951–57*. London: Bloomsbury.
Maddison, A. (1991). *Dynamic forces in capitalist development*. Oxford: Oxford University Press.
Malone, M. S. (1995). *The microprocessor: A biography*. New York: Springer.
Marr, A. (2009). *A history of modern Britain*. London: Macmillan.
McWhan, D. (2012). *Sand and silicon*. Oxford: Oxford University Press.
Miller, S. (Ed.). (1994). *Photocopiers and fax machines: A market sector overview*. Hampton: Keynote Publications.
Mitchell, B. R. (1988). *British historical statistics*. Cambridge: Cambridge University Press.
Morris, P. R. (1990). *A history of the world semiconductor industry*. London: Peregrinus.
Morrison, L. (Ed.). (1998). *Photocopiers and fax machines: 1998 market report*. Hampton: Keynote Publications.
O'Hare, G. A. (1998). *Towards the 1990s: Developments in the British video market*. London: Goodall Alexander O'Hare.

Oxford Dictionary of National Biography, online edition.
Price, C. (2007). Depression and recovery. In F. Carnevali & J.-M. Strange (Eds.), *20th century Britain*. Harlow: Pearson Education.
Proudfoot, W. B. (1972). *The origin of stencil duplicating*. London: Hutchinson.
Pugh, M. (2008). *We danced all night*. London: Bodley Head.
Rhodes, B., & Streeter, W. W. (1999). *Before photocopying: The art and history of mechanical copying* (pp. 178–938). Northampton, MA: Heraldry Bindery.
Riordan, M., & Hoddeson, L. (1998). *Crystal fire*. New York: Norton.
Rogers, E. M. (1983). *Diffusion of innovations*. London: Collier Macmillan.
Rowe, A. P. (1948). *One story of radar*. Cambridge: Cambridge University Press.
Seidenberg, P. (1997). From germanium to silicon—A history of change in the technology of the semiconductors. In A. Goldstein & W. Aspary (Eds.), *Facets: New perspectives on the history of semiconductors* (pp. 34–74). New Brunswick: IEEE Center for the History of Electrical Engineering.
Smith, M. (1998). *Station X: The codebreakers of Bletchley Park*. London: Channel 4 Books.
Smith, M. R., & Marx, L. (1994). *Does technology drive history? The dilemma of technological determinism*. Cambridge, MA: MIT Press.
Taylor, A. J. P. (1977). *English history 1914–1945*. Oxford: Oxford University Press.
Thompson, R. J. (2007). *Crystal clear: The struggle for reliable communication technology in World War II*. Hoboken, NJ: Wiley-Interscience.
Tyne, G. F. J. (1994). *Saga of the vacuum tube*. Indianapolis: Prompt Publications.
Uglow, J. (2002). *The lunar men*. London: Faber & Faber.
Verne, J. (1876). *From the earth to the moon*. London: Ward, Lock.
Weightman, G. (2011). *Children of light: How electrification changed Britain forever*. London: Atlantic.
Wheen, A. (2011). *Dot-Dash to Dot. Com: How modern telecommunications evolved from the telegraph to the Internet*. New York: Springer.
Whiting, V. (1998). *DVD video: European market assessment and forecast, 1998–2000*. London: Screen Digest.
Williams, R. (1992). *The long revolution*. London: Hogarth.
Wood, L. (1986). *British films 1927–1939*. London: BFI Publishing.
Zimmerman, D. (2010). *Britain's shield: RADAR and the defeat of the Luftwaffe*. Stroud: Amberley.

Index

J.B. Williams, *The Electronics Revolution*, Springer Praxis Books,
DOI 10.1007/978-3-319-49088-5

Printed in the United States
By Bookmasters